A guide to the vegetation of
Britain and Europe

A guide to the vegetation of Britain and Europe

Oleg Polunin and Martin Walters

OXFORD UNIVERSITY PRESS
1985

Oxford University Press, Walton Street, Oxford OX2 6DP
Oxford New York Toronto·
Delhi Bombay Calcutta Madras Karachi
Kuala Lumpur Singapore Hong Kong Tokyo
Nairobi Dar es Salaam Cape Town
Melbourne Auckland
and associated companies in
Beirut Berlin Ibadan Mexico City Nicosia

Oxford is a trade mark of Oxford University Press

Published in the United States
by Oxford University Press, New York

British Library Cataloguing in Publication Data
Polunin, Oleg
A guide to the vegetation of Britain and Europe.
1. Botany—Europe 2. Plants—Identification
I. Title II. Walters, Martin
581.94   QK281
ISBN 0-19-217713-3

Photosetting by Cotswold Typesetting Ltd.
Printed in Great Britain by
St Edmundsbury Press Ltd.
Bury St Edmunds, Suffolk

Preface

This is the first attempt, as far as is known, to describe in outline the natural and semi-natural vegetation cover of Europe (as far as the Russian border), at a level which is comprehensible to the layman. We have attempted to make the account as 'visual' as possible, with photographs of most of the different types of vegetation and with drawings of the most characteristic species in each community. This is, in a sense, a field guide to European vegetation and will, we hope, enable the naturalist—whether he or she be interested in plants or animals—to decide where to go to find the best examples of natural and semi-natural vegetation and ecosystems persisting to the present day.

The aim is primarily to familiarize the reader with the most important plant communities to be found in Europe. By important we mean those communities which are likely to be seen by the traveller, or those communities which are especially characteristic in particular regions or environments. This book does not attempt to describe all the different communities which have been recognized by ecologists. Because we are writing primarily for the interested layman and traveller we have concentrated on the communities which are most distinctive; thus woodland communities are more fully covered than, for example, grasslands. These latter require considerable botanical knowledge to understand the different types, while most people can distinguish, for example, the difference between oak and beech woods.

The study of plant communities has been underway for a considerable time, but it received a great impetus from the work of Braun-Blanquet in the 1920s, and subsequently a classification of plant communities has been evolved mainly by the so-called Central European school of phytosociology. This classification is not easily comprehensible or available to the layman as it requires a considerable knowledge of plant taxonomy; also much of the work is published in relatively obscure journals in foreign languages. This book aims to bridge the gap between the scientific approach to vegetation and that of the layman-naturalist in the hope that the latter will be better able to comprehend the importance of vegetation, to identify the main types whilst travelling in Europe, and also to understand something of how plant communities have developed in different environments to reach relatively stable climax communities. Also he will be better able to understand the extent of man's influence on the vegetation and the changes and destruction that he has brought about. Such considerations are outlined in the introductory chapter but are largely excluded from the 'guide' to the main types of vegetation in the subsequent chapters.

Only superficial accounts of a limited selection of communities can be included in a single volume, but reference is made to the literature where further information can be obtained for each country in Europe—albeit in different languages.

In describing each community it has only been possible to list the characteristic species present, with little reference to their interrelationships with each other and with the environment. Again it has been very difficult to select representative examples of each type of plant community. There are so many local variations and no two communities are exactly similar; thus, in general, contrasting examples from different countries or regions have been selected to give some indication of the range and composition of each community.

About 100 different communities in Europe have been briefly described in this volume. The majority are zonal and are described in the different climatic regions: Arctic, Boreal, Atlantic, Central European, Mediterranean, and Pannonic. Other communities are azonal and occur in a wide range of localities and climates in Europe and have many similarities of composition and species. Examples are alpine, wetland, and coastal communities.

In the description of each community, it has unfortunately not been possible to give more than a passing mention of the species of ferns, mosses, liverworts, and lichens that occur, while the fungi, algae, and micro-organisms have not been included; neither have the animals that are part of each ecosystem been included.

The important process of succession, which is the sequence of changes through which every mature community evolves, in a series of stages (seres) in the closest association with the environment, to arrive at a more or less stable climax community, is only briefly mentioned in some cases. But succession is a very important process, which can readily be deflected by the activities of man and his animals, resulting in what is termed semi-natural vegetation. This semi-natural vegetation, which may remain relatively stable for long periods so long as these activities continue, is composed largely of native species and it covers the greater part of Europe at the present day—land which is not otherwise intensively cultivated for crops for man and his animals.

The realization that vegetation communities are of prime importance to the survival of life on earth, at a time when the natural and semi-natural communities are being destroyed at an ever-increasing pace by contemporary man, has resulted during the last half-century in the establishment in the countries of Europe of Nature Reserves and National Parks. Though they are but very small samples of the natural and semi-natural ecosystems, nevertheless this is a beginning and natural communities are being conserved, where the plant and animal life can live relatively undisturbed. The most important Reserves and National Parks in Europe are listed in this volume, and in these it is possible to observe and study these natural ecosystems.

This is a most important step forward in the preservation of what remains of our natural wild life, and it is hoped that in the next half century these Reserves and National Parks will be considerably extended. The conservation of the rarer and decreasing species of plants and animals is only possible in their natural environments where they have become adapted to living under certain environmental conditions, and where they can hold their own in competition with other species. Individual species cannot survive in isolation. Consequently it is the whole ecosystem that requires protection and recently reserves have been established largely with this aim is view.

Godalming and Cambridge
February 1983

O. P.
M. W.

Acknowledgements

This volume would not have been possible without the intensive work on the literature and study of vegetation undertaken over a period of two years by Martin Walters. He has selected the examples of vegetation and has planned the layout of the major plant communities of Europe, and they have subsequently been re-edited and ordered by myself.

Both authors have travelled considerably to visit and observe the different types of vegetation in Europe—from northern Scandinavia to Turkey-in-Europe, and from Portugal to Bulgaria—and to obtain photographic examples of most of the natural and semi-natural communities in Europe. Additional photographs have kindly been loaned by Professor Hugo Sjörs from Uppsala, Sweden, numbers 67, 71, 72, 75; Dr Manfred Fischer, Botanisches Institut und Botanisches Garten, Vienna, Austria, numbers 92, 93, 94, 100, 101, 102, 103, 133, 134, 135, 136, 137, 138, 139, 140, 144, 148, 149; Dr Čedomil Šilič, Zemalski Muzej, Sarajevo, Yugoslavia, numbers 95, 112; Dr Klaus Kaplan, Ruhr Universität, Bochum, W. Germany, numbers 28, 41, 42, 43, 45, 46; Dr J. Akeroyd, Department of Botany, University of Reading, number 37; Jan Jerzy Karpiński, Poland, number 18.

The most characteristic species occurring in each different type of community have been drawn to help the reader to identify the different communities as he or she travels through Europe. We are greatly indebted to the two botanical artists for their fine and accurate work made from herbarium specimens. Ann Farrer has drawn the plates of the grasses, sedges, etc., and those of mosses, lichens, and ferns, while Rosemary Wise has drawn the large number of plates of flowering plants, trees, and shrubs. The skill of both these artists is much appreciated, and it is hoped that the layman will be able to identify many of the plants he finds in the different communities. I would also like to thank Serena Marner of the Herbarium at Oxford, for her help in providing large numbers of specimens for Rosemary Wise's drawings; also Michael Wilkinson of the Forest Herbarium. Rosemary Wise has also prepared the diagrams under the direction of Martin Walters and Oxford University Press, and Christine Quartley has prepared the glossary; we are grateful for their help and participation.

Once again I am greatly indebted to Kathleen Cooke for her excellent typescript, and to my wife who has accompanied me with great pleasure on our journeys to remote parts of Europe in search of photographs of vegetation. She has also prepared the two indexes.

O. P.

I should also like to thank Sarah, Rachel, and Katherine for their encouragement, advice, and help.

M. W.

Contents

Plates 61–170 form a separate colour section at the end of the text

PART I
General introduction

1 Some ecological concepts and terms

The general term *vegetation* refers to groups of plants which live in close proximity to each other, and which interact with each other and with the environment in which they live. Plants only occasionally live in nature as isolated individuals. Vegetation has been described primarily in relation to the species which make up the plant communities—that is, its floristic composition. Considered on a regional basis, vegetation may be regarded as consisting of all the different plant communities occurring within a region both on land and in water.

During the last 200 years or so, attempts have been made to classify the different types of vegetation found covering the earth's surface. The main criteria used in the classification of vegetation are the floristic composition, the dominance and relationship of species to each other, the structure of the community, the general appearance (physiognomy), and the periodicity of development and maturity of the community.

In this book, where a very wide range of types of vegetation are recorded, the floristic composition of the community takes precedence over the other criteria. It is on this basis that we have listed the main types of vegetation occurring in Europe.

The more detailed and advanced quantitative descriptive approach, involving the precise measurement of vegetational features such as density of population, cover, frequency, height, biomass, age, structure, etc., as well as soil type, is not appropriate for such large vegetational units as we are recording, nor in many cases is the knowledge available.

In describing the vegetation of the earth's surface, a number of important ecological terms and concepts must be outlined. A few of the most widely accepted are listed below.

The *ecosystem* refers to the whole community comprising all the plants including fungi, bacteria, etc., and all the animals living within the community, all of which are largely interdependent on each other. In general, the plants are the 'producers', building up organic matter in the ecoystem; the animals are the 'consumers', living on the organic matter in the ecosystem; while the bacteria and fungi are the 'decomposers', breaking down the organic matter in the ecosystem and releasing mineral salts and carbon dioxide. These latter are again quickly assimilated by the 'producers' in an actively growing community.

All this organic matter within the ecosystem, whether living or dead, is referred to as the *biomass*.

The concept of the *plant community* has many interpretations. In general, it is a distinctive assemblage of plants found growing together and interacting with each other—not a random aggregation of plants. Different environments have different plant communities, each composed of different assemblages of species. Each community has a distinctive layering, degree of cover, and relative abundance exhibited by each species, and is thus usually clearly distinguishable. However there are also many cases where there is a gradual merging of one community with another, often with the presence of distinctive intermediate stages.

Within each community there are usually one or more species which are described as *dominant*. They have a marked influence on the community and on other species of the community. For example, the extent and density of the leaf canopy of the dominant tree species in a forest profoundly affects the composition of the subordinate communities which grow in the lowest layers of the forest.

Competition between species is another very important concept, particularly in relation to the existing environmental resources. Competition for light is one of the most important factors in the development of such a complex community as a forest. The amount of light entering the forest has a profound effect on the species that are able to colonize and survive in the lower layers. Lack of light may exclude the presence of certain species or may reduce the reproductive capacity of other species present, while increase of light, when a tree dies or is removed, may result in colonization by different light-loving species, or in rapid recolonization by the dominant trees. Again there may be intense competition between species in the community for the existing mineral resources in the soil, or for soil water, etc. Thus competition between species is a very important aspect of community development and stabilization.

Other species, however, may be *complementary* within the community and occupy niches which are not otherwise competed for; as for example the growth of mosses and lichens on tree trunks, or the presence of certain pre-vernal species in the field layer which flower before the leaf canopy of the deciduous trees in the forest develops.

There are other relationships between species within a community. Some species are dependent on the presence of other plants for their existence within the community. For example, mosses and ferns, which can grow in the lower light intensity and higher humidity on the forest floor, will quickly disappear when the forest is removed. Or again, different fungal species growing on humus from different sources are entirely dependent on the plants from which the humus is originally derived. Other relationships include those of the parasitic species which obtain their nutriment directly from other species; and symbiotic species which live in close association with other species, to their mutual advantage. Or again there are epiphytic species which by obtaining support from other species, live in distinctive microhabitats as, for example, epiphytic ferns growing on trees.

Plant communities are not static. There is a continuous change in the composition and proportion of species occurring in each community as it matures and ages within a stable environment. In their development towards maturity, plant communities pass through many stages. This process of change is known as *succession* and it is an important ecological process. No integrated communities of organisms occur spontaneously. Each community gradually develops stage by stage as the plants, the animals, and the environment come more and more closely into contact and balance, and interact with each other. In the succession of a community there is a series of stages of development each with its own distinctive plant communities and these are known as *seres* or *seral stages*. Thus the successive plant communities developing in water, as a result of the accumulative changes brought about by the action of the communities on the environment, are known as *hydroseres*, while the successive communities developing on rocks are known as *lithoseres*.

Each vegetational succession has its distinctive seral stages,

Some ecological concepts and terms

which may for periods of time remain relatively stable, and which can usually be identified by their species composition. However, in reality there is a continuous if irregular development, in which each sere gradually supersedes the preceding sere in the process of succession. In some cases there are intermediate stages which may be distinctive and are known as *ecotones*. They may be of particular interest, for not only may they contain species from the two seres, but they may also have additional species which can survive in these intermediate communities.

Thus there is often an 'edge effect' where the two different communities lie in close proximity to each other, and the intermediate zone may have some different species which are not present in either community. For example, where forest and grassland converge certain additional shrubs and herbaceous species may be found which do not otherwise occur in either community. The semi-natural vegetation of hedges which line meadows are often examples of this 'edge effect' where unusual species can occur.

Another important ecological concept is that of the *climax* community. This is the ultimate and most complex community which develops in relation to a particular environment of an area or region. The climax community is, in contrast to the seral stages which have preceded it, a relatively stable community which can be considered to be in closest balance with that particular environment. In general it can remain relatively stable as long as the environment continues unchanged. But where there are local changes of soil or climate, as for example from locality to locality, different climax communities may occur. These are described as sub-climax communities by some ecologists, or as components of the *polyclimax* by others. For example in cases where mature coniferous forests occur on certain soils or in local climates, where elsewhere the climatic climax is deciduous forest, these examples are commonly referred to as sub-climax communities. By contrast, in the northern taiga it is the coniferous forests which are the climatic climax communities. Man may also be a main factor in the creation of sub-climax communities; for example, many of the northern heathlands of the British Isles have resulted from man's influence or that of his domestic animals. There seems to be little likelihood of their ever developing to what theoretically would be considered as the climatic climax, that of coniferous or deciduous woodlands, such as an acid oak-birch community.

2 Soils of Europe

Soils are highly complex structures which usually take many years to reach maturity and to form a relatively stable structure. They develop as a result of a complex interrelationship between the climate, the rocks or substrates on which, and from which, they are formed, the vegetation cover, and the soil fauna and flora. Each plays a vital and closely integrated part in the formation of soils, as indicated below.

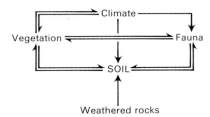

Thus the vegetation, as a result of the decomposition of dead leaves and branches, etc. by fungi and bacteria, first produces humus and then ultimately releases mineral salts and carbon dioxide into the soil. The climate determines, among other things, the water movements and hence the transport of soluble materials in the soil, the oxygen concentration in the soil, and the temperature of the soil. The fauna further breaks down the organic matter and affects the oxygen and carbon dioxide concentration in the soil. The type of weathered rock from which the soil develops, whether quartz, silicate, or limestone, largely affects not only the mineral content of the soil but also the soil acidity or alkalinity, both of which are very important in influencing the type of vegetation that develops.

Another factor in soil development is relief, for the slope of the ground not only affects the amount of solar radiation reaching the vegetation but also the drainage systems. Thus vegetation on the northern and southern slopes of a mountain may be very different and distinctive soils may be developed. Again water drainage down slopes into valleys may influence the development of vegetation on the slopes and in the valley bottom. These are but a few of the complex interrelationships which occur in the development of soils.

Certain characteristics of soils will now be briefly outlined before the major types of soil in Europe are described.

The *soil profile* refers to the distinct layering which occurs in many soils as they develop and reach maturity. Three horizons are widely recognized in many soils. These horizons develop largely as a result of the downward movement of water as it percolates through the soil. This transports soluble mineral salts and humus downwards, and deposits them at lower levels.

Thus the *A horizon* is the upper horizon in which humus accumulates, and out of which the soluble materials and humus are removed, or *leached*, and carried downwards by water to the lower levels. The *B horizon* is the horizon in which humus and mineral salts are deposited, often in distinct layers. The *C horizon* is the relatively unchanged rock or substrate from which the soil develops. Many soils do not show this clear-cut profile; in some cases only two horizons are developed, while in others, mostly immature soils, there are no distinct horizons. In saline soils of eastern Europe, by contrast,

net water movement and transport of minerals is upwards and, consequently, as a result of evaporation of water at the surface, salts are deposited in the uppermost soil layers.

Another very important soil factor is that of acidity and alkalinity. This affects to a large extent the type of vegetation that develops on a soil. Certain families of plants, notably the heath family, *Ericaceae*, are better adapted to live on acid soils—they are termed *calcifuge*. Other plants are *calcicole*, that is to say that they tend to grow only on calcareous soils, as for example many species of orchids. Most plants however have a wider range of tolerance of acidity or alkalinity, though in the more extremely acid or alkaline soils they may not be competitive and thus rarely occur in natural communities on such soils.

The measurement of soil acidity is in terms of the concentration of hydrogen ions (H^+) in the soil, and is indicated by pH. The greater the concentration of ions of hydrogen (expressed by a low pH number) the higher the acidity. Acid soils have hydrogen ion concentration, or pH's, ranging from 4 to less than 7, while alkaline soils have pH's of above 7 up to 9; a pH of 7 indicates a neutral soil.

The mineral particles formed as a result of the physical and chemical weathering of rocks can be grouped according to average particle size. This materially determines the texture of the soil, whether for example it is sandy, loamy, silt, or clay.

Size	Diameter
Stones and gravel	over 2.0 mm
Coarse sand	2.0–0.2 mm
Fine sand	0.2–0.02 mm
Silt	0.02–0.002 mm
Clay	smaller than 0.002mm

Humus is a most important constituent of soils. It is formed from the partial decomposition of organic remains of plants, and in many soils it is mixed with mineral matter. Under normal soil conditions humus is rapidly broken down by the activity of soil organisms, with the ultimate release of soluble mineral salts and carbon dioxide. These products are quickly reassimilated in actively growing vegetation. Thus there is an annual or seasonal cycle of breakdown and build-up of organic material in every plant community. Humus rich in mineral salts and usually alkaline, is known as mild humus, or *mull*. By contrast raw humus, or *mor*, contains few nutrients and is acid.

In conditions of low oxygen content, as in waterlogged soils, or in highly acid soils, the numbers and activity of the decomposing fungi, bacteria, and fauna are much reduced, and in consequence they bring about the partial decomposition of only a small amount of the organic matter. Consequently this partly decomposed organic matter accumulates season by season as *peat*. The accumulation of peat is a very important factor in determining the type of vegetation that develops.

The general term *mire* refers to communities developed on peat-rich substrates. Mires show all gradations between communities developed in alkaline peat or *fens*, and those developed on acid peat or *bogs*.

Nitrogen is present only in small quantities in soil but it is very important for plant growth. Areas of local nitrogen enrichment are rare in nature but common where human and animal interference is strong. Certain communities such as those with nettles (*Urtica*), oraches (*Atriplex*), and goosefoots (*Chenopodium*) are examples. Nitrogen-fixing species such as members of the pea family, *Leguminosae,* also increase the nitrogen content of the soil.

Serpentine soils are rich in magnesium and have a characteristic vegetation, and are particularly widespread in southeastern Europe. Saline soils also have a very distinctive influence on the dominating plant communities.

An outline of the main types of soils in Europe now follows.

Brown-earth soils

These are developed over most types of rock and particularly those rich in clay, with a neutral or mildly acid reaction. They are characteristically developed under deciduous forests, in Central Europe and the drier parts of the Atlantic region, but they are not found north of latitude 60°. They are developed in relatively humid and temperate climates. Most brown-earth soils are fertile and are now largely cleared of forest and used for cultivating crops.

Brown-earths are so called because they contain ferric oxide and are in general brown in colour. The leaf litter from the forest vegetation forms a weak mull humus, which with the soil particles forms a loam with a crumb structure. The A horizon is consequently dark brown, by contrast with the yellow-to-reddish-brown iron-rich B horizon. Clay is also removed from the upper layers to some extent, and is deposited in the B horizon, giving the lower layers a somewhat heavier texture. The two horizons are not clearly differentiated however and merge one into the other. Also, as a result of leaching, calcium ions are largely removed from the A horizon and this is consequently mildly acid with a pH below 6.5.

Acid brown-earth soils occur in cooler and moister climates, and are usually developed under mixed deciduous–coniferous forests, or under mountain grasslands. Because of the higher rainfall they are more completely leached and have a higher acidity, while the upper soil layers are typically sandy loams. A mean annual rainfall of above about 1000 mm is roughly the transition point between brown-earth and acid brown-earth soil regions.

Black-earth soils or Chernozems

These soils develop largely under grasslands in gently undulating plains in Central and Eastern Europe. They are usually very fertile and at the present day much of this grassland is under cultivation. Black-earths are formed from *loess,* which is a fine dust originating from glacial deposits exposed after the ice retreated, or from the outwash of glacial streams. This dust was carried by winds over large areas of land surface and deposited to a greater or lesser depth in different localities.

Black-earths are developed in climates with cold winters and are commonly frozen to a depth of 60–80 cm by temperatures of −7°C to −10°C. A low rainfall or snowfall of about 550 mm in the year is also required and in summer the temperatures are relatively high, 19–21°C, usually with a period of drought in the late summer.

These soils have a deep and rich A horizon, often to a depth of 80–100 cm, which has an alkaline mull humus and a well developed crumb structure. The pH is neutral or slightly acid, primarily as a result of leaching by snow-melt water in spring and the early summer rains. During late summer, by contrast, because of the excess of evaporation of soil moisture over rainfall, there is an upward movement of soil water, resulting in the accumulation of calcium concretions and nodules in the lower layers of the soil profile at depths of about 90–180 cm.

The rapid growth of grasses and herbaceous plants in spring results in a rich yearly addition of fresh organic matter as the plants die down each summer. With the late summer drought and cold winters, decomposition of humus is retarded and there is a gradual increase in the accumulation of fertile mineral-rich humus, thus giving a black colour to the soil, and hence its name.

Rendzina soils

These are relatively shallow soils developed over calcareous rocks, such as limestones and chalk, in rather humid conditions. They develop under grasslands and deciduous woodlands. Only the A and C horizons are developed. The A horizon is leached by rainwater of soluble bicarbonates, and only the non-soluble soil particles remain, and these with a mull-type humus form a thin but mineral-rich alkaline layer usually only 10–20 cm in depth. This A horizon is commonly dark brown to blackish. The lower layers of the A horizon may contain limestone fragments. Below this lies the more or less unchanged calcareous rock. Some limestones may have a thin cover of *drift* transported there by wind or glaciers and these may weather into reddish-brown calcareous soils. Such soils have a similar profile to brown-earth soils but the lower horizons are alkaline.

Southern dry soils

Both soils described here are 'fossil soils'—that is to say they were formed in a previous age under different climatic conditions. Both *terra rossa* and *terra fusca* are neutral or acid and were mainly formed in the Tertiary period.

Terra rossa, or *red Mediterranean soil,* is a dry reddish soil, developed in southern Europe, under Mediterranean coniferous or evergreen woods, or under the sub-climax shrub vegetation of maquis or garigue which now largely replace the woods. They are bright-red clay soils, developed commonly over limestone, calcareous marls, karst, or less commonly over sandstones and shales. They occur in climates with mild winters averaging 5–15°C, with mostly winter and spring rainfall of about 500 mm, and hot dry summers with temperatures between 25 and 30°C. The length of the summer drought is important in the formation of these soils. Erosion is another important factor, resulting in the redistribution of soil which is transported down into the valleys where it accumulates, leaving bare rocky hills above.

The A horizon of terra rossa soils has a layer of mull humus; below this is the B horizon in which some clay accumulates, and also in some cases iron, thus further reddening the horizon. Below lies the C horizon of the unchanged bedrock or substrate, while at the junction of the B and C horizons may be a layer mixed with fragmented particles of the bedrock.

Terra fusca or *brown Mediterranean soils* are the southern equivalents of the brown-earth soils. Some are formed on non-calcareous rocks and have a rich friable humus layer, brown in colour. The B horizon has a denser but less friable clay layer. On soils developed over calcareous rocks, the lower A horizon is leached of calcium, and clay has been carried down into the B horizon. This clay has been redeposited in the B horizon and in fissures in the C horizon, often with the accumulation of iron, thus giving a reddish colour to the lower layers.

Podsols

These soils occur in areas of high rainfall, largely in the Boreal and Atlantic regions, but also in the mountainous regions of Europe. They are developed under coniferous forest or heath-land, and the organic matter which is deposited by these communities is slowly and only partially decomposed to produce acid *mor* humus. At the same time the heavy rainfall results in strongly leached upper layers, so that soluble minerals and humus are carried downwards and deposited in the lower soil layers, thus producing a distinctive soil profile with well marked horizons (See Fig. 14, p. 57).

In podsols the main horizons are

L   A layer of leaf-litter on the soil surface;

Ao   A fibrous layer with partially decomposed litter and containing roots of living plants;

Al   A dark usually black layer of mor humus;

A2   A layer from which humus, iron, and clay have been leached, leaving a bleached layer mainly of insoluble silica. This process is known as *podsolization;*

B1   A layer in which humus is deposited and is consequently black or dark-brown;

B2   A layer in which iron oxides accumulate and which is rust-brown in colour. This layer may become concreted into a hard impervious iron-pan. Clay may also be deposited in this layer;

C   The partly weathered rock or substrate;

D   The unaltered rock or substrate.

Not all of these horizons may develop under different conditions of climate, vegetation, and rocks. For example, in moorlands dominated by grasses, rushes, and bog mosses, no distinctive B1 horizon of humus is found, but it does have a well developed B2 horizon with a distinctive rust-coloured iron-pan. This is sometimes referred to as an iron-podsol, in comparison with a humus-podsol when the B1 horizon only is distinctive, or an iron-humus podsol when both B1 and B2 layers are well formed.

Gley soils

These are developed on valley bottoms, or on poorly drained sites, where there are changes in the water level from season to season. They are commonly developed under damp meadows, river-valley woodlands, or swamps. Owing to the lack of oxygen in the water-logged layers of the soil, ferric iron salts are reduced to ferrous salts thus giving a grey colour throughout the soil profile. This is known as *gleying.* In soils in which the levels of waterlogging change periodically, both reduction and oxidation of iron salts occur, and the soil layers subjected to these changes have an overall grey colour but are mottled with orange-to-rust-brown patches of ferric iron.

Peat soils

These are termed organic soils, for they are developed primarily from organic matter and contain none of the soil particles of the substrate over which they are formed. Peat is the partly decomposed remains of vegetation from heaths, mires, and swamps, where waterlogged conditions retard or prevent the oxidation by the decomposing soil organisms of the organic matter. Peat forms where the water table lies close to, or above the ground surface for at least much of the year. Peat consequently accumulates year by year and gradually increases in depth. In peat bogs it may accumulate to a depth of 6 m or more.

Peat may be formed in acid, neutral, or alkaline waterlogged terrain. *Bog peat* is acid in reaction, and is formed largely from bog mosses, *Sphagnum*; sedges; and cotton grasses, *Eriophorum*, in conditions where rainfall, which is deficient in mineral salts, is the main source of water. It is developed in cool wet climates, and in some cases, as in raised bogs, it may bring about the actual raising of the water level.

Fen peat by contrast is neutral or alkaline, and as it is largely flooded by ground-water it is relatively rich in mineral salts. Fen peat is largely derived from sedges, reeds, and other herbaceous species as well as from willows and alders. Again, the height of the water table, at or above the ground level, is a crucial factor in the development of fen peat. As in bogs, the anaerobic conditions reduce the activity of the soil organisms and prevent the complete decomposition of the plant remains; in consequence, fen peat accumulates from season to season. There are many gradations between bog and fen peats; collectively these are called *mires.* The main source of the water supply is crucial in determining the type of mire that developes. Rainwater is largely mineral-deficient and hence the peat that develops in heavy rainfall areas is usually acid. Drainage water on the other hand can be either mineral-rich or mineral-deficient depending on the rock or substrate from which it drains; consequently, either fens, intermediate mires, or bogs may develop in areas of less rainfall.

Saline soils

Maritime saline soils occur in close proximity to the sea, where, for example, there is periodic flooding at high tides, or where wind-blown sea water carries salt inland, or where under-water sediments become exposed above the sea level. Only a small number of species, from families of plants such as the goosefoot family, *Chenopodiaceae*, and some grasses and rushes have become adapted to live competitively in such high-salt-content soils—these are termed *halophytes.* They form the distinctive plant communities of salt marshes, mud flats, and sand dunes.

Inland saline soils commonly develop in plains and valleys where there are low rainfall and high summer temperatures; they occur largely in Central and Eastern Europe. As a result of the high summer temperatures and minimal rainfall, evaporation from the soil surface takes place, and soil water is drawn upwards, carrying salts towards the surface from the lower soil horizons. As the water evaporates, the salts are deposited in the upper soil layers and gradually accumulate producing a saline soil. Such inland saline soils carry a highly characteristic vegetation, composed of a mixture of some maritime halophytes and some distinctive inland halophytes.

Tundra soils

These are developed in the far north of Europe in low snow and rainfall areas, over a wide range of rocks and substrates. During the arctic summer, the upper soil layers thaw out but usually remain saturated with water, due to the impermeable, permanently frozen, *permafrost* zone beneath. As a result, wherever the substrate slopes, there is a gradual downward movement of the saturated soil—this is termed *solifluction*. The finer soil particles are transported further downwards, while the larger rock fragments remain in the upper part of the slope; thus there is a gradual separation of the different-sized soil particles, often in ridges or polygons.

Tundra soils develop under dwarf shrubs, largely of the heath family, *Ericaceae,* or under lichen or moss communities. Only in more sheltered or deeper soil sites do communities of low shrubs and bushes occur, largely of willows and birches.

Soil horizons do not develop in tundra soils as leaching is very limited and, as a result of the alternate freezing and thawing, there is a process of 'churning'. The upper layers of unfrozen soil freeze first with a fall in temperature before the lower unfrozen soil layers. An expansion occurs in the process of freezing and pressures are built up which eventually rupture the surface, with the consequence that the lower unfrozen layers may be forced up and spill over the upper frozen layers.

Raw soils

These are young soils which are little changed from the original rock or substrate. They show no regular accumulation of humus and no layering, and they are referred to as azonal soils and are not characteristic of any particular climatic zone. Some are commonly found on steep rocky slopes, particularly of calcareous rocks, in the montane regions. Some raw soils remain young by the constant removal of the upper layers of accumulating humus by rain, melt-water, or wind. Other raw soils are found on sands and sand dunes where excessive leaching of the pervious substrate results in the removal of humus and salts. Thus, though superficial horizons may quickly develop, they may not persist owing to soil mobility. In other cases they may show multiple profiles as a result of fresh sand being periodically deposited over the surface.

Alluvial soils also come under the heading of azonal soils. They are deposited by rivers and are periodically flooded with the addition of more soil material so that they are often layered. Such alluvial soils may form river terraces or flood plains; these can be very fertile and are in most cases under cultivation. In general alluvial soils vary considerably in texture, drainage, and state of maturity.

Mountain soils are also azonal. They are shallow soils developed over the parent rock, and they are often rich in organic matter. But because of soil erosion, or *slip,* on mountain slopes, they do not develop distinctive profiles. Under climates with heavy cloud and mist or rain, peat may develop; elsewhere the soils may be gleyed with mottled areas of ferrous and ferric iron. Other important factors in the formation of mountain soils include aspect in relation to the sun, altitude, rock formation, slope, and soil drainage. The vegetation associated with mountain soils is very distinctive. It is composed of many unique species that are only adapted to grow in the mountains and in mountain climates. Mountains also form refuges for many species which had a wider distribution in the past. Some mountain species are also found in the northern regions, and these are descibed as arctic–alpine species.

3 Climates of Europe

Climate has a profound effect on the development of vegetation over the earth's surface. The tolerance of species to temperature, rainfall, and seasonal change varies enormously, and so do the plant communities which grow under these different climatic conditions.

In consequence of this we have described most of the types of vegetation found in Europe primarily in their climatic zones or regions. These are

Arctic
Boreal
Atlantic
Central European
Mediterranean
Pannonic.

Each has a very different climate and distinctive plant communities. However there are many cases where one type of plant community may occur in more than one of the climatic

zones; also there are regions where intermediate climates and communities occur.

In other cases, certain types of vegetation occur where the local environmental conditions have precedence over the main climates of each climatic zone. The vegetation of the alpine regions, of the wetlands, and of the coastal regions in Europe are examples. These are referred to as azonal communities and show many similarities over wide geographical and climatic ranges.

Climate consists primarily of the prevailing weather conditions as a result of the interplay of temperature, barometric pressure, winds, sunshine, rainfall, cloud, and humidity. Other important factors determining climate are latitude, altitude, and distance from the sea.

In Europe the climatic range is exceptional and unlike any other region of the globe. This is largely due to the North Atlantic Drift which carries warm waters northwards to the western coasts of Europe. Nowhere else on the globe do warm

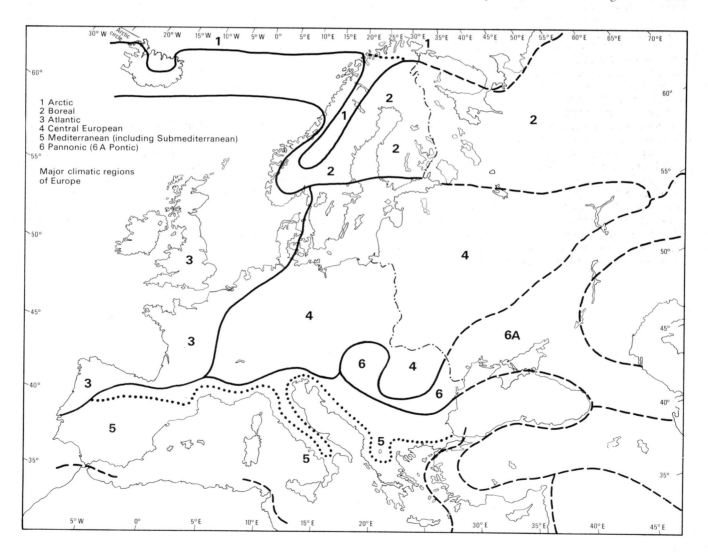

1 Arctic
2 Boreal
3 Atlantic
4 Central European
5 Mediterranean (including Submediterranean)
6 Pannonic (6 A Pontic)

Major climatic regions
of Europe

Climates of Europe

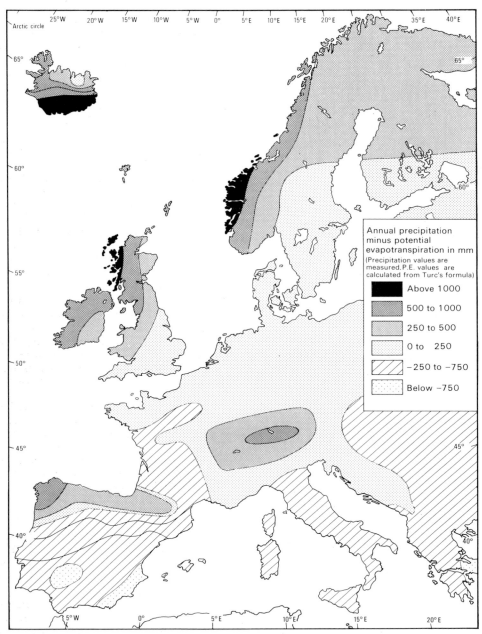

Annual precipitation minus potential evapotranspiration in mm
(Precipitation values are measured. P.E. values are calculated from Turc's formula)

Above 1000
500 to 1000
250 to 500
0 to 250
−250 to −750
Below −750

currents penetrate so near to the North Pole and maintain unfrozen seas so far north of the Arctic Circle as in northern Scandinavia. Also, in the northern hemisphere the prevailing winds are westerly, and they carry inland relatively warm moist air from the Atlantic ocean eastwards across Europe in a series of depressions. As there are no large mountain ranges running north–south in most of Western Europe, which would otherwise act as barriers to the passage of these depressions, they frequently penetrate deep into Central Europe. While the Mediterranean climate is again quite different, yet it has parallels with other regions of the world.

Largely as a result of these oceanic conditions in Western Europe, the equable climate is unique in having an average temperature difference of less than 11.5°C (20°F) between the average winter and summer temperatures. In contrast, in Eastern Europe, which is under the influence of the continental high-pressure zone, this temperature difference may be as much as 24°C (43°F), as, for example, in Belgrade.

Many depressions—on average 100 a year—travel inland from the Atlantic into Central Europe at all seasons of the year. They cause sudden changes in weather, bringing in clouds, wind, and rainfall but under relatively mild conditions.

In winter these depresssions are strong and cause very variable weather in most countries of Western Europe. There are, on average, up to 20 depressions passing in from the Atlantic in January, while in summer the depressions are weaker and occur, on average, two to three times each month.

In Eastern Europe, in contrast, the high-pressure system prevails and the winters are cold with one or more months of freezing, while the summers are relatively hot and dry.

In the Mediterranean region the climate is quite different. During the winter months frequent depressions travel inland from the Atlantic passing eastwards, and these bring to the Mediterranean region variable weather which is mild and humid with considerable rainfall. In summer the contrast is dramatic. The few weak depressions from the Atlantic pass to the north or to the south of the Mediterranean, thus having little influence on the Mediterranean itself, so that lands bordering the sea have hot dry summers with several months of drought and with maximum sunshine.

Thus in general, the winter isotherms run from south to north in most of Europe, with temperatures declining strongly on passing inland from the Atlantic seaboard to Central and Eastern Europe. In winter the January isotherm of 4.4°C (40°F) passes through the northern Adriatic and Aegean seas in the Mediterranean; yet in Western Europe the same isotherm runs north of the Shetland Islands. The −1.1°C (30°F) isotherm runs south of the Danube in Romania and northwards up through western Scandinavia and crosses Iceland. In the winter, maritime air is unable to penetrate deeply into continental Europe, being blocked by the continental high-pressure system of Central and Eastern Europe with its cold, dry, easterly air streams. In the summer months the reverse is the case, and the Atlantic depressions are able to penetrate far into Central Europe.

In contrast to the south-to-north running winter isotherms, the summer isotherms run from west to east. Thus on proceeding northwards in summer, the average temperatures become progressively lower. The warmest parts of Europe in summer, where temperatures average over 26.7°C (80°F), are the inland areas of southern Spain, Italy, and Greece. In the extreme north of Scandinavia, the summer average falls below 10°C (50°F); this underlies the great contrast between the summer of the Mediterranean region and that of the Arctic and this contrast has a profound effect on vegetation.

Arctic climate

In general terms, this is a climate in which the summer months have an average temperature of less than 10°C (50°F). There is continuous daylight for over two months with little or no freezing during this period, though the subsoil remains permanently frozen—the *permafrost* zone. During winter there are months of permanent freezing and snow cover and for two months there is no sunlight. Rainfall is low, averaging usually about 250 mm (10 inches) during the year, and much of it occurs as snow. The cold winter winds are strong. Such an arctic climate occurs in the extreme northern part of Scandinavia, northern Iceland, and northern USSR, while a similar climate also occurs in the arctic–alpine mountain regions of Scandinavia. The growing season for plants is consequently very limited and, in general terms, growth of herbaceous plants only occurs when the daily temperature exceeds about 6°C (43°F).

Boreal climate

This is a climate that is profoundly influenced by the polar air masses of northern Eurasia, and it is relatively little affected by the warmer moisture-laden Atlantic winds. The Scandinavian mountains, running from north to south, prevent the penetration inland of this Atlantic influence.

In consequence, the climate of the Boreal region is very different from that of the Atlantic region lying only a hundred or so miles to the west. The Boreal winters are cold to very cold, with average temperatures of less than −7°C (21°F). The whole of Finland, except the coastal strip, is covered with snow for nearly six months, while the Gulf of Bothnia is frozen for four to six months. In northern Sweden snow cover lies for four to six months or even up to seven months in some localities.

Summers by contrast are relatively warm, considering the latitude. The average temperatures of the warmest months may be as high as 15.6°C (60°F). Rainfall is relatively low and rarely exceeds 500 mm (20 inches), except in the mountainous regions, and it occurs largely in late summer.

The growing period for plants is thus very limited in the Boreal region. For example, in northern Sweden there may be only up to 100 days in the year when the average temperature rises above about 6°C (43°F) when most herbaceous plants are able to commence growth. Thus only perennial frost-hardy species can survive and be competitive. In central Sweden the growth period may be 140–160 days, while in the extreme south, due largely to the Atlantic influence, it may be up to 200 days or more. Here some Atlantic and continental plant communities may become dominant.

Atlantic climate

The Atlantic climate of Europe is quite distinctive. Not only is it one of the cloudiest regions of the world, but the cool, moist, windy conditions along with low temperature-fluctuations result in a climate strongly contrasted with the more continental climates of much of Europe.

A succession of depressions bringing rain and cloud passes inland from the Atlantic throughout the year. Thus the coastal and western land masses are strongly influenced by the relatively cool temperature of the sea in the summer, and the relatively warm sea in the winter. This influence is felt inland as far east as southern Sweden, western Poland, and the western Alps throughout the year. Only in summer do the depressions travel as far eastwards as the USSR. The mean temperature range in the western Atlantic is less than in any other region of Europe; thus the average temperature difference between winter and summer is only 8°C (14.5°F) in western Ireland, 11°C (20°F) in eastern England, and 14°C (25°F) in eastern France and Germany.

The strong westerly winds from the Atlantic carry inland

cloudy, damp air with fog and mist; in consequence there is little snowfall and no freezing of the sea as far north as the North Cape in Scandinavia. This contrasts strongly with the frozen seas of the Gulf of Bothnia and the Baltic Sea much further south.

In summer the average temperatures are relatively low owing to the changeable weather. Thus the July averages are: for Bergen 15°C (59°F); London 16°C (60.9°F); Paris 18.1°C (64.6°F); Santiago 18°C (64.3°F). The January means for these localities are: Bergen 1.2°C (34.2°F); London 3.8°C (39°F); Paris 2.4°C (36.7°F); Santiago 7.3°C (45.1°F).

Rainfall occurs throughout the year, with minor maxima in winter and autumn, but the amount of rainfall varies considerably from place to place. Thus in the lowlands in general it may range from an average of 580 mm (23 inches) in Paris, 600 mm (24 inches) in London, 780 mm (30.7 inches) in Bordeaux, 787 mm (31 inches) in Trondheim, to as much as 2134 mm (84 inches) in Bergen and 1651 mm (65.2 inches) in Santiago. As evaporation is relatively slow due to low soil and air temperatures, there is usually ample soil moisture for active and rich plant growth throughout the spring and summer. In certain mountainous areas however the rainfall is very heavy, as for example in the mountains of Scotland and Scandinavia. Here rainfall may exceed 3048 mm (120 inches) a year on the western side, while in the lee of the mountains on the eastern side the rainfall may decrease rapidly.

In consequence of this variable, humid western Atlantic climate the amount of sunshine received in Western Europe is relatively low, with a mean of about 1700 hours a year in southern England for example.

Central European climate

Central Europe is largely dominated by the high-pressure system of Eastern Europe; thus the general climate is considerably colder in winter and warmer in summer than in the Atlantic region. The western parts of Central Europe are somewhat under the influence of the westerly oceanic winds, for in the summer a few weak depressions may penetrate as far east as western USSR. But, at most other times of the year, the continental high-pressure system blocks depressions from the west, and a truly continental climate prevails.

Thus in winter East Germany may have prolonged periods of bitterly cold, crisp, clear, calm air, while at the same latitudes in France and England a succession of depressions carries warm, moist air with much rain inland from the Atlantic.

The January mean temperatures of some Central European localities are as follows: Warsaw −3°C (26.5°F); Berlin −5°C (23°F); Vienna −2°C (28.5°F); Belgrade −1.7°C (29°F); while the lowest winter temperatures of southern Europe, −3.9°C (25°F), occur in the Danubian plains. There may be one to three months of continuous freezing and snow in much of Central Europe and many of the rivers are frozen over for a month or more.

The summers by contrast are relatively hot, with July temperatures averaging: in Warsaw 18.5°C (65°F); Berlin 18°C (64°F); Vienna 19°C (66°F); and in Belgrade 22.2°C (72°F); with maximum summer averages in the Danubian plains of 26.7°C (80°F). Thus the average temperature difference between winter and summer is usually over 19°C (35°F).

The average rainfall in Central Europe is low and decreases eastwards. It is rarely more than 635 mm (25 inches), with again the smallest amount of rainfall occurring in the Danube plains—about 380 mm (15 inches). Precipitation occurs throughout the year, but largely as snow in winter, and there is a small maximum of rain in the summer; there is no really dry season, except in the Danube plains. By contrast, in the mountain ranges, the rainfall is more than 1400 mm (55 inches) and it may be as much as 3000 mm (120 inches) locally in the year.

Pannonic climate

This is the driest region of Europe, and it has a large winter–summer temperature range of over 19°C (35°F). The total rainfall is low, averaging under 450 mm (18 inches), and most of the rain occurs in May and June, largely from thunderstorms. As a result of these heavy rainstorms, water run-off is rapid and relatively little penetrates the dry hard ground. At the same time, owing to the high summer temperatures, evaporation takes precedence over precipitation and, in consequence, salts are brought up to the surface and deposited there as the soil water evaporates. This results in the accumulation of salts in the surface soil, producing a salinity which has a profound effect on the natural and semi-natural vegetation which occurs in the pannonic plains.

July and August are the driest months; there is a summer drought and consequently plant life is dormant or inactive at this season. Plant growth largely occurs in the spring, or in the autumn before the onset of the prolonged winter freeze. The average July temperature in Bucharest is 22.8°C (73°F), while in January the average temperature falls as low as −3.9°C (25°F).

Mediterranean climate

This climate is the most distinctive and clear-cut of all European climates, though there are considerable local variations. The winters are mild and moist, with average temperatures of 6°C (43°F), while the summers are sunny and cloudless with average temperatures of 21°C (70°F), and with one or more months without rain. In consequence of these warm rainless summers and mild winters, the natural vegetation is quite distinct. It is largely dominated by evergreen trees and shrubs, and has many herbaceous species which die down during the summer, and a rich assortment of annuals flowering in the autumn and spring. Thus, unlike any other region of Europe, the main periods for plant growth are the autumn with the beginning of the rains and the spring. The Iberian and Balkan peninsulas have a greater number of species of native flowering plants than any of the other regions of Europe.

The temperature range and rainfall of different Mediterranean localities are shown at the top of the next page.

	January	July	Rainfall
Gibraltar	12.8°C (55°F)	22.8°C (73°F)	907mm (35.7in)
Marseilles	6.7°C (44°F)	22.2°C (72°F)	574mm (22.6in)
Rome	7.2°C (45°F)	24.4°C (76°F)	831mm (32.7in)
Athens	8.9°C (48°F)	26.7°C (80°F)	394mm (15.5in)

The change of climate begins in the western Mediterranean at the end of September, when depressions from the Atlantic penetrate deep into the Mediterranean and bring variable cooler weather with clouds and rain. Temperatures gradually fall as the season progresses and the relatively warm waters of the Mediterranean begin to cool. Thus in Naples the mean October temperature is 18.3°C (65°F), while in March the mean temperature falls to 11.1°C (52°F). Depressions enter the eastern Mediterranean later in the year, and are not so pronounced or numerous as in the west. Over a distance of about 2000 miles from west to east, the average winter temperatures differ from about 5–10°C (41–50°F) in the west, to 13.3°C (56°F) in the east.

Rain falls throughout the autumn, winter, and spring, but during relatively few days—a total of rarely more than 100 days in the year. The rainfall is irregular, with many short heavy storms, followed by periods of clear weather. Rainfall decreases as you go eastwards in the Mediterranean, while the driest parts of the Mediterranean region are the interiors of the land masses, such as the meseta of Spain and some of the inland regions of Yugoslavia and Greece.

The three peninsulas of Iberia, Italy, and Greece, with their mountain ranges lying inland from the Mediterranean sea, have an effect on the rainfall locally. One of the heaviest rainfall areas in Europe is the southern part of the Dalmation coast of Yugoslavia, where there may be as much as 4500 mm (183 inches) of rain or more in the year, while opposite at the same latitude on the Adriatic coast of Italy, the rainfall is as low as 508 mm (20 inches). The mountain ranges act as barriers to the prevailing westerly winds and much rain carried by the depressions is deposited on the western mountain slopes, thus causing a rain shadow on the eastern side where there is markedly less rainfall. For example, in Skopje in Macedonia the rainfall averages 480 mm (19 inches), and occurs on relatively few days in the year.

The summer, by contrast, has an entirely different climate. It is distinguished by bright clear days with a minimum of cloud and maximum sunshine and by the almost complete absence of rain for several months. The subtropical anticyclone of the Azores penetrates deeply into the western Mediterranean over the relatively cool waters, and it deflects any depressions coming in from the Atlantic to the north or south of the Mediterranean sea. A few depressions may enter the west but none penetrates as far as the eastern Mediterranean, which is influenced by the high-pressure air mass from further east. Thus the summer climate is almost entirely continental.

In general, in the summer, the winds over the whole area blow from the north and they may reach considerable force and constancy. They blow from mid-May to mid-October, and they are known as the Etesian winds. Locally in the eastern Mediterranean they are known as the *meltemi* and they blow steadily from the north during the daytime at speeds of 10–30 miles per hour, or sometimes up to 45 miles per hour. Other local winds are the *mistral* which blows from southern France seawards for an average of 103 days in the year, and the *bora* which blows down from the karst of Yugoslavia across the Adriatic, sometimes attaining speeds of up to 100 miles per hour. Hot winds also blow from the south from North Africa, such as the *sirocco* from Algeria, and the *leveche* northwards across Spain. They are hot dry dusty winds which may blow for up to 50 days in the year and which have a pronounced effect on the vegetation.

Sunshine is nearly continuous during the summer. Southern Spain, southern Italy, and southern Greece may have a total of up to 2500 hours of sunshine in a year. Such a contrasting and unique type of climate has favoured types of vegetation that occur nowhere else in Europe. At the same time man's early colonization of the Mediterranean shores has resulted in his prolonged and intensive influence on the vegetation of the region. Consequently, at the present day little remains of the natural plant communities.

The influence of mountain ranges in Europe

These have a considerable effect on the local climates of Europe and, in consequence, on the vegetation. Thus the long north–south mountain ranges of Scandinavia not only result in the tundra vegetation penetrating further south than elsewhere, but they also allow the Atlantic climate of the North Sea to penetrate further north than anywhere else in the world. The Alps have a very marked effect on the climates of Central Europe and the diagram (see Fig. 19, page 72) shows the characteristic vegetation at the different altitude zones. These zones are as follows:

Planar. Lowlands, largely flat country with originally climax woodlands such as oak–hornbeam. They are now mostly well populated and converted into agricultural land.

Colline. The hill zone, originally with natural woodlands of mixed oak woods and, on drier south-facing slopes, steppe-heaths. Agricultural land predominates, particularly with viticulture, to an altitude of between 300 and 500 m.

Submontane. The lower montane zone dominated typically by beech woods with some oak–hornbeam and Scots pine in the central continental Alps, and with planted sweet chestnut in the southern region. Cultivation is still important to the upper level of about 800 m.

Montane. This zone contains primarily beech woods mixed with silver fir, spruce, and larch. Fir and spruce increase at the higher levels and meadows and pastures become more abundant. Cultivation is still profitable at lower levels, and the general range of this zone is 1600–2000 m.

Sub-alpine. This zone is typified by the zone of dwarf shrubs which occur above the tree line and are protected by snow cover in the winter. They may also occur lower down in the forest zone where the forest has been cleared, or on exposed slopes. The range of this zone lies between 2000 and 2300 m.

Alpine. The zone of alpine grasslands and heaths ranging from about 2400 to 3260 m.

Climates of Europe

Nival. This is above the snow line where there are limited habitats where a few plants can survive—particularly mosses and lichens.

Note: the sub-alpine and alpine zones are described in the Alpine section.

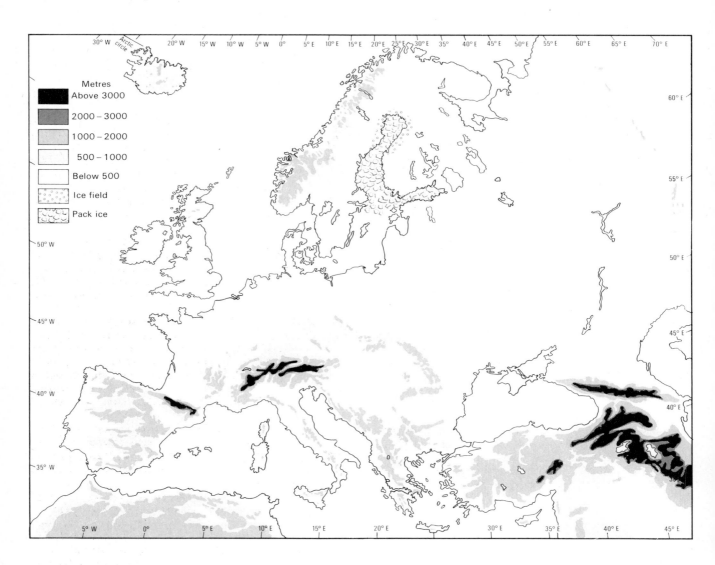

4 Recent history of the development of vegetation in Europe

Our knowledge of the vegetation of the past is based primarily on two types of evidence; that from macroscopic fossil remains, and that from pollen analysis. Macroscopic remains are very localized and fragmentary, and are commonly laid down only under certain very restricted conditions. However they are usually located where the animals and plants lived and consequently give us a potentially clearer insight into the local communities of the past. Pollen grains by contrast can be transported much further afield and are more dispersed. On the other hand, they not only enable us to identify the species sometimes, but pollen counts also give the relative frequency of species. Pollen grains have a very resistant cuticle which decays very slowly, and the cuticles of different species often have very distinctive sculpturing so that the pollen grains can be positively identified and related to present-day living species. Thus reliable information can be obtained, not only about the species present in past communities, but also about their relative abundance. Pollen analysis—palynology—has been the discipline most responsible for revealing the composition of vegetation over something like the last two million years.

Another important technique is radio-carbon dating. This operates over a span of something like the last 70 000 years. The proportion of radioactive carbon present in a given sample of organic matter can be measured and, since its decay rate is known, a comparison of this proportion with the equivalent proportions in atmospheric carbon dioxide can give an estimate of the age of the sample. Thus an accurate estimate of the time at which the organic material was deposited can be made within this period.

Evidence from other sources, such as changes in sea level, and identification of insects, all help to build up the information on vegetation cover of the past, and this becomes progressively more detailed as we come nearer to the present day.

Our knowledge of the vegetation of the Tertiary period, about 65 to about 2 million years ago, is very fragmentary. The earliest known fossil remains of plants from this period in Europe were similar in some ways to plants found today in the tropical lowlands of the Indo-Malaya region. They included palms, etc. Another type of vegetation was of woody plants characteristic of the warm-temperate and subtropical climates of today, which included trees like *Magnolia*.

Following these tropical and subtropical periods in Europe, there was a progressive cooling with distinct changes in the vegetation.

Fossil evidence from Poland from the late-Tertiary period includes many species which are characteristic of the modern European flora, as well as many other species present today in the mountains of western China, eastern Tibet, and Japan. These include such trees as spruce, *Picea;* hornbeam, *Carpinus;* wing-nut, *Pterocarya;* tulip-tree, *Liriodendron;* and hemlock, *Tsuga*. Many of these trees later became extinct in Europe during the further climatic changes which followed at the end of the Tertiary period, though they are still present in North America and Asia. Later still there were further changes

towards the development of the contemporary European flora, with the presence of oak and coniferous forests with pine, spruce, and fir, and with such trees of warmer temperate climates as hickory, *Carya;* and walnut, *Juglans,* as well as *Pterocarya* and *Magnolia*. By the end of the Tertiary period most eastern Asiatic species had disappeared from Europe, and by the beginning of the Quaternary period there were few genera present which do not now occur in the modern European flora.

The Quaternary period commenced about one to two million years ago. Fossil deposits show that at the beginning there was a temperate vegetation with mixed deciduous and coniferous woods with oak, hemlock, and alder, and pine and spruce. During cooler periods heaths with ericaceous species were present, while in the cold glacial phases tundra communities with dwarf willows, dwarf birch, and with herbaceous species such as saxifrages and mountain sorrel, *Oxyria,* etc, occurred.

As the climate improved, birch and pine woods appeared. During the warmest periods, oak, elm, and ash formed woods, commonly with hazel. During the Quaternary period there were probably many ice ages, which profoundly affected the vegetation of Europe; the last three of these were very pronounced.

The most severe of these developed ice sheets as far south as Kiev in Russia and the Severn and Rhine estuaries, and also in the Alps and Central European mountains. Following this there were warmer periods and it was during these periods that the first flint artefacts appeared, indicating the presence of Palaeolithic man. Such woody plants present as ivy, *Hedera;* yew, *Taxus;* and holly, *Ilex,* indicated a warmer interglacial period.

The second to last major glaciation commenced about 150 000 years ago and, at its height, a distinctive tundra vegetation characterized by mountain avens, *Dryas octopetala,* covered much of Europe, with the elimination to a large extent of birch and poplar.

The final major glaciation known as the Weichselian, or in the Alps the Würm, commenced about 110 000 years ago and lasted to about 10 000 years ago with its maximum extension occurring about 20 000 years ago. The ice sheet covered parts of northern Germany but did not extend as far south as the Netherlands and only occurred in eastern Britain north of the Thames, with ice-free areas in northern Britain, Scotland, and south-western Ireland. In the Alps the ice extended downwards in glaciers to the European plain, to the level of Munich and Grenoble. At the same time the sea-level became 90–100 metres lower than it is at the present day, as a result of the water being 'locked-up' in the polar and alpine ice sheets. The North Sea and the Irish Sea were areas of land at that time, so that the vegetation of Britain and Ireland was in direct contact with the vegetation of continental Europe and there were no barriers to the recolonization of the northern areas as the ice retreated.

Thus during the maximum extension of the ice, the tundra-like vegetation of southern Europe was composed largely of

Recent history of development of vegetation in Europe

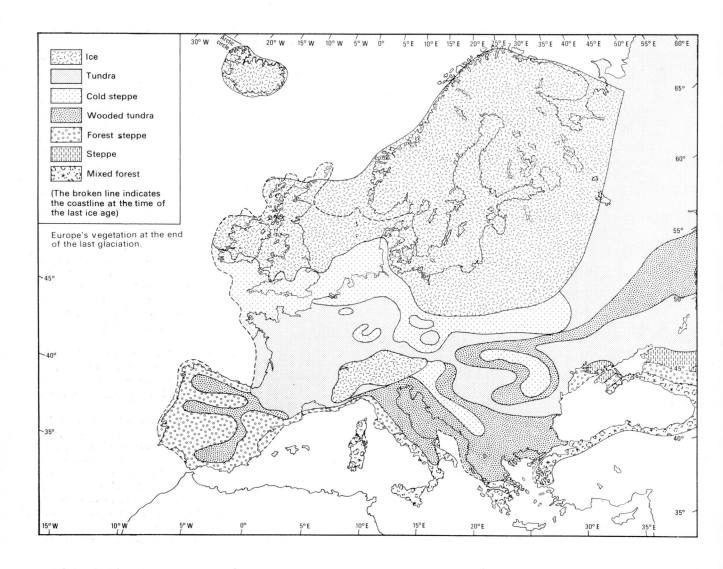

Legend:
- Ice
- Tundra
- Cold steppe
- Wooded tundra
- Forest steppe
- Steppe
- Mixed forest

(The broken line indicates the coastline at the time of the last ice age)

Europe's vegetation at the end of the last glaciation.

grasses and sedges, with dwarf birch, *Betula nana;* dwarf willows such as *Salix herbacea* and *Salix polaris,* and juniper. Also many present-day arctic−alpine species were widespread, such as: purple saxifrage, *Saxifraga oppositifolia;* alpine meadow-rue, *Thalictrum alpinum;* and mountain avens, *Dryas octopetala,* and, in saline areas, such plants as the sea pink, *Armeria maritima;* sea-blight, *Suaeda maritima;* and other salt-tolerant species. The northern Mediterranean region was at this time covered with dry treeless steppe. A later glaciation occurred about 15 000–13 000 years ago, and this was succeeded by the period from the final retreat of the ice about 10 000 years ago to the present day. The main post-glacial periods occurring during the last 10 000 years approximately are

Pre-Boreal	10 000 to 9500 years ago
Boreal	9500 to 7500 years ago
Atlantic	7500 to 5000 years ago
Sub-Boreal	5000 to 2500 years ago
Sub-Atlantic	2500 to present day.

During the pre-Boreal period the ice was rapidly melting and retreating, and as a result, the level of the sea was rising. First the North Sea was flooded and Ireland became separated from Britain, and then about 7500 years ago (at the end of the Boreal period) Britain was separated from the continent by the flooding of the English Channel.

The isolation of these land masses profoundly influenced the re-establishment of continental species in the vegetation of Ireland and Britain as the climate ameliorated. This largely accounts for the relatively small numbers of species which are native in these two countries, by comparison with the number of native species present in the adjacent European countries.

Subsequent to this land separation, a small number of species were able to migrate across the sea barriers, but many more were introduced by man, either accidentally or intentionally.

Further south in Europe, mountain masses such as the Pyrenees, Alps, and Carpathians acted as partial barriers to the migration of species northwards, but there were in some places lowland corridors through which migration took place, as for example south of the Pyrenees and through France, and in the east through the Danube valley and Poland.

During the pre-Boreal period, pollen analysis shows a rapid increase of juniper followed by birch and later an influx of pine from the warmer south. Light woods developed in the northern parts of Europe.

In the Boreal period warmer-loving deciduous trees and bushes, such as oaks and elms, and in particular hazel, *Corylus avellana*, became abundant in northwestern Europe. Forests of pine and birch and thickets of hazel became widespread in the early-Boreal period.

The Atlantic period was the warmest and most humid of the post-glacial periods. Oaks and wych elm, *Ulmus glabra*, became abundant and formed forests, and at the same time there was a marked increase in alder, *Alnus glutinosa*. Limes, *Tilia* species, had their widest distribution in England at this time, indicating the warmer climate, while in Scandinavia there was a marked increase of pollen of ivy, *Hedera helix;* and the great fen-sedge, *Cladium mariscus*, in Sweden. The distinctive increase of peat deposits and raised bogs also indicated heavier rainfall and greater humidity.

Mixed oak forests, with elms and limes became the climax forests and covered large areas of Europe. They were well established in Britain for example, to altitudes of at least 750 m, while species of open habitats became much less evident. Coastal plants like sea buckthorn, *Hippophae rhamnoides;* thrift, *Armeria maritima;* and sea campion, *Silene vulgaris* ssp. *maritima*, survived in local refuge areas. Similarly in the mountains, such widespread late-glacial and pre-Boreal species as the dwarf birch, *Betula nana;* mountain avens, *Dryas octopetala;* spring gentian, *Gentiana verna;* and alpine meadow-rue, *Thalictrum alpinum*, continue to survive in 'refuges' to this present day, far south of their present main distribution. Also such open-habitat species as docks, *Rumex* species, and wormwoods, *Artemisia* species, showed much reduced pollen counts at this period, only to return abundantly later when man-made habitats began to develop.

The sub-Boreal period commenced with a steep fall in the percentage of elm pollen in northwestern Europe. At the same time pollen counts indicate a notable increase of many herbaceous species, including ribwort plantain, *Plantago lanceolata;* chickweeds, *Cerastium* species; knotgrasses, *Polygonum* species; docks, *Rumex* species; wormwoods *Artemisia* species; members of the goosefoot family, *Chenopodiaceae;* and bracken, *Pteridium aquilinum*. Also there was an increase of such pioneering trees as ash, *Fraxinus excelsior;* and birch,

Betula species, which were usually among the first trees to recolonize areas cleared of forest.

All these changes were probably due to the activities of Neolithic and early Bronze Age man, who by this time was keeping livestock and carrying out shifting cultivation. Forests were cleared locally, and for several years crops were grown in these clearings. At the same time invasion of weed species occurred. However after a few years the soil fertility became much reduced and these cleared areas were abandoned. They were quickly recolonized by such trees as ash and birch, to develop in time into the mixed climax forest. This clearance occurred particularly on lighter soils, and there was a marked increase of such spiny bushes as hawthorn, *Crataegus;* and gorse, *Ulex*. The expansion of grasslands and heaths also occurred at this time.

The sub-Atlantic period more or less coincided with the Iron Age in northwestern Europe, when for the first time permanent fields were established. There was at this time a marked (real or apparent) decline in ash, *Fraxinus excelsior;* and limes, *Tilia* species, and this is thought at least in part, to be due to the process of pollarding, when branches of these trees were regularly cut for fodder for man's domestic animals. Beech and hornbeam, by contrast, increased at this time, but there is no clear indication as to why this occurred.

In general the climate was cooler and wetter in this sub-Atlantic period than in the preceding period. There was a considerable increase in peat-covered areas, and forest was replaced by bogs and fens, particularly in the more northerly parts of Europe.

This brings us to the present day when little, if any, of the natural vegetation remains, largely due to the activities of man and his livestock. Only in remote or special sites, such as mountains, lakes, marshes, etc. can we find anything like the natural vegetation which once covered Europe before the influence of man. His influence began to be felt in Neolithic times about 5000 years ago. About this time man developed, at least in northern Europe, the polished stone-axe which enabled him to cut down forest trees and clear areas for cultivation, and this marks the emergence of man as the potentially dominant factor in the ecosystem.

During the last 100 years or so sophisticated agricultural processes using modern machines, and involving mass production of many agricultural products, have further rapidly decreased the natural and semi-natural vegetation of Europe.

In the last few decades man has at last begun to realize the destruction that he has brought about, and in a small way he is beginning to think of how some of the last remaining tracts of natural and semi-natural vegetation can be saved from destruction.

The establishment of National Parks and Nature Reserves in the countries in Europe is a small step towards conserving what remains of the natural plant communities, on which all living organisms ultimately depend.

PART II
Plant communities of Europe

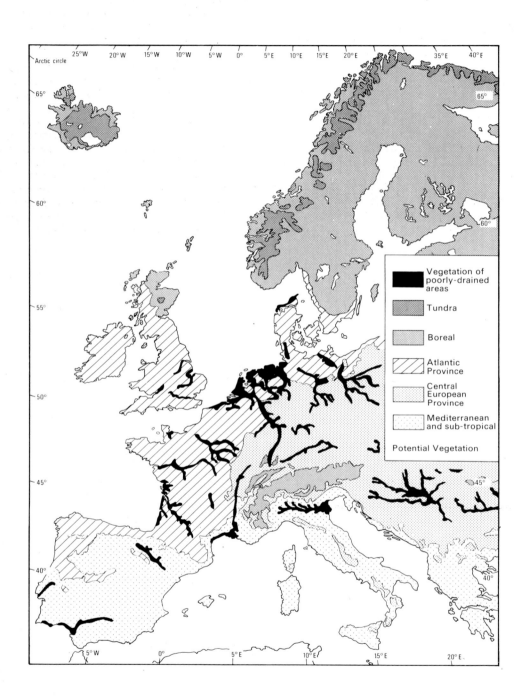

Vegetation of poorly-drained areas

Tundra

Boreal

Atlantic Province

Central European Province

Mediterranean and sub-tropical

Potential Vegetation

5 Introduction

This book is organized in regions, based largely on climates of Europe. Each region has a characteristic vegetation, e.g. Arctic tundra, Boreal taiga, Atlantic heaths, Central European deciduous forests, Mediterranean evergreen forests, Pannonic steppes, etc., and these divisions, although to some extent arbitrary, are logical as far as differences in the 'natural' vegetation are concerned. These vegetational types are termed zonal. However, types of vegetation may be common in more than one region, e.g. beech woods are highly characteristic of the Central European region but also occur in the Atlantic and montane–Mediterranean regions.

Certain other types of vegetation however have such considerable similarities over wide areas in Europe, that it is more logical to consider these in separate categories, such as alpine, wetlands, and coastal communities, though again there are considerable regional variations. These communities are termed azonal.

The map (p. 20) shows the whole of Europe divided ʋ into these regions, and indicates the natural vegetation that each area would support. This is the potential natural vegetation, and represents what the dominant vegetation would be without human interference.

The more extensive this human interference has been, the more the actual vegetation we encounter departs from the natural state. These six regions result from a convenient splitting-up of Europe into areas with significantly differing climates and hence vegetation. The true Arctic climate, for example, carries highly characteristic plant communities, whilst the moist Atlantic climate affects plant growth from the west of Norway right down to north-west Spain and Portugal and many species extend over wide areas in this region.

In general, the extent of human destruction of the natural vegetation increases as one moves southward, or descends from the mountains. Thus natural and semi-natural communities are widespread in northern Scandinavia, but almost non-existent in the Mediterranean region.

6 Arctic vegetation (Plates 1–3; Plates 61–68)

Arctic vegetation is quite distinctive. It is found north of the forested Boreal zone and is largely composed of dwarf evergreen shrubs mostly of the heath family, and of sedges and rushes, mosses and lichens, with grasses being of lesser importance. These plants are able to adapt themselves to the very adverse northern growing conditions, which include low temperatures, short or very short growing season, snow cover for long or short periods, and often extreme exposure to cold winds, as well as frozen soil or *permafrost*.

Arctic vegetation occupies a narrow zone in northern Norway, but further east in northern Russia this zone becomes much more extensive. It also extends down the mountain ranges of Scandinavia, and includes the high plateaux of Iceland (by many, these would be termed the sub-arctic zones) (Map 1). The main types of vegetation are tundra, arctic-alpine communities, arctic woodlands, arctic mires, and arctic grasslands.

Tundra. These are the treeless undulating plains, which are covered in snow for up to six months or more, with short but relatively mild summers of 60–120 days where the July mean is less than 10°C. Permafrost, the permanently frozen soil layers, predominates and only the upper soil layers thaw and dry out to a greater or lesser extent in the summer months.

The main types of tundra are moss-tundra, lichen-tundra, dwarf-heath tundra, and arctic–alpine tundra. So often these different tundra types occur as a mosaic, dependent on local variation in soil, soil moisture, exposure, retreat of snow cover, etc. (Plate 61).

Moss-tundra is little developed in Europe. It occurs in damp areas, often forming peaty mounds interspersed with puddles of water. The mosses include species of *Dicranum, Hylocomium, Hypnum, Polytrichum, Racomitrium,* and some liverworts and lichens.

Lichen-tundra occurs on drier and shallower soils, where thin layers of peat accumulate. The *fruticose* lichens (Plate 62)—those in which the plant is attached by a single basal stem and which is usually copiously branched above—often occur as a compact carpet, and give a greyish or whitish appearance to the landscape. This is particularly characteristic on the plateaux of northern Norway. The following lichens are typical.

> *Coelocaulon divergens*
> *Alectoria nigricans*
> *Alectoria ochroleuca*
> *Cetraria cucullata*
> *Cetraria ericetorum*
> *Cetraria islandica*
> *Cladonia rangiferina*, reindeer moss
> *Cladonia stellaris*
> *Sphaerophorus globosus*

Often dwarf shrubs and some herbaceous plants such as grasses and rushes are found growing between the lichen masses. On dry ground swept clear of snow, the encrusting lichen, *Ochrolechia tartarea*, is common.

Dwarf-heath tundra (Plate 63) is composed of stunted or creeping shrublets, with their winter resting-buds placed just above ground, or in the surface soil. They are mostly 10–20 cm high but may reach 30 cm. They are usually found on sandy and quartz soils which tend to dry out in summer. These soils are often very poor in nutrients, and a layer of raw humus accumulates at the surface and thus has some water-retaining capacity.

The shrublets are mostly evergreen, with small thick leaves, and with the ability to retain water during dry periods. Others, like the willows and dwarf birch, are deciduous. Characteristic species of the dwarf-heath tundra are

	Salix glauca
Woolly willow	*Salix lanata*
Cloudberry	*Rubus chamaemorus*
Stone bramble	*Rubus saxatilis*
Diapensia	*Diapensia lapponica*
	Rhododendron lapponicum
Labrador-tea	*Ledum palustre*
Trailing azalea	*Loiseleuria procumbens*
Blue heath	*Phyllodoce caerulea*
Alpine bearberry	*Arctostaphylos alpinus*
Bearberry	*Arctostaphylos uva-ursi*
Bog bilberry	*Vaccinium uliginosum*
Cowberry	*Vaccinium vitis-idaea.*

By contrast, on neutral or alkaline soils, heathlands dominated by the mountain avens, *Dryas octopetala*—known as Dryas-heaths—are characteristic and have a richer association of species including

> *Draba nivalis*
> *Potentilla nivea*
> *Oxytropis lapponica*
> *Rhododendron lapponicum*
> *Campanula uniflora*
> *Arnica alpina.*

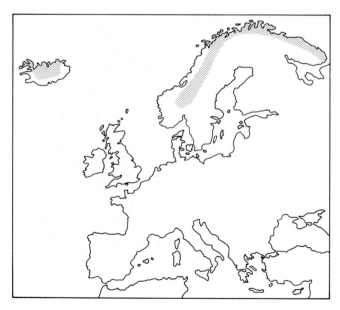

Map 1. Arctic region.

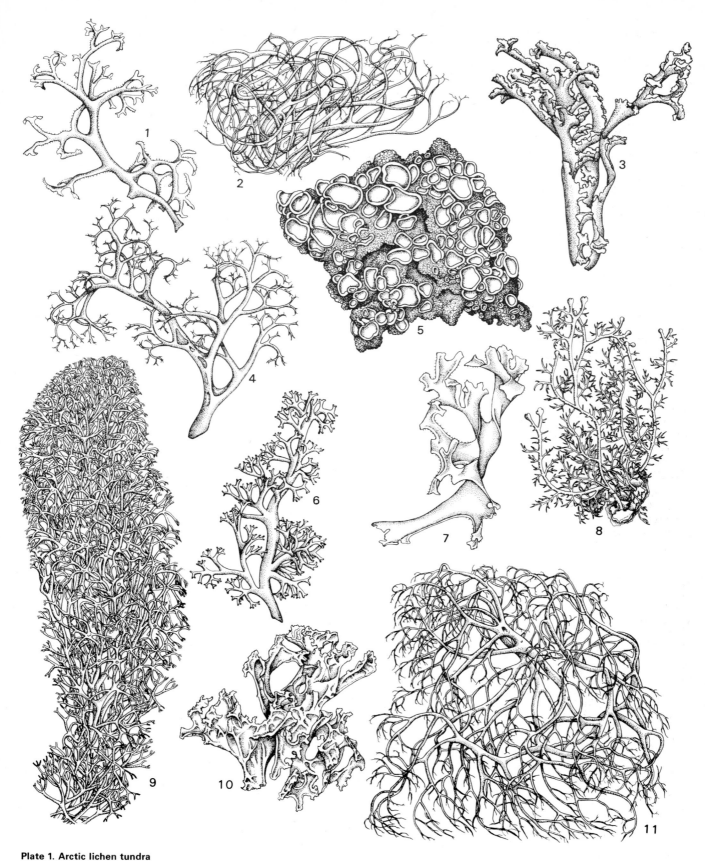

Plate 1. Arctic lichen tundra

1. *Cetraria ericetorum*; **2**. *Alectoria nigricans*; **3**. *Cetraria cucullata*; **4**. *Coelocaulon divergens*; **5**. *Ochrolechia tartarea*; **6**. *Cladonia rangiferina*; **7**. *Cetraria islandica*; **8**. *Sphaerophorus globosus*; **9**. *Cladonia stellaris*; **10**. *Cetraria nivalis*; **11**. *Alectoria ochroleuca*.

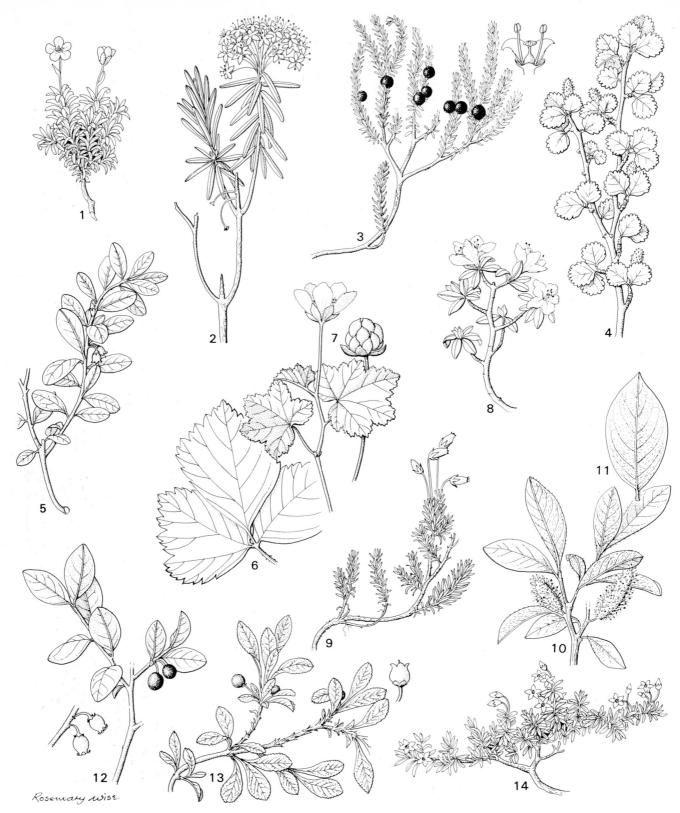

Plate 2. Arctic dwarf-heath tundra

1. *Diapensia, Diapensia lapponica*;   **2**. Labrador-tea, *Ledum palustre*;   **3**. Crowberry, *Empetrum nigrum* (inset×6);   **4**. Dwarf birch, *Betula nana*;   **5**. Cowberry, *Vaccinium vitis-idaea*;   **6**. Stone bramble, *Rubus saxatilis* (leaf);   **7**. Cloudberry, *Rubus chamaemorus*;   **8**. *Rhododendron lapponicum*;   **9**. Blue heath, *Phyllodoce caerulea*;   **10**. *Salix glauca*;   **11**. Woolly willow, *Salix lanata* (leaf);   **12**. Bog bilberry, *Vaccinium uliginosum* (inset×2);   **13**. Alpine bearberry, *Arctostaphylos alpinus*;   **14**. Trailing azalea, *Loiseleuria procumbens*.

24

Plate 3. Arctic arctic–alpine tundra

1. Alpine saw-wort, *Saussurea alpina*;   **2**. Mountain avens, *Dryas octopetala*;   **3**. *Cassiope hypnoides*;   **4**. *Cassiope tetragona*;   **5**. *Salix polaris* (leaf);   **6**. Net-leaved willow, *Salix reticulata*;   **7**. Dwarf willow, *Salix herbacea* (leaf);   **8**. Starry saxifrage, *Saxifraga stellaris*;   **9**. Dwarf cudweed, *Omalotheca supina*;   **10**. Glacier crowfoot, *Ranunculus glacialis*;   **11**. Alpine bartsia, *Bartsia alpina*;   **12**. Purple saxifrage, *Saxifraga oppositifolia*;   **13**. *Pedicularis lapponica*;   **14**. Moss campion, *Silene acaulis*;   **15**. *Potentilla nivea*.

Arctic vegetation

The following may also occur

Net-leaved willow	*Salix reticulata*
Moss campion	*Silene acaulis*
Purple saxifrage	*Saxifraga oppositifolia*
The sedge	*Kobresia myosuroides*
Black Alpine-sedge	*Carex atrata.*

Arctic–alpine tundra (Plate 64) is found on higher ground, and spreads southwards along the mountain backbone of Scandinavia; it also occurs on the high plateaux of Iceland. The vegetation is locally very varied, and is often dominated by a single distinctive species, usually one of the following: three-leaved rush, *Juncus trifidus*; diapensia, *Diapensia lapponica*; mountain avens, *Dryas octopetala*; moss campion, *Silene acaulis*; glacier crowfoot, *Ranunculus glacialis*; stiff sedge, *Carex bigelowii*. Other widespread plants of the middle arctic–alpine belt, occurring above about 1000 m are

	Cassiope hypnoides
	Cassiope tetragona
Curved wood-rush	*Luzula arcuata*
Sheep's-fescue	*Festuca ovina*
Alpine meadow-grass	*Poa alpina*
	Poa alpigena
	Poa arctica
Alpine hair-grass	*Deschampsia alpina*
Alpine holy-grass	*Hierochloe alpina*
Lapland small-reed	*Calamagrostis lapponica.*

The shrubby plants growing at the highest altitudes are usually members of the heath family such as the alpine bearberry, *Arctostaphylos alpinus*; blue heath, *Phyllodoce caerulea*; trailing azalea, *Loiseleuria procumbens*; and the closely related crowberry, *Empetrum nigrum*. The length of snow-lie is very important in determining the plants which can establish themselves in these arctic–alpine conditions. So also is the composition of the bedrock, which can either be acidic on quartz or sandstone, or neutral, or alkaline from limestones or dolomites, thus resulting in the considerable variety of plant communities which often occur as a patchwork. The direction and degree of slope, as well as the variation in water retention are other important local factors (see Fig. 1).

Some arctic–alpine communities in relation to length of snow-cover

	Acidic bedrock	Calcareous bedrock
Almost snow-free, wind-exposed	Crowberry Three-leaved rush	Mountain avens
Early snow-free but protected	Bilberry	Mountain avens with *Cassiope tetragona*
Late snow-free	Alpine hair-grass, Sweet vernal-grass	Net-leaved willow, Alpine meadow-grass
Very late snow-free	Dwarf willow *Salix herbacea*	'Polar willow', *Salix polaris*
Extremely late snow-free	Mosses including: *Polytrichum norvegicum*	Mosses including: *Distichium capillaceum*

Forest-tundra (Plate 65) occurs in sheltered valleys with more fertile soils, and is dominated largely by the northern form of the downy birch, *Betula pubescens* subsp. *tortuosa*. In mountainous regions it may grow up to altitudes of 600 m, as in northern Lapland, while further south it may occur as high as 800 m.

In exposed situations, the trees commonly have several crooked trunks arising from the base and are characteristically bare of branches in the lower part, as a result of being covered by snow to a depth of 1–1.5 m during the long winters. Elsewhere this downy birch has a single erect trunk. The following trees are commonly present, often in distinctive northern varieties

Goat willow	*Salix caprea*
Aspen	*Populus tremula*
Grey alder	*Alnus incana*
Rowan	*Sorbus aucuparia*
Bird cherry	*Prunus padus.*

An example of a birch wood in Iceland has the following associated species

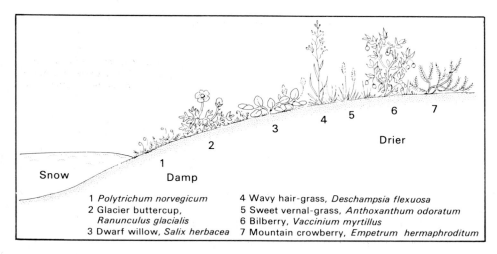

Fig. 1. Snow bed in the Arctic–Alpine region (acid soil). These are some of the plants associated with snow patches. The duration of snow cover is important in determining which species occur.

Snow   Damp   Drier

1 *Polytrichum norvegicum*
2 Glacier buttercup, *Ranunculus glacialis*
3 Dwarf willow, *Salix herbacea*
4 Wavy hair-grass, *Deschampsia flexuosa*
5 Sweet vernal-grass, *Anthoxanthum odoratum*
6 Bilberry, *Vaccinium myrtillus*
7 Mountain crowberry, *Empetrum hermaphroditum*

Shrub layer

Woolly willow	*Salix lanata*
Tea-leaved willow	*Salix phylicifolia*

Field layer

Alpine bistort	*Polygonum viviparum*
Alpine meadow-rue	*Thalictrum alpinum*
Stone bramble	*Rubus saxatilis*
Water avens	*Geum rivale*
Wild strawberry	*Fragaria vesca*
White clover	*Trifolium repens*
Wood crane's-bill	*Geranium sylvaticum*
Bearberry	*Arctostaphylos uva-ursi*
Heather	*Calluna vulgaris*
Bilberry	*Vaccinium myrtillus*
Bog bilberry	*Vaccinium uliginosum*
Crowberry	*Empetrum nigrum*
Northern bedstraw	*Galium boreale*
Slender bedstraw	*Galium pumilum*
Self-heal	*Prunella vulgaris*
Red fescue	*Festuca rubra*
Wavy hair-grass	*Deschampsia flexuosa*
Sweet vernal-grass	*Anthoxanthum odoratum*
Brown bent	*Agrostis canina*
Common bent	*Agrostis capillaris*
	Carex begelowii

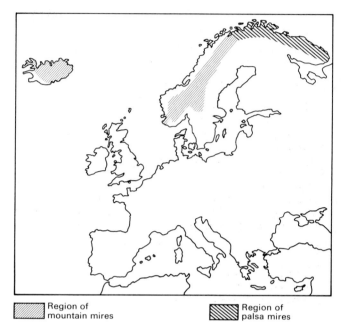

Region of mountain mires

Region of palsa mires

Map 2. Regions of palsa mires and mountain mires.

Arctic mires (Plate 61) are formed where peat accumulates in acid and moist conditions (Map 2), and where there are permanently frozen layers of subsoil, the *permafrost*, below the surface. Typical of these are *palsa mires* (see Fig. 2) which occur in the Arctic region where there is permanent or intermittent permafrost. The upper layers of peat only melt in the summer months, and this results in the formation of a series of hummocks which build up on an otherwise relatively flat mire or bog. Each hummock (Plates 67 and 68) may be up to 20–35 m long and 10–15 m wide, and usually up to 7 m high, but in some cases they may be raised to 50 m. Snow collects in the watery hollows surrounding the hummocks thus giving the underlying peat some protection from freezing. By contrast, the hummocks become more exposed to colder conditions, and this results in ice-layers developing year by year inside the hummocks, thus gradually pushing up the hummocks as the ice-layers increase, and so bringing about increased exposure. During the summer, the outer layers of peat on the hummocks melt and are thus able to support an active mire vegetation which is quite distinct from that growing in the watery hollows. With increasing height the hummocks become drier, resulting in the decay of the vegetation, and consequently erosion ultimately sets in, thus limiting the upward growth of the hummocks.

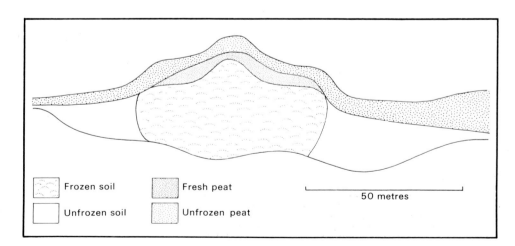

Fig. 2. Palsa mire—cross-section in summer. Frozen soil pushes up the peat to form these characteristic Arctic mires.

Frozen soil		Fresh peat	
Unfrozen soil		Unfrozen peat	

50 metres

Arctic vegetation

Typical plants of palsa mire hummocks (Plate 68) are

Dwarf birch	*Betula nana*
Labrador-tea	*Ledum palustre*
Cowberry	*Vaccinium vitis-idaea*
Crowberry	*Empetrum nigrum*
Cloudberry	*Rubus chamaemorus.*

Mosses including *Sphagnum, Polytrichum, Pleurozium,* and *Dicranum* are common, as well as lichens and, in particular, *Cladonia* species. The bases of the hummocks are usually dominated by the dwarf birch, *Betula nana.* The surrounding watery hollows have the species

Marsh cinquefoil	*Potentilla palustris*
Bogbean	*Menyanthes trifoliata*

Common cottongrass	*Eriophorum angustifolium*
	Eriophorum russeolum
Water sedge	*Carex aquatilis*
Mountain bog-sedge	*Carex rariflora*
	Carex rotundata.

South of the permafrost zone other mires like the *aapa mires* and the *raised bogs* occur. These are well developed in the Boreal region and are described there.

Likewise, the Arctic grasslands (Plate 66) differ little from those of the Boreal grasslands, though they are poorer in species; these also are described in the Boreal grasslands section.

7 Boreal vegetation (Plates 4–10; Plates 69–76)

The distinctive vegetation of the Boreal region (Map 3) is the evergreen coniferous forest, commonly called the *taiga*. The dominant trees have needle-like leaves which may remain on the twigs for several years. They are able to withstand several months of freezing and snow-cover and transpiration water loss is much reduced during dry or cold periods by the thick cuticles of these leaves. Some deciduous trees occur, particularly birches, and they may form sub-alpine woods above the coniferous zone. Other deciduous trees such as alders and willows are largely associated with wetter soils.

In mature coniferous forests, the shrub and field layers are often absent or poorly developed, but in more open stands, which are common as a result of man's tree-felling activities, members of the heath family and related species form dwarf shrub layers, or alternatively there is moss or lichen coverage.

In localities where drainage is impeded, mires occur over wide areas, with the accumulation of peat layers at the surface. These mires are dominated by dwarf shrubs, sedges, rushes, cottongrasses, etc., with sphagnum mosses and lichens. Different types of mires develop, ranging from acid bogs to alkaline fens, the latter populated by alders and willows and a rich assortment of herbaceous species.

Grasslands are primarily the result of man's activities, though the native species predominate. Cattle and sheep grazing and cutting for hay are the most important activities which maintain these grasslands.

Heaths on the other hand are natural and occur mainly in the montane and alpine regions. They are largely composed of dwarf ericaceous shrubs, dwarf willows, etc.

Taiga

The dark coniferous forests of the Boreal region are widely known as the *taiga*. They are dominated by the Norway spruce, *Picea abies*; and the Scots pine, *Pinus sylvestris* (Fig. 3). Norway spruce is the most important and widespread climax tree, forming dense forests on most types of soil, except on the wettest where mires prevail. In addition to Scots pine, the downy birch, *Betula pubescens*; rowan, *Sorbus aucuparia*; and the grey alder, *Alnus incana*, in damper areas, are the only trees of any importance in the spruce forests. The shrub layer is equally sparse in species, being largely composed of dwarf shrubs of the heath family, such as bilberry, cowberry, and heather.

The field and moss layers are richer in species, and the composition depends primarily on the soil-type and moisture, as well as the development of the tree layer.

Scots pine becomes the dominant forest-former in many of the peripheral areas, for example, on drier coarser soils, rocky ground, sandy and gravelly soils, and the edges of mires. It also colonizes cleared areas more rapidly than spruce—at least for the first 20 years or so—and is more resistant to forest fires (Plate 71). Where man's activities are strong, i.e. timber-felling, livestock-grazing, charcoal production, fires, etc., Scots pine may be much increased, to the detriment of spruce. In the northern taiga, Norway spruce forests occur only on well drained soils near the drainage influence of rivers, while mires become the dominant feature of the landscape. Here Scots pine forests are common on the sandy river terraces and round the mire margins.

The commonest type of Norway spruce forest (Plate 69) is found on richer soils with average water content. It is distinguished by the presence of the wood-sorrel, *Oxalis acetosella*, with the bilberry, *Vaccinium myrtillus*, and the buckler-fern, *Dryopteris assimilis*. Other characteristic herbaceous species include

Chickweed wintergreen	*Trientalis europaea*
May lily	*Maianthemum bifolium*
Hairy wood-rush	*Luzula pilosa*
Wavy hair-grass	*Deschampsia flexuosa.*

Mosses include

Barbilophozia lycopodioides
Dicranum majus
Hylocomium splendens
Hylocomium umbratum
Plagiochila asplenioides
Pleurozium schreberi
Polytrichum formosum
Ptilium cristacastrensis
Rhytidiadelphus loreus
Sphagnum quinquefarium.

On wet soils, the dominant mosses are species of bogmoss, *Sphagnum*, with both the bilberry, *Vaccinium myrtillus*, and the cowberry, *Vaccinium vitis-idaea*, in the shrub layer. The herb layer may include

Cloudberry	*Rubus chamaemorus*
Chickweed wintergreen	*Trientalis europaea*
The small-reed	*Calamagrostis purpurea*

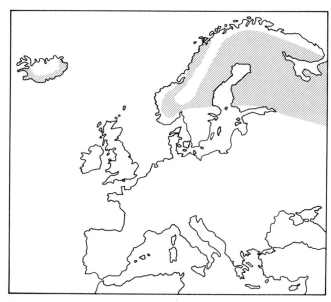

Map 3. Boreal region.

Boreal vegetation

Sheathed sedge	*Carex vaginata*
	Carex brunnescens
Lesser twayblade	*Listera cordata*
Wood horsetail	*Equisetum sylvaticum*

Man's interference, for example by forest clearance, livestock grazing, and fires, has tended to encourage the introduction of other species including

Sheep's sorrel	*Rumex acetosella*
Raspberry	*Rubus idaeus*
Rosebay willowherb	*Epilobium angustifolium*
Hairy wood-rush	*Luzula pilosa*
Wavy hair-grass	*Deschampsia flexuosa*
Bent species	*Agrostis* sp.
	Calamagrostis arundinacea
Bracken	*Pteridium aquilinum*

and, in wetter sites

Meadowsweet	*Filipendula ulmaria*
Wild angelica	*Angelica sylvestris*
Tufted hair-grass	*Deschampsia cespitosa.*

In the northern drier regions of the taiga the lichen species of *Cladonia* and *Cetraria* become more abundant in the spruce forests. In the southern taiga the spruce forests have a much richer ground flora which includes such herbaceous plants as

Wood anemone	*Anemone nemorosa*
Hepatica	*Hepatica nobilis*
Wild strawberry	*Fragaria vesca*
Stone bramble	*Rubus saxatilis*
Common dog-violet	*Viola riviniana*
Round-leaved wintergreen	*Pyrola rotundifolia*
Serrated wintergreen	*Orthilia secunda*

One-flowered wintergreen	*Moneses uniflora*
Small cow-wheat	*Melampyrum sylvaticum*
Hawkweed species	*Hieracium* sp.
Fingered sedge	*Carex digitata.*

The moss layer includes

Hylocomium splendens
Plagiomnium affine
Mnium spinosum
Pleurozium schreberi
Rhytidiadelphus triquetrus.

Scots pine woods, often mixed with Norway spruce, are developed on dry sandy plains with lower rainfall. The bilberry and cowberry, dwarf birch, *Betula nana*; and Labrador-tea, *Ledum palustre*, often form a very well developed dwarf-shrub layer, while there is a general increase of lichens in the ground layer further north (Fig. 4).

Characteristic plants of the field layer are

The clubmoss	*Diphasium companulatum*
	Pyrola chlorantha
Creeping lady's-tresses	*Goodyera repens.*

Mosses and lichens include

Dicranum fuscescens
Dicranum scoparium
Dicranum undulatum
Hylocomium splendens
Pleurozium schreberi
Cladonia arbuscula
Cladonia rangiferina
Peltigera aphthosa.

Fig. 3. Boreal pine and spruce forest near the Otra river, S. Norway.

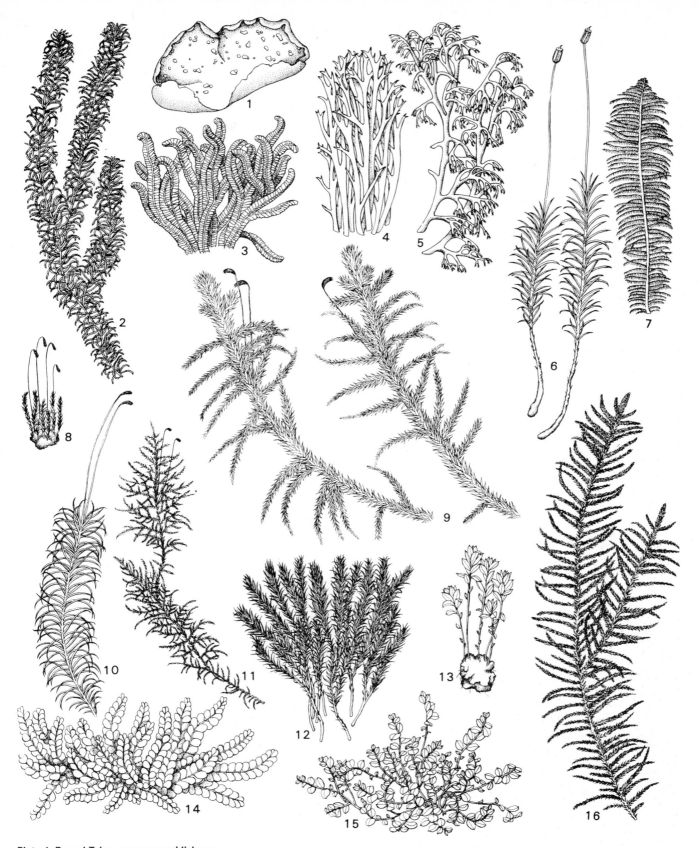

Plate 4. Boreal Taiga—mosses and lichens
1. *Peltigera aphthosa*; **2**. *Sphagnum capillifolium*; **3**. *Barbilophozia lycopodioides*; **4**. *Cladonia uncialis*; **5**. *Cladonia arbuscula*; **6**. *Polytrichum formosum*; **7**. *Ptilium cristacastrensis*; **8**. *Pohlia nutans*; 9. *Rhytidiadelphus triquetrus*; **10**. *Dicranum majus*; **11**. *Hylocomium splendens*; **12**. *Dicranum spurium*; **13**. *Mnium spinosum*; **14**. *Plagiochila asplenioides*; **15**. *Plagiomnium affine*; **16**. *Pleurozium schreberi*.

Plate 5. Boreal herbaceous species of taiga

1. Round-leaved wintergreen, *Pyrola rotundifolia*; **2**. *Pyrola chlorantha* (flowers); **3**. Bilberry, *Vaccinium myrtillus (inset×1); **4**. Lesser twayblade, *Listera cordata*; **5**. One-flowered wintergreen, *Moneses uniflora*; **6**. Common dog-violet, *Viola riviniana* (×1½); **7**. May lily, *Maianthemum bifolium*; **8**. Wood-sorrel, *Oxalis acetosella*; **9**. Hairy wood-rush, *Luzula pilosa*; **10**. Small cow-wheat, *Melampyrum sylvaticum* (inset×3); **11**. Chickweed wintergreen, *Trientalis europaea*; **12**. Creeping lady's-tresses, *Goodyera repens*; **13**. *Dryopteris assimilis*; **14**. *Diphasium complanatum*; **15**. Wood horsetail, *Equisetum sylvaticum*.

The Scots pine woods on the margins of the mires contain many characteristic mire species, such as Labrador-tea, bog rosemary, crowberry, cranberry, cloudberry, and cottongrass, while, in drier areas, heather and cross-leaved heath are typical, particularly in the more oceanic regions.

Pine forests (Plate 70) on the shallower drier soils are low and stunted, usually lichen-rich with

Cladonia arbuscula
Cladonia rangiferina
Cladonia stellaris
Cladonia uncialis.

The lichens colonize the moss cushions of *Racomitrium lanuginosum*, and form mats which easily become detached and are blown away. Similar mats of lichens may form in the dwarf shrubs of heather and bearberry.

Mires

These are the second most extensive plant communities in the Boreal region and are characteristic of it. In Europe they

Fig. 4. A relatively dry scots pine forest with lichens *Cladonia* species in the ground layer.

occupy 5.5 per cent of the whole surface (an area the size of France). A mire can be either a moss, herbaceous, or shrub community, which develops over peat accumulating at or just above the water-level. The peat is built up, layer upon layer, as a result of the annual partial decay of the vegetation, under conditions of low oxygen content, in either acid, neutral, or alkaline water. Mires which are poor in nutrients are known as *oligotrophic*, and support a bog community. Such bogs develop where the supply of water is largely from rain or snow, and consequently are deficient in mineral salts. Or bogs are formed where drainage water comes from acid substrates which are poor in minerals. Only certain species of plants can grow vigorously in such mineral-deficient waters, while slow and partial decomposition of their annual die-back brings about an accumulation of acid peat at the rate of about 1 mm per year.

By contrast, fen communities are developed on peat which is rich in mineral salts; they are spoken of as *eutrophic*. The peat is formed by the partial decay of plants, in conditions of low oxygen content and with drainage water from streams or springs which contain ample mineral salts washed out of mineral-rich substrates. Peat accumulates, but at a considerably slower rate than in acid bogs, and the vegetation of fens is quite distinct and has a much richer flora. There are many intermediate stages between acid and alkaline mires; these are often given local names in different countries. The classification of mires is complicated; different research workers have given different priorities to the structural, topographical, chemical, nutritional, and floristic composition of mires in their classification. The following are some of the main types of mires in Europe. Each will be described in the region of Europe in which it is most extensively developed.

Palsa mires. These occupy some 50 per cent of the land surface in northern Fennoscandia and are described in the Arctic region.

Aapa mires. Such mires are characteristic of the Boreal region, south of the permafrost zone. They are described below.

Raised bogs. These are distinguished by their convex central area, bordered by a slope, with drainage channels (rills), and marginal streams (laggs). They are best developed in the northern part of Central Europe.

Blanket bogs. These occur in heavy rainfall areas, covering hillsides, or on plateaux or saddles. These are described in the Atlantic region.

Transitional mires. These form a mosaic of bog and fen, or occur where bog or fen succeed each other.

Intermediate mires. These are so named where the plant communities are intermediate between those of bogs and those of fens.

Fen communities. These occur where drainage is restricted, on peat which is rich in mineral salts supplied by ground water, and not primarily dependent on rainfall; thus they may occur outside the humid areas. Fens can be further divided into marsh types, lake and river-verge types, and spring or hanging-fen types.

Boreal vegetation

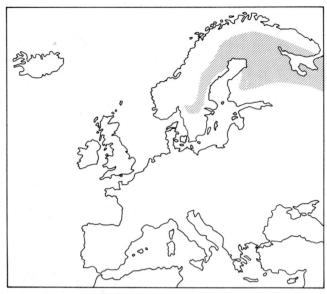

Map 4. Region of aapa mires.

Aapa mires (Fig. 5, Map 4, Plate 72), or string mires, are widely developed in the Boreal region. They are distinguished by the series of ridges or hummocks, with alternating depressions or 'flarks', which develop in the mire. These ridges run at right-angles to the general slope of the ground, forming a series of very shallow terraces one above the other, like a 'paddy-field'. The ridges arise partly as a result of the lateral pushing effect of the ice covering the depressions during the winter. These watery flarks are relatively rich in minerals, while the ridges, which are largely fed by rainwater, are poor in minerals. Consequently different distinctive species colonize the ridges, while other species occur in the depressions.

Commonly present in the wetter parts of the flarks are the white beaked-sedge, *Rhynchospora alba*; and oblong-leaved sundew, *Drosera intermedia*, while in drier parts of the depressions are found

Round-leaved sundew	*Drosera rotundifolia*
Bog rosemary	*Andromeda polifolia*
Cranberry	*Vaccinium oxycoccos*
Deergrass	*Scirpus cespitosus*
Hare's-tail cottongrass	*Eriphorum vaginatum.*

The commonest and most important bogmoss, *Sphagnum cuspidatum*, is the main peat-former in the aapa mire depressions. Also the greyish 'fluffy' lichen, *Cladonia squamosa*, is often present, together with species of liverwort.
   Quite different plants grow on the ridges; these include

Dwarf birch	*Betula nana*
Cloudberry	*Rubus chamaemorus*
Heather	*Calluna vulgaris*
Crowberry	*Empetrum nigrum*
Hare's-tail cottongrass	*Eriophorum vaginatum.*

While the reddish bogmosses, *Sphagnum magellanicum* and *S. capillifolium* (Plate 73), are the most important peat-formers of the ridges, *S. imbricatum* and *S. fuscum* are also common. As the hummocks age, they are invaded by lichens such as *Cladonia rangiferina*, *C. portentosa*, *C. arbuscula*, and *C. stellaris.*
   In mires with a somewhat richer mineral content, the following plants commonly occur

Bogbean	*Menyanthes trifoliata*
Intermediate bladderwort	*Utricularia intermedia*
Lesser bladderwort	*Utricularia minor*
Rannoch-rush	*Scheuchzeria palustris*
Bog asphodel	*Narthecium ossifragum*

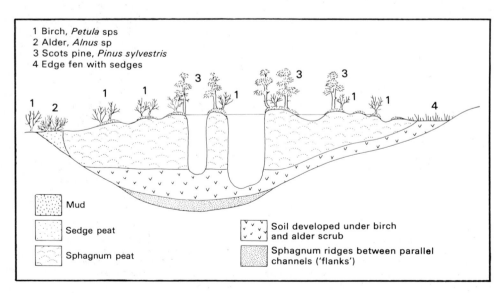

1 Birch, *Petula* sps
2 Alder, *Alnus* sp
3 Scots pine, *Pinus sylvestris*
4 Edge fen with sedges

Mud

Sedge peat

Sphagnum peat

Soil developed under birch and alder scrub

Sphagnum ridges between parallel channels ('flanks')

Fig. 5. Here we see the cross-section of a raised bog showing the string surface structure found on many boreal Aapa mires. (Redrawn from Ellenberg 1982.)

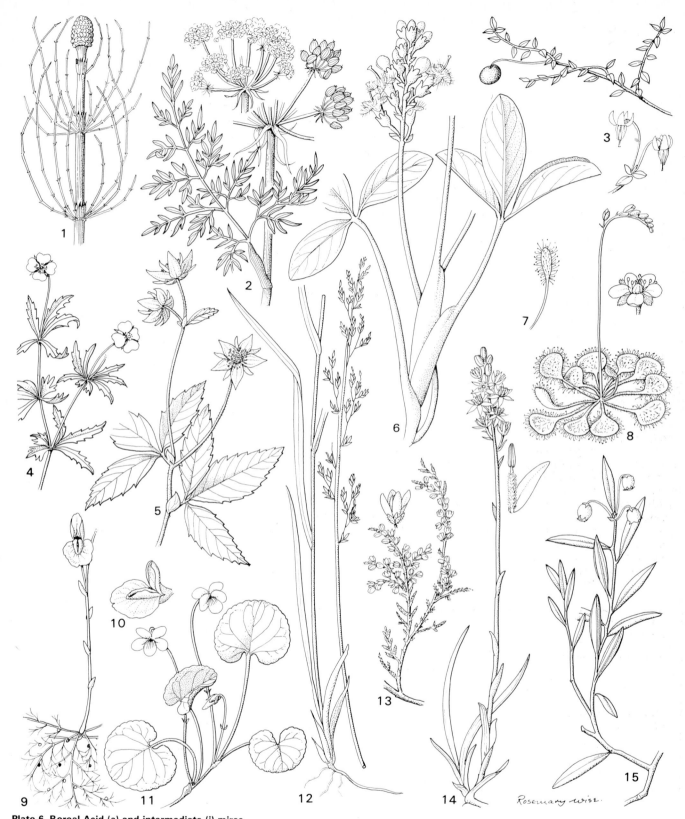

Rosemary Wise

Plate 6. Boreal Acid (a) and intermediate (i) mires

1. Water horsetail, *Equisetum fluviatile* (i); **2**. Milk-parsley, *Peucedanum palustre* (i); **3**. Cranberry, *Vaccinium oxycoccos* (a) (inset×2); **4**. Tormentil, *Potentilla erecta* (i); **5**. Marsh cinquefoil, *Potentilla palustris* (i); **6**. Bogbean, *Menyanthes trifoliata* (i); **7**. Oblong-leaved sundew, *Drosera intermedia* (leaf) (a); **8**. Round-leaved sundew, *Drosera rotundifolia* (a); **9**. Intermediate bladderwort, *Utricularia intermedia* (i); **10**. Lesser bladderwort, *Utricularia minor* (i) (flower×4); **11**. Marsh violet, *Viola palustris* (i); **12**. Purple moor-grass, *Molinia caerulea* (i); **13**. Heather, *Calluna vulgaris* (a) (inset×6); **14**. Bog asphodel, *Narthecium ossifragum* (i) (inset×4); **15**. Bog rosemary, *Andromeda polifolia* (a).

Boreal vegetation

Purple moor-grass	*Molinia caerulea*
	Eriophorum russeolum
Hare's-tail cottongrass	*Eriophorum vaginatum*
Slender sedge	*Carex lasiocarpa*
Bog sedge	*Carex limosa*
	Carex magellanica
Few-flowered sedge	*Carex pauciflora*
	Carex rotundata
Bottle sedge	*Carex rostrata*
Water horsetail	*Equisetum fluviatile.*

With a richer supply of mineral soil water communities more like poor fens have such distinctive species as

Tormentil	*Potentilla erecta*
Marsh cinquefoil	*Potentilla palustris*
Marsh violet	*Viola palustris*
Milk-parsley	*Peucedanum palustre*
Common marsh-bedstraw	*Galium palustre*
Bulbous rush	*Juncus bulbosus*
Brown bent	*Agrosis canina*
White sedge	*Carex curta*
Star sedge	*Carex echinata*
Common sedge	*Carex nigra.*

Wooded bog vegetation (Plate 75) is distinctive and it covers large areas of mires in southeastern Sweden. Scots pine, *Pinus sylvestris,* is the most important tree, often occurring with the downy birch, *Betula pubescens.* The dwarf shrubs commonly associated with these woods include

Dwarf birch	*Betula nana*
Cloudberry	*Rubus chamaemorus*
Heather	*Calluna vulgaris*
Labrador-tea	*Ledum palustre*
	Chamaedaphne calyculata
Bog bilberry	*Vaccinium uliginosum*
Crowberry	*Empetrum nigrum.*

Hare's-tail cottongrass, *Eriophorum vaginatum,* is very common. Bogmosses, *Sphagnum* species, dominate in the moss layer.

Intermediate mires may have the following characteristic assemblage of species

Marsh willowherb	*Epilobium palustre*
Marsh lousewort	*Pedicularis palustris*
Bog orchid	*Hammarbya paludosa*
Purple moor-grass	*Molinia caerulea*
Slender cottongrass	*Eriophorum gracile*
Hare's-tail cottongrass	*Eriophorum vaginatum*
Dioecious sedge	*Carex dioica*
Few-flowered sedge	*Carex pauciflora.*

Fen communities rich in mineral salts may have the following species

Arctic rush	*Juncus arcticus*
Chestnut rush	*Juncus castaneus*
Hard rush	*Juncus inflexus*
Blunt-flowered rush	*Juncus subnodulosus*

Brown bog-rush	*Schoenus ferrugineus*
Black bog-rush	*Schoenus nigricans*
False sedge	*Kobresia simpliciuscula*
Scorched alpine-sedge	*Carex atrofusca*
Hair sedge	*Carex capillaris*
	Carex capitata
Long-stalked yellow-sedge	*Carex lepidocarpa*
Bristle sedge	*Carex microglochin*
	Carex parallela
Narrow-leaved marsh-orchid	*Dactylorhiza traunsteineri.*

Mosses, particularly bogmosses, *Sphagnum* species, have different assortments of species in the poor and intermediate mires, and rich fens.

Marsh-fens are found near the margins of the mineral-rich lakes and pools. Here the most distinctive species are

Purple loosestrife	*Lythrum salicaria*
Yellow loosestrife	*Lysimachia vulgaris*
Gipsywort	*Lycopus europaeus*
Great fen-sedge	*Cladium mariscus*
Slender tufted-sedge	*Carex acuta*
Water sedge	*Carex aquatilis*
Tufted-sedge	*Carex elata*
	Carex juncella
Greater pond-sedge	*Carex riparia.*

Wooded or *bush-fens* occur on sites over fen peat which have a lower water table, usually below the soil surface. They are often called *carr.* Here alders, birches, willows, and spruce are common members of the tree layer. Both the grey alder, *Alnus incana,* and the common alder, *Alnus glutinosa,* occur, commonly with the downy birch, *Betula pubescens,* and willow species. Under denser tree cover in the wooded fens, the following tall herbaceous species are common

Meadowsweet	*Filipendula ulmaria*
Water avens	*Geum rivale*
Marsh hawk's-beard	*Crepis paludosa*
Narrow small-reed	*Calamagrostis neglecta*
Fibrous tussock-sedge	*Carex appropinquata*
	Carex cespitosa.

In *wooded-fens* with a sparser tree cover the following may also occur

Marsh-marigold	*Caltha palustris*
Grass-of-Parnassus	*Parnassia palustris*
Fen bedstraw	*Galium uliginosum*
Marsh helleborine	*Epipactis palustris*
Broad-leaved cottongrass	*Eriophorum latifolium*
Star sedge	*Carex echinata*
Tawny sedge	*Carex hostiana*
Common sedge	*Carex nigra*
Carnation sedge	*Carex panicea.*

Boreal deciduous woodland

Birchwoods. The most widely distributed deciduous tree in the Boreal region is the downy birch, *Betula pubescens,* while the silver birch, *Betula pendula,* only occurs in the south of the

R. Wise.

Plate 7. Boreal mineral-rich mires and wooded fens

1. Purple loosestrife, *Lythrum salicaria*; **2**. Narrow-leaved marsh-orchid, *Dactylorhiza traunsteineri*; **3**. Fen bedstraw, *Galium uliginosum*; **4**. Yellow loosestrife, *Lysimachia vulgaris*; **5**. Meadowsweet, *Filipendula ulmaria*; **6**. Bog orchid, *Hammarbya paludosa* (inset×3); **7**. Marsh hawk's-beard, *Crepis paludosa*; **8**. Grass-of-Parnassus, *Parnassia palustris*; **9**. Great fen-sedge, *Cladium mariscus*; **10**. Marsh lousewort, *Pedicularis palustris*; **11**. Gipsywort, *Lycopus europaeus*; **12**. Greater pond-sedge, *Carex riparia*; **13**. Water avens, *Geum rivale*; **14**. Marsh helleborine, *Epipactis palustris*; **15**. Marsh willowherb, *Epilobium palustre*.

Plate 8. Boreal mires—sedges and rushes
1. Broad-leaved cottongrass, *Eriophorum latifolium*;   **2**. Common cottongrass, *Eriophorum angustifolium*;   **3**. Rannoch-rush, *Scheuchzeria palustris*;   **4**. Brown bog-rush, *Schoenus ferrugineus*;   **5**. Hard rush, *Juncus inflexus*;   **6**. Bog sedge, *Carex limosa*;   **7**. *Carex parallela*;   **8**. White beak-sedge, *Rhynchospora alba*;   **9**. *Eriophorum russeolum*;   **10**. Few-flowered sedge, *Carex pauciflora*;   **11**. Water sedge, *Carex aquatilis*;   **12**. *Carex rotundata*;   **13** Chestnut rush, *Juncus castaneus*;   **14**. Arctic rush, *Juncus arcticus*.

region. Downy birch woods are well developed in the sub-alpine zone of the Scandinavian mountains, above the coniferous belt (Fig. 6). It is also common in the taiga forests and rapidly colonizes areas after clearing or burning. (Fig. 7). The most northern examples of downy birch woods occur in the forest-tundra, and these have been described previously. Lowland birch woods growing on dry acid soils have a characteristic dwarf-shrub layer composed largely of bilberry, *Vaccinium myrtillus;* and crowberry, *Empetrum nigrum.* The following herbaceous species are commonly present on drier sites

Tormentil	*Potentilla erecta*
Common cow-wheat	*Melampyrum pratense*
Small cow-wheat	*Melampyrum sylvaticum*
Golden-rod	*Solidago virgaurea*
Wavy hair-grass	*Deschampsia flexuosa.*

On wetter sites, one frequently finds

Globeflower	*Trollius europaeus*
Wood crane's-bill	*Geranium sylvaticum*
Yellow wood-violet	*Viola biflora*
Dwarf cornel	*Cornus suecica*
Highland cudweed	*Omalotheca norvegica.*

Another type, described as *Meadow birch wood,* occurs on sloping ground with mobile ground-water, on soil which is usually either calcareous, neutral, or slightly acid. The downy birch is dominant and the shrub layer consists of the northern form of the bird cherry, *Prunus padus;* the downy currant, *Ribes spicatum;* and the grey alder, *Alnus incana,* with the shrub mezereon, *Daphne mezereum.* A lush and rich herbaceous layer may include

Globeflower	*Trollius europaeus*
Northern monkshood	*Aconitum septentrionale*
	Corydalis intermedia
Meadowsweet	*Filipendula ulmaria*
Wood-sorrel	*Oxalis acetosella*
Wood crane's-bill	*Geranium sylvaticum*
Wood forget-me-not	*Myosotis sylvatica*
Alpine sow-thistle	*Cicerbita alpina*
Whorled solomon's-seal	*Polygonatum verticillatum*
Wood millet	*Milium effusum*

and the ferns

Lady-fern	*Athyrium filix-femina*
Broad buckler-fern	*Dryopteris dilatata*
	Matteuccia struthiopteris.

Wet birch woods contain willow species and the dwarf birch, *Betula nana,* with a number of sedges, cottongrasses, horsetails, and grasses.

Willow thickets are well developed by Boreal rivers on sand and mud banks; the characteristic willow species are the almond willow, *Salix triandra,* and the northern *Salix daphnoides.* They are small trees growing up to 10 m in height; and commonly associated with them are such herbaceous plants as

Creeping buttercup	*Ranunculus repens*
Corn mint	*Mentha arvensis*
Creeping bent	*Agrostis stolonifera*
Reed canary-grass	*Phalaris arundinacea*
Field horsetail	*Equisetum arvense.*

Fig. 6. Montane birch wood well developed above the coniferous forest zone in the mountains of Scandinavia—here at about 450 m.

Boreal vegetation

Small stands of willows occurring on wet peat, often flanked by Norway spruce, contain the willow species

Eared willow	*Salix aurita*
	Salix borealis
Grey willow	*Salix cinerea*
Bay willow	*Salix pentandra*
Tea-leaved willow	*Salix phylicifolia.*

Alder woods, or *carrs,* are widely distributed in the Boreal region. The grey alder, *Alnus incana,* is the commoner of the two alders in the north and eastern parts of the Boreal region, where it fringes the Bothnian shores and spreads inland along river valleys into the mountainous areas (Fig. 8). It occurs on drier soils than the common alder, and the community is rich in tall herbaceous species which are often grazed. A typical riverside alder *carr* in the Boreal region, has, in addition to the dominant grey alder, *Alnus incana,* the bird cherry, *Prunus padus,* in the tree layer. The downy currant, *Ribes spicatum,* occurs in the shrub layer. The field layer includes

Wood anemone	*Anemone nemorosa*
Common valerian	*Valeriana officinalis*
Herb-Paris	*Paris quadrifolia*
Bearded couch	*Elymus caninus*
The fern	*Matteuccia struthiopteris*
Shady horsetail	*Equisetum pratense*
The moss	*Cirriphyllum piliferum.*

In the valley-woods on more fertile soils, grey alder and ash may be co-dominant. In such woods additional distinctive herbaceous species may occur, such as

Goldilocks buttercup	*Ranunculus auricomus*
Lesser celandine	*Rancunculus ficaria*
Wood stitchwort	*Stellaria nemorum*

Giant bellflower	*Campanula latifolia*
Yellow star-of-Bethlehem	*Gagea lutea.*

The common alder, *Alnus glutinosa,* also forms woods on the shores of the Bothnian gulf, and it may be mixed with the fluttering elm, *Ulmus laevis;* and with the black currant, *Ribes nigrum,* common in the shrub layer. Herbaceous species include

Marsh marigold	*Caltha palustris*
Meadowsweet	*Filipendula ulmaria*
Milk-parsley	*Peucedanum palustre*
Yellow loosestrife	*Lysimachia vulgaris*
Gipsywort	*Lycopus europaeus*
Bittersweet	*Solanum dulcamara*
Field horsetail	*Equisetum arvense.*

Both common and grey alders may grow together and form woods, often with the downy birch. In these mixed woods such plants as the bog arum, *Calla palustris;* tufted loosestrife, *Lysimachia thyrsiflora;* large bittercress, *Cardamine amara;* and elongated sedge, *Carex elongata,* may be found.

Other trees, in addition to ash, are found occasionally in woodlands in the south, but rarely form pure woods in the Boreal region. These include oak, elm, maple, aspen, and lime. There are however elm stands in western Norway, and oak–lime–maple communities in southeastern Sweden.

Grasslands

Northern grasslands (Plate 76) occur in the Boreal and sub-Arctic regions but, with the exception of the Alpine grasslands, they are usually associated with grazing and hay-cutting in the vicinity of human habitations. They are best developed on well-drained morainic soils, with a southern aspect, which

Fig. 7. Land left bare by retreating glaciers colonized by birch—here with a field layer of heather.

Plate 9. Boreal deciduous woods

1. Downy birch, *Betula pubescens* ssp. *tortuosa*;   **2**. Bird cherry, *Prunus padus*;   **3**. Wood forget-me-not, *Myosotis sylvatica*;   **4**. Alpine sow-thistle, *Cicerbita alpina*;   **5**. Common cow-wheat, *Melampyrum pratense*;   **6**. Whorled Solomon's-seal, *Polygonatum verticillatum*;   **7**. Goldenrod, *Solidago virgaurea*;   **8**. Globeflower, *Trollius europaeus*;   **9**. Wood Crane's-bill, *Geranium sylvaticum*;   **10**. Northern monkshood, *Aconitum septentrionale*;   **11**. *Corydalis intermedia*;   **12**. Highland cudweed, *Omalotheca norvegica*;   **13**. Dwarf cornel, *Cornus suecica*;   **14**. Yellow wood violet, *Viola biflora*.

Plate 10. Boreal deciduous woods—wet woods

1. Marsh-marigold, *Caltha palustris*;   **2**. Downy currant, *Ribes spicatum*;   **3**. Tufted loosestrife, *Lysimachia thyrsiflora*;   **4**. Goldilocks, *Ranunculus auricomus*;   **5**. Bay willow, *Salix pentandra* (leaf);   **6**. Giant bellflower, *Campanula latifolia*;   **7**. Wood stitchwort, *Stellaria nemorum*;   **8**. Almond willow, *Salix triandra*;   **9**. Violet willow, *Salix daphnoides* (leaf);   **10**. Shady horsetail, *Equisetum pratense*;   **11**. Bog arum, *Calla palustris*;   **12**. Reed canary-grass, *Phalaris arundinacea*;   **13**. Large bittercress, *Cardamine amara*;   **14**. Common alder, *Alnus glutinosa*;   **15**. Grey alder, *Alnus incana* (leaf);   **16**. Bittersweet, *Solanum dulcamara*.

retain their moisture during the spring but dry out in the summer. In Iceland these grasslands are composed of the grasses

Sheep's-fescue	*Festuca ovina*
Red fescue	*Festuca rubra*
Smooth meadow-grass	*Poa pratensis*
Tufted hair-grass	*Deschampsia cespitosa*
Brown bent	*Agrostis canina*
Common bent	*Agrostis capillaris*

with the sheathed sedge, *Carex vaginata*, and stiff sedge, *C. bigelowii*. Distinctive flowering species include cuckoo flower, *Cardamine pratensis;* marsh violet, *Viola palustris;* and dandelions *Taraxacum* species. Manured hay meadows have a richer flora and may have such additional species as

Common sorrel	*Rumex acetosa*
Meadow buttercup	*Ranunculus acris*
Creeping buttercup	*Ranunculus repens*
Tufted vetch	*Vicia cracca*
White clover	*Trifolium repens*
Wood crane's-bill	*Geranium sylvaticum*
Wild pansy	*Viola tricolor*
Yellow rattle	*Rhinanthus minor.*

The grasslands of Scandinavia show considerable variation but the following species are often typical

White clover	*Trifolium repens*
Self-heal	*Prunella vulgaris*
Yarrow	*Achillea millefolium*
Heath wood-rush	*Luzula multiflora*
Red fescue	*Festuca rubra*
Common bent	*Agrostis capillaris*
Mat-grass	*Nardus stricta.*

In northeastern Sweden the extensive semi-natural grasslands, which are regularly cut for hay crops and which are usually found along river verges and sea shores, are often dominated by the tufted hair-grass, *Deschampsia cespitosa*, with the water sedge, *Carex aquatilis*, and water horsetail, *Equisetum fluviatile*.

Heaths (Plate 74)

Most of the heaths of the northern Boreal region are found in the montane or alpine zones, but to the west, under the influence of the Atlantic, they may reach lower altitudes, as in Norway and Iceland. Examples of montane heaths in Iceland, occurring at altitudes of about 350 m, have the following ericaceous shrublets: heather, *Calluna vulgaris;* Bilberry, *Vaccinium myrtillus;* bog bilberry, *Vaccinium uliginosum;* bearberry, *Arctostaphylos uva-ursi; Cassiope hypnoides;* trailing azalea, *Loiseleuria procumbens;* and the crowberry, *Empetrum nigrum*. The dwarf birch, *Betula nana*, and the downy birch, *B. pubescens*, and the willows: *Salix glauca;* woolly willow, *S. lanata;* tea-leaved willow, *S. phylicifolia;* with the juniper, *Juniperus communis*, also occur in these heaths. Other herbaceous species include

Alpine bistort	*Polygonum alpinum*
Alpine mouse-ear	*Cerastium alpinum*
Mountain avens	*Dryas octopetala*
Slender bedstraw	*Galium pumilum*
The false-sedge	*Kobresia myosuroides.*

The majority of lowland heaths are found in southern Scandinavia; these will be described in the following Atlantic region.

Fig. 8. Grey alder *Alnus incana* is widespread in the Boreal region. A typical damp wood with a rich fern undergrowth.

8 Atlantic vegetation (Plates 11–21; Plates 77–90)

There is much evidence to show that the Atlantic region (Map 5) was dominated by deciduous forests in the middle of the post-glacial period. The dominant trees were oaks, ash, beech, birch, with other trees such as elm, maple, hornbeam, and willows in local environments. These natural and extensive forests were progressively decimated by man as he colonized the region, and today it is no longer possible to find natural unchanged woods. What remain today are mostly semi-natural woodlands strongly influenced by the activities of man, but with most of the native plants still present. Much of the land originally covered by forest has become agricultural land but, in the western Atlantic region in particular, the forests have been replaced by heathlands which are now typical of the region. A great range of distinctive heathlands now occur from Scandinavia to northwestern Spain and inland to West Germany.

Natural coniferous woods are restricted to montane and some maritime areas in the south, but elsewhere in the lowlands they occur as a result of man's original introduction.

Mires are widespread, particularly in the more humid oceanic western part of the Atlantic region. There are many types of mire, ranging from blanket bogs on highly acid soils, to fens on mineral-rich alkaline soils. Again these have been much restricted by man through drainage, peat removal, and clearance.

Grasslands are widespread in the region, occurring on a wide variety of soils ranging from acid to highly alkaline, but in almost every case they are semi-natural and maintained as grassland by man or his animals. Recently entirely artificial grasslands have been sown by man. Only in a very few cases, such as some montane grasslands and some western maritime grasslands, are they in any sense natural.

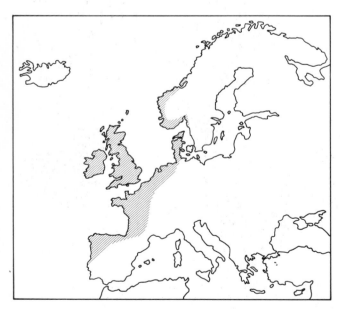

Map 5. Atlantic region.

Atlantic deciduous woodlands

Deciduous woodlands are multi-layered, usually with distinctive shrub, field, and moss layers, and often with a lower tree layer as well. They require a climate which has a growth period of four to six months, and a cool but mild winter period of three to four months, when their leaves are shed.

Oak woods

The oak woods of the Atlantic region are dominated by two oaks, the sessile oak, *Quercus petraea*, and the pedunculate oak, *Quercus robur*, both of which are widespread in the western and central European regions and occupy a wide range of soils. A third species, the Pyrenean oak, *Quercus pyrenaica*, forms local woodlands in southwestern Europe.

The sessile oak favours drier acid siliceous soils, and occurs mainly in the north and west of the Atlantic region, while the pedunculate oak grows on heavier richer, often wetter, soils and usually requires a more continental climate, though it penetrates further north and east than the former species. The Pyrenean oak is restricted largely to the Iberian Peninsula and southwestern France and favours acid soils.

In general oak woods have well-developed shrub and field layers. In Britain there occur characteristic oak–hazel coppices (Fig.9) with widely-spaced oaks, and hazel shrubs which are regularly cut. High oak forests with a closed canopy of oak trees are commoner in other parts of Atlantic Europe.

Sessile oak woods occurring in the extreme west of Ireland in a humid and mild Atlantic climate have the following composition (c.f. Fig. 10)

Tree and shrub layer

Yew	*Taxus baccata*
Silver birch	*Betula pendula*
Downy birch	*Betula pubescens*
Sessile oak	*Quercus petraea*
Rowan	*Sorbus aucuparia*
Holly	*Ilex aquifolium*

Field layer

Wood sorrel	*Oxalis acetosella*
Heather	*Calluna vulgaris*
Bilberry	*Vaccinium myrtillus*
Great wood-rush	*Luzula sylvatica*
Hay-scented buckler-fern	*Dryopteris aemula*
Hard fern	*Blechnum spicant*
Bracken	*Pteridium aquilinum.*

The moss layer is very rich, particularly covering boulders, and includes *Hylocomium brevirostre*, *Rhytidiadelphus loreus*, *Thuidium tamariscinum*, with many liverworts, and Wilson's filmy-fern, *Hymenophyllum wilsonii*, and Tunbridge filmy-fern, *H. tunbrigense*. Rather similar woods occur in northwestern Iberia.

Sessile oak woods (Plate 77) mixed with birch are frequent in montane northern Britain and southwestern Norway, often occurring with the rowan, *Sorbus aucuparia*; bird cherry, *Prunus padus*; honeysuckle, *Lonicera periclymenum*; and

Plate 11. Atlantic deciduous woods—mosses and ferns

1. Marsh fern, *Thelypteris palustris*; **2**. Lemon-scented fern, *Thelypteris limbosperma*; **3**. Hard fern, *Blechnum spicant*; **4**. Lady-fern, *Athyrium filix-femina*; **5**. Wilson's filmy-fern, *Hymenophyllum wilsonii*; **6**. Hay-scented buckler-fern, *Dryopteris aemula*; **7**. Bracken, *Pteridium aquilinum*; **8**. Broad buckler-fern, *Dryopteris dilatata*; **9**. Royal fern, *Osmunda regalis*; **10**. *Isothecium myosuroides*; **11**. *Thuidium tamariscinum*; **12**. *Sphagnum squarrosum*; **13**. *Rhytidiadelphus loreus*; **14**. *Hylocomium brevirostre*; **15**. *Atrichum undulatum*.

eared willow, *Salix aurita*. One type of wood has a field layer rich in such grasses as

Sweet vernal-grass	*Anthoxanthum odoratum*
Yorkshire-fog	*Holcus lanatus*
Common bent	*Agrostis capillaris*

with the bluebell, *Hyacinthoides non-scripta*, and the lemon-scented fern, *Thelypteris limbosperma*. Another type of sessile oak wood characteristically has the bilberry, *Vaccinium myrtillus*, with

Juniper	*Juniperus communis*
Hazel	*Corylus avellana*
Cowberry	*Vaccinium vitis-idaea*
Great wood-rush	*Luzula sylvatica*
Common bent	*Agrostis capillaris*
Bracken	*Pteridium aquilinum*
Hard fern	*Blechnum spicant.*

In some areas, mixed woods with both sessile and pedunculate oaks may occur, largely on acid soils, and these usually have poorly developed shrub and field layers. Sherwood Forest, England, has such a mixed wood, with a tree layer of

Sessile oak	*Quercus petraea*
Pedunculate oak	*Quercus robur*
Downy birch	*Betula pubescens*
Crab apple	*Malus sylvestris*
Rowan	*Sorbus aucuparia.*

The shrub layer is sparse, with hawthorns, *Crataegus* sps., and elder, *Sambucus nigra*. The field layer contains

Wavy hair-grass	*Deschampsia flexuosa*
Creeping soft-grass	*Holcus mollis*

Common bent	*Agrostis capillaris*
Bracken	*Pteridium aquilinum.*

Pedunculate oak woods usually occur on alkaline or neutral soils which are richer in minerals; they are particularly well developed on damp clay soils. They are often rich in both woody and herbaceous species and also have a rich and characteristic ground flora. The activities of man such as timber-felling, coppice-cutting, and selective planting, have had a long and lasting effect on these woods and they show all stages of semi-natural composition and structure, yet the majority of species present are in all probability native species that would occur naturally in these woods. Coppicing of ash, hazel, field maple, hawthorn, and willows all take place to a greater or lesser extent. In southeastern England the pedunculate oak woods (Plate 78) commonly have the ash, *Fraxinus excelsior*, in the upper tree layer, and in the lower tree layer the field maple, *Acer campestre*. Other trees may include

Grey poplar	*Populus canescens*
Aspen	*Populus tremula*
Silver birch	*Betula pendula*
Downy birch	*Betula pubescens*
Alder	*Alnus glutinosa*
Hornbeam	*Carpinus betulus*
Wych elm	*Ulmus glabra*
Crab apple	*Malus sylvestris*
Wild cherry	*Prunus avium*
Holly	*Ilex aquifolium.*

The shrub layer includes

Grey willow	*Salix atrocinerea*
Eared willow	*Salix aurita*

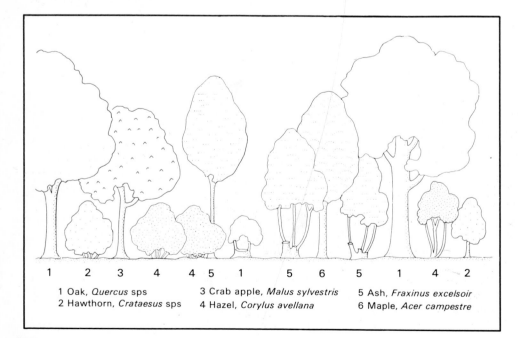

Fig. 9. Atlantic oak coppice. A damp oak woodland under traditional coppice management. Note standard oaks. The left side was felled the previous winter; the right is ready for felling after five seasons' growth. (Redrawn from Rackham 1975.)

1 2 3 4 4 5 1 5 6 5 1 4 2

1 Oak, *Quercus* sps
2 Hawthorn, *Crataesus* sps
3 Crab apple, *Malus sylvestris*
4 Hazel, *Corylus avellana*
5 Ash, *Fraxinus excelsoir*
6 Maple, *Acer campestre*

Goat willow	*Salix caprea*
Hazel	*Corylus avellana*
Hawthorn	*Crataegus monogyna*
Blackthorn	*Prunus spinosa*
Guelder-rose	*Viburnum opulus.*

On soils rich in lime or other basic minerals, shrubs such as

Spindle	*Euonymus europaeus*
Buckthorn	*Rhamnus catharticus*
Dogwood	*Cornus sanguinea*
Privet	*Ligustrum vulgare*
Elder	*Sambucus nigra*

commonly occur. Roses, *Rosa* sps., and brambles, *Rubus* sps., are ubiquitous in the lower shrub layer, while climbing species include the ivy, *Hedera helix*, and the honeysuckle, *Lonicera periclymenum*. On mineral-rich and calcareous soils, the shrub and field layers of oak woods are very rich in species, contrasting with oak woods on acid siliceous soils, which have a limited shrub and field layer.

Many plants flower before the leaf-cover of the dominant trees and shrubs develop; this is known as the *pre-vernal* aspect. These early-flowering pre-vernal species include

Wood anemone	*Anemone nemorosa*
Celandine	*Ranunculus ficaria*
Wild strawberry	*Fragaria vesca*
Common dog-violet	*Viola riviniana*
Dog's mercury	*Mercurialis perennis*
Primrose	*Primula vulgaris*
Ground-ivy	*Glechoma hederacea.*

Vernal species are those which flower at approximately the same time as the deciduous tree and shrub layers break into leaf. They commonly include

Greater stitchwort	*Stellaria holostea*
Red campion	*Silene dioica*
Wood spurge	*Euphorbia amygdaloides*
Sanicle	*Sanicula europaea*
Bugle	*Ajuga reptans*
Yellow archangel	*Lamiastrum galeobdolon*
Germander speedwell	*Veronica chamaedrys*
Wood speedwell	*Veronica montana*
Bluebell	*Hyacinthoides non-scripta*
Lords-and-ladies	*Arum maculatum*

and many other less widespread species.

Aestival species are those which flower in summer, after the foliage of the trees and shrubs is fully developed. They include

Wood avens	*Geum urbanum*
Enchanter's-nightshade	*Circaea lutetiana*
Rosebay willowherb	*Epilobium angustifolium*
Broad-leaved willowherb	*Epilobium montanum*
Hedge woundwort	*Stachys sylvatica*
Foxglove	*Digitalis purpurea*
Nettle-leaved bellflower	*Campanula trachelium*
Burdock	*Arctium minus.*

There are often many other species present, including grasses. The commonest mosses in the ground layer in damp oak woods are *Atrichum undulatum* and *Thuidium tamariscinum*.

Pedunculate oak woods, growing on dry porous soils, have a very different but characteristic field layer which includes bracken, *Pteridium aquilinum;* blackberry, *Rubus fruticosus;* rosebay willowherb, *Epilobium angustifolium;* and bluebell, *Hyacinthoides non-scripta*, with the grasses common bent, *Agrostis capillaris;* and creeping soft-grass, *Holcus mollis*.

Pedunculate oak woods on wet or damp soils are, by contrast, distinguished by such moisture-loving species as the

Fig. 10. Atlantic oak wood with holly *Ilex aquifolium*. Yarner's wood, S. Devon, UK.

Rosemary Wise.

Plate 12. Atlantic deciduous wood; trees and shrubs

1. Blackthorn, *Prunus spinosa* (inset×1);   **2**. Hazel, *Corylus avellana*;   **3**. Ash, *Fraxinus excelsior*;   **4**. Hawthorn, *Crataegus monogyna* (inset×1½);   **5**. Crab apple, *Malus sylvestris*;   **6**. Eared willow, *Salix aurita* (leaf);   **7**. Goat willow, *Salix caprea*;   **8**. Yew, *Taxus baccata*;   **9**. Pedunculate oak, *Quercus robur*;   **10**. Holly, *Ilex aquifolium*;   **11**. Field maple, *Acer campestre*;   **12**. Honeysuckle, *Lonicera periclymenum*;   **13**. Rowan, *Sorbus aucuparia*.

Plate 13. Atlantic oak woods—herbaceous plants

1. Common dog-violet, *Viola riviniana*;   **2**. Celandine, *Ranunculus ficaria*;   **3**. Red Campion, *Silene dioica*;   **4**. Enchanter's nightshade, *Circaea lutetiana* (inset×1½);   **5**. Dog's mercury, *Mercurialis perennis* (inset×2);   **6**. Bluebell, *Hyacinthoides non-scripta*;   **7**. Primrose, *Primula vulgaris*;   **8**. Wood anemone, *Anemone nemorosa*;   **9**. Ground-ivy, *Glechoma hederacea*;   **10**. Yellow archangel, *Lamiastrum galeobdolon*;   **11**. Lords-and-ladies, *Arum maculatum* (×⅓);   **12**. Bugle, *Ajuga reptans*;   **13**. Herb-Paris, *Paris quadrifolia*;   **14**. Sanicle, *Sanicula europaea* (inset×2);   **15**. Greater stitchwort, *Stellaria holostea*.

Atlantic vegetation

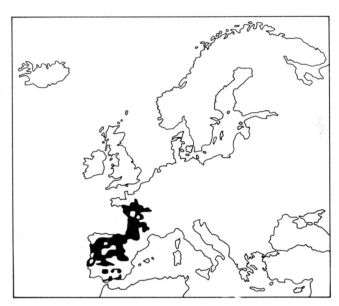

Map 6. Pyrenean oak *Quercus pyrenaica*.

common nettle, *Urtica dioica;* creeping buttercup, *Ranunculus repens;* and the tufted hair-grass, *Deschampsia cespitosa,* with various rushes and sedges. In addition such distinctive tall herbaceous plants as the hedge woundwort, *Stachys sylvatica;* water figwort, *Scrophularia auriculata;* and the marsh thistle, *Cirsium palustre,* are common.

The Pyrenean oak, *Quercus pyrenaica* (Map 6), prefers a mildly oceanic climate, and well developed woods are found on the western side of the Iberian peninsula, particularly along river valleys. It is also common in central Spain, and extends along the western seaboard of France.

Pyrenean oak woods are commonly found on acid soils, and have many acid-loving species in the lower layers. Extensive grazing often results in the destruction of these woods. As a result the heather-gorse heaths take their place in the more oceanic areas, while cistus–lavender heaths occur in drier sites at lower altitudes. The Pyrenean oak (Plate 79) also forms distinct woodland communities along the verges of rivers in northwestern Spain and in west Portugal; often mixed with this oak is the 'Iberian' willow, *Salix salvifolia.*

In general, Pyrenean oak woods have few other trees present, the commonest being the sessile oak, *Quercus petraea.* The shrub layer contains

Rose species	*Rosa* sps
Hawthorn	*Crataegus monogyna*
Blackthorn	*Prunus spinosa*
Broom	*Cytisus scoparius*
	Genista florida
Tree heath	*Erica arborea*
Bell heather	*Erica cinerea*
Heather	*Calluna vulgaris*
Honeysuckle	*Lonicera periclymenum.*

Common field-layer species include

Wild strawberry	*Fragaria vesca*
Common dog-violet	*Viola riviniana*
Wild basil	*Clinopodium vulgare*
Wood sage	*Teucrium scorodonia*
Large self-heal	*Prunella grandiflora*
Marjoram species	*Origanum* sps
Germander speedwell	*Veronica chamaedrys*
Betony	*Stachys officinalis*
Common cow-wheat	*Melampyrum pratense*
Lady's bedstraw	*Galium verum*
Hawkweed species	*Hieracium* sp.
Southern wood-rush	*Luzula forsteri*
Cock's-foot	*Dactylis glomerata*
False-brome species	*Brachypodium* sps.

An example of a Pyrenean oak wood at Macizo, Spain, at a height of 850–1100 m, has the dominant oak trees 10–14 m high and a shrub layer with

Pyrenean oak	*Quercus pyrenaica*
Rose species	*Rosa* sps
Hawthorn	*Crataegus monogyna*
Alder buckthorn	*Frangula alnus*
Tree heath	*Erica arborea.*

The field layer includes

Greater stitchwort	*Stellaria holostea*
Stinking hellebore	*Helleborus foetidus*
Hepatica	*Hepatica nobilis*
	Geum sylvaticum
Barren strawberry	*Potentilla sterilis*
Hairy violet	*Viola hirta*
Narrow-leaved lungwort	*Pulmonaria longifolia.*

Other examples of pyrenean oak woods from southern Madrona, Spain, have the wild service tree, *Sorbus torminalis,* and the narrow-leaved ash, *Fraxinus angustifolia,* in the tree layer, and honeysuckle in the shrub layer. The field layer includes

Three-nerved sandwort	*Moehringia trinervia*
Wood avens	*Geum urbanum*
Tutsan	*Hypericum androsaemum*
Enchanter's nightshade	*Circaea lutetiana*
Sanicle	*Sanicula europaea*
Common vincetoxicum	*Vincetoxicum hirundinaria*
Wood sage	*Teucrium scorodonia*
	Tanacetum corymbosum
	Scilla lilio-hyacinthus
Golden-scaled male-fern	*Dryopteris borreri*
Hard fern	*Blechnum spicant*
Bracken	*Pteridium aquilinum*

Pyrenean oaks are mixed with stone pine, *Pinus pinea,* in central Catalonia, with Scots pine in the Montes de Granada, and with the semi-evergreen Portuguese oak, *Quercus faginea,* and the evergreen holm oak, *Q. ilex,* in the Sierra Morena. Pyrenean oak woods are also well developed in the northern Iberian Mountains.

Atlantic beech woods (Plate 80)

The main domain of beech woods is in Central Europe, but well developed woods are widely established in the Atlantic region, particularly in northwest and central France, southeast England, and Belgium, while they occur as far north as southern Sweden. They also occur to a lesser extent in the montane region of northern Spain. Beech woods are found on shallow dry porous acid, neutral or alkaline soils. They cannot tolerate waterlogged soils, and cannot compete with oaks on the richer deeper soils. In most European beech woods, selective felling, to supply valuable timber, has resulted in very uniform stands of similar-aged trees, with straight grey trunks branched only above, and with a dense closed leaf canopy. The very low light intensity under the leaf canopy—only up to 2 per cent of light penetrates the shrub and field layers in summer—and the heavy leaf-fall with slow decomposition, results in a poorly developed shrub layer and a poor, but often distinctive, field layer. On shallow chalk soils, particularly on escarpments in England, the beech, *Fagus sylvatica,* is dominant, with ash, *Fraxinus excelsior;* common whitebeam, *Sorbus aria;* and wild cherry, *Prunus avium,* frequently present in the tree layer. The evergreens, yew, *Taxus baccata* and holly, *Ilex aquifolium,* are often common in the lower tree layer.

The shrub layer has

Hazel	*Corylus avellana*
Field maple	*Acer campestre*
Spindle	*Euonymus europaeus*
Box	*Buxus sempervirens*
Elder	*Sambucus nigra.*

Common field layer species are

Wood anemone	*Anemone nemorosa*
Dog's mercury	*Mercurialis perennis*
Sanicle	*Sanicula europaea*
Lords-and-ladies	*Arum maculatum*
Wood meadow-grass	*Poa nemoralis*
Hairy-brome	*Bromus ramosus*
False brome	*Brachypodium sylvaticum,*

while characteristic, but less common field layer species are

Green hellebore	*Helleborus viridis*
Columbine	*Aquilegia vulgaris*
Spurge-laurel	*Daphne laureola*
Yellow bird's-nest	*Monotropa hypopitys*
Solomon's-seal	*Polygonatum multiflorum*
Broad-leaved helleborine	*Epipactis helleborine*
White helleborine	*Cephalanthera damasonium*
Bird's-nest orchid	*Neottia nidus-avis.*

In northwest France and Belgium additional field-layer species may commonly occur in beech woods; these include

Stinking hellebore	*Helleborus foetidus*
Cowslip	*Primula veris*
Common vincetoxicum	*Vincetoxicum officinale*
Wood sage	*Teucrium scorodonia*
Honeysuckle	*Lonicera periclymenum*
Angular Solomon's-seal	*Polygonatum odoratum*
Black bryony	*Tamus communis*
Glaucous sedge	*Carex flacca.*

The shrub layer often includes

Hazel	*Corylus avellana*
Holly	*Ilex aquifolium*
Cornelian cherry	*Cornus mas*
Privet	*Ligustrum vulgare*
Wayfaring-tree	*Viburnum lantana,*

Fig. 11. Beechwood on acid soil with sparse field layer. Holland.

Plate 14. Atlantic Beech woods

1. Wild cherry, *Prunus avium*; **2**. Beech, *Fagus sylvatica*; **3**. Bramble, *Rubus fruticosus*; **4**. Solomon's-seal, *Polygonatum multiflorum*; **5**. Box, *Buxus sempervirens*; **6**. Broad-leaved helleborine, *Epipactis helleborine*; **7**. Yellow bird's-nest, *Monotropa hypopitys*; **8**. Spurge-laurel, *Daphne laureola*; **9**. White helleborine, *Cephalanthera damasonium*; **10**. Bird's-nest orchid, *Neottia nidus-avis*; **11**. Spindle, *Euonymus europaeus* (fruits); **12**. Cowslip, *Primula veris*; **13**. Stinking hellebore, *Helleborus foetidus*; **14**. Elder, *Sambucus nigra* (inset×4).

while oaks, ash, and lime may occur in the tree layer.

On poor acid soils the almost pure beech woods have a sparser field layer (Fig. 11), commonly including

Wood sorrel	*Oxalis acetosella*
Slender St. John's-wort	*Hypericum pulchrum*
Heather	*Calluna vulgaris*
Common cow-wheat	*Melampyrum pratense*
Hairy wood-rush	*Luzula pilosa*
Wavy hair-grass	*Deschampsia flexuosa*
Pill sedge	*Carex pilulifera*
Bracken	*Pteridium aquilinum*
Narrow buckler-fern	*Dryopteris carthusiana*
Hard fern	*Blechnum spicant.*

On deeper, well-drained soils, beech is co-dominant with the pedunculate oak, *Quercus robur,* or the ash, *Fraxinus excelsior,* or the hornbeam, *Carpinus betulus.* The shrub layer is usually richer—with holly, bramble, ivy, honeysuckle, etc. The field layer contains many of the oak wood species.

Ash woods

These are largely found north of the range of beech in the British Isles where they form woods on limestone rocks or screes. Ash woods also occur as a stage in the succession to the climax beech woods further south. They also commonly form wet ash woods in river valleys. Owing to the relatively open foliage of ash woods which allows more light to penetrate to the lower layers, both the scrub and field layers are well developed and rich in species (Fig. 12).

In addition to the dominant ash, *Fraxinus excelsior;* wych elm, *Ulmus glabra;* yew, *Taxus baccata;* common whitebeam, *Sorbus aria;* and small-leaved lime, *Tilia cordata,* are frequent in the tree layer, particularly in the west. The shrub layer comprises

Hazel	*Corylus avellana*
Traveller's-joy	*Clematis vitalba*
Hawthorn	*Crataegus monogyna*
Blackthorn	*Prunus spinosa*
Spindle	*Euonymus europaeus*
Buckthorn	*Rhamnus catharticus*
Dogwood	*Cornus sanguinea*
Privet	*Ligustrum vulgare*
Elder	*Sambucus nigra*
Wayfaring-tree	*Viburnum lantana.*

The field layer includes

Celandine	*Ranunculus ficaria*
Columbine	*Aquilegia vulgaris*
Dog's mercury	*Mercurialis perennis*
Woodruff	*Galium odoratum*
Wood forget-me-not	*Myosotis sylvatica*
Moschatel	*Adoxa moschatellina*
Giant bellflower	*Campanula latifolia*
Ramsons	*Allium ursinum*
Lords-and-ladies	*Arum maculatum*
Greater butterfly orchid	*Platanthera chlorantha.*

In the north of the Atlantic region the globeflower, *Trollius europaeus;* and Jacob's ladder, *Polemonium caeruleum,* are distinctive in the field layer.

On drier soils, different herbaceous species are found in the field layer including

Stone bramble	*Rubus saxatilis*
Hairy St. John's-wort	*Hypericum hirsutum*
Wood sage	*Teucrium scorodonia*
Ground-ivy	*Glechoma hederacea*
Lily-of-the-valley	*Convallaria majalis.*

Fig. 12. Atlantic ash wood showing dense shrub layer.

Ash woods on the northern coastal region of Spain have the narrow-leaved ash, *Fraxinus angustifolia*, dominant with the common alder, *Alnus glutinosa;* and hazel, *Corylus avellana*, in the shrub layer. The characteristic field layer species include

Wood spurge	*Euphorbia amygdaloides*
Primrose	*Primula vulgaris*
Bugle	*Ajuga reptans*
Betony	*Stachys officinalis*
False brome	*Brachypodium sylvaticum*
Remote sedge	*Carex remota*
Thin-spiked wood-sedge	*Carex strigosa*
Lady-fern	*Athyrium filix-femina.*

Oak–hornbeam woods

These woods are characteristic of Central Europe but some occur in southwest England and northwest France. They are distinguished by the presence of the wild daffodil, *Narcissus pseudonarcissus;* bluebell, *Hyacinthoides non-scripta;* honey-suckle, *Lonicera periclymenum;* greater stitchwort, *Stellaria holostea;* barren strawberry, *Potentilla sterilis;* and great wood-rush, *Luzula sylvatica*. In southwest France additional species such as butcher's-broom, *Ruscus aculeatus;* and southern wood-rush, *Luzula forsteri*, have a high constancy.

Wet woodlands (Plate 82)

These occur in valleys, by rivers, and on lake-sides on more or less waterlogged soils, which are rich in minerals. They usually develop on wet fen peat where drainage is impeded (Fig. 13). As they mature and the water-table drops, they dry out and may ultimately develop into wet oak woods (see Fig. 17, p. 66). Alder woods, or *carr*, which are dominated by *Alnus glutinosa*, are found throughout the Atlantic region, from southwestern Norway to northern Spain, usually occurring in small stands. They are also well developed locally in the Central European region. Commonly associated trees are the ash, *Fraxinus excelsior*, and the downy birch, *Betula pubescens*. The rich shrub layer contains

Grey willow	*Salix cinerea*
Black currant	*Ribes nigrum*
Red currant	*Ribes rubrum*
Gooseberry	*Ribes uva-crispa*
Hawthorn	*Crataegus monogyna*
Spindle	*Euonymus europaeus*
Buckthorn	*Rhamnus catharticus*
Alder buckthorn	*Frangula alnus*
Privet	*Ligustrum vulgare*
Guelder-rose	*Viburnum opulus.*

The field layer includes

Hop	*Humulus lupulus*
Common nettle	*Urtica dioica*
Meadowsweet	*Filipendula ulmaria*
Ivy	*Hedera helix*
Hedge bindweed	*Calystegia sepium*
Comfrey	*Symphytum officinale*
Bittersweet	*Solanum dulcamara*
Hemp agrimony	*Eupatorium cannabinum*
Yellow iris	*Iris pseudacorus*
Greater tussock-sedge	*Carex paniculata*
Marsh fern	*Thelypteris palustris*

and other sedge species.

Alder woods also occur on acid soils in Ireland and northern Spain. In the latter country, alder woods contain such additional field layer species as

Cornish money-wort	*Sibthorpia europaea*
Lesser skullcap	*Scutellaria minor*
Royal fern	*Osmunda regalis*
Broad buckler-fern	*Dryopteris dilatata*
Hard fern	*Blechnum spicant*

and the bogmoss, *Sphagnum squarrosum.*

Fig. 13. Wet birch community with fen undergrowth. Steinhuder Meer, W. Germany.

Plate 15. Atlantic wet woodlands

1. Hedge bindweed, *Calystegia sepium*;   **2**. Guelder-rose, *Viburnum opulus*;   **3**. Hop, *Humulus lupulus*;   **4**. Alder buckthorn, *Frangula alnus*;   **5**. Comfrey, *Symphytum officinale*;   **6**. Greater tussock-sedge, *Carex paniculata*;   **7**. Black currant, *Ribes nigrum*;   **8**. Grey willow, *Salix cinerea*;   **9**. Buckthorn, *Rhamnus catharticus*;   **10**. Ivy, *Hedera helix*;   **11**. Red currant, *Ribes rubrum* (leaf);   **12**. Hemp agrimony, *Eupatorium cannabinum*;   **13**. Privet, *Ligustrum vulgare*;   **14**. Common nettle, *Urtica dioica*.

Atlantic vegetation

Birch woods

These are most frequent in the northern and montane zones of the Atlantic region. They occur from sea level up to an altitude of about 600 m in Scotland and southern Scandinavia, are well developed above the oak woods or pine woods, and form the highest forest communities. They are often open in structure, with widely spaced rather uniform trees, and the field and shrub layers are commonly highly grazed, particularly in northern Scotland.

There are two main types of birch wood—those dominated by grasses, and those which are bilberry-rich. The former has such common grasses in the field layer as the sweet vernal-grass, *Anthoxanthum odoratum;* common bent, *Agrostis capillaris;* the Yorkshire-fog, *Holcus lanatus,* with the bluebell, *Hyacinthoides non-scripta;* and the lemon-scented fern, *Thelypteris limbosperma.* The bilberry-rich birch woods have, in addition to the bilberry, *Vaccinium myrtillus,* the cowberry, *Vaccinium vitis-idaea;* great wood-rush, *Luzula sylvatica;* common bent, *Agrostis capillaris;* and the hard fern, *Blechnum spicant.* The hazel, *Corylus avellana;* and the juniper, *Juniperus communis,* occur frequently in the shrub layer.

Atlantic coniferous woods

The Atlantic climate does not on the whole favour the evolution of trees and shrubs with small thick narrow evergreen leaves. These have largely been evolved to reduce water loss during unfavourable seasons of the year and they are more prominent in drier or harsher climates, such as in the Mediterranean or montane regions. The two most important pines of the Atlantic regions are the Scots pine, *Pinus sylvestris,* and the maritime pine, *Pinus pinaster,* of southwestern France, northwestern Spain, and Portugal. Another coniferous evergreen tree, which forms very limited natural woods in northwestern France, southern Britain, and northern Portugal, is the yew, *Taxus baccata* (Plate 81).

The Scots pine is native in Scandinavia and Scotland, while further south in the Atlantic region it has been extensively planted and has become naturalized and subspontaneous. Further east in Central Europe it is native. Boreal pine woods have already been discussed. Natural Scottish pine woods are found particularly on mountain slopes in the central and eastern Highlands, on drier shallower sandy or gravelly soils poor in mineral salts. They have the following composition

Tree layer

Scots pine	*Pinus sylvestris*
Downy birch	*Betula pubescens*
Common alder	*Alnus glutinosa*
Aspen	*Populus tremula*
Rowan	*Sorbus aucuparia*

Shrub layer

Juniper	*Juniperus communis*
Heather	*Calluna vulgaris*
Bilberry	*Vaccinium myrtillus*
Cowberry	*Vaccinium vitis-idaea*

Field layer

Tormentil	*Potentilla erecta*
Chickweed wintergreen	*Trientalis europaea*
Round-leaved wintergreen	*Pyrola rotundifolia*
Intermediate wintergreen	*Pyrola media*
Common wintergreen	*Pyrola minor*
Serrated wintergreen	*Orthilia secunda*
One-flowered wintergreen	*Moneses uniflora*
Heath bedstraw	*Galium saxatile*
Hard fern	*Blechnum spicant.*

The Moss layer includes

Hylocomium splendens
Rhytidiadelphus loreus
Rhytidiadelphus triquetrus
Pleurozium schreberi
Ptilium cristacastrensis.

Scots pine woods occurring at the present day in southern England and western France are in all probability not native, though in the Boreal period, 10 000 to 9500 years ago, they were probably dominant on lighter drier soils. In the milder Atlantic period that followed, these Scots pine woods were replaced by deciduous woods which became dominant. Present day scots pine woods of the south, with their characteristic heath shrub and field layers, are almost certainly subspontaneous—that is to say that they are derived from trees which were originally planted, possibly in Roman times.

Maritime pine woods (Plate 83) dominated by *Pinus pinaster,* develop naturally on sandy soils and dunes, in a climate which is strongly influenced by the Atlantic. They are well developed in Les Landes, bordering the Bay of Biscay, in France but they have been very largely extended by planting in the nineteenth century. Semi-natural maritime pine woods may have oaks such as the pedunculate oak *Quercus robur,* the Pyrenean oak, *Quercus pyrenaica,* and the holm oak, *Quercus ilex,* associated with the pine. The shrub and herb layers are composed of heathland plants typical of southern Atlantic heaths.

Deciduous bush communities (Plate 84)

These are largely stages in the natural succession of vegetation where grassland is gradually superseded by the climax woodland. However, locally bush communities are quite common in the Atlantic region, particularly where grazing is relaxed, or cultivation has ceased. Closely associated and rather similar are the bush communities of hedges, roadside verges, commons, and greens, where similar woody species grow in close proximity and form thickets. Typical shrubs of these communities are

Juniper	*Juniperus communis*
Hazel	*Corylus avellana*
Bramble	*Rubus fruticosus*
Field rose	*Rosa arvensis*
Common whitebeam	*Sorbus aria*
Hawthorn	*Crataegus monogyna*
Blackthorn	*Prunus spinosa*

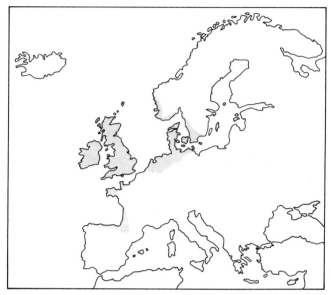

Map 7. Main areas in which lowland heaths occur. (After Gimingham (1972).)

Spindle	*Euonymus europaeus*
Buckthorn	*Rhamnus catharticus*
Dogwood	*Cornus sanguinea*
Privet	*Ligustrum vulgare*
Wayfaring-tree	*Viburnum lantana*

with the climbers

Traveller's-joy	*Clematis vitalba*
Ivy	*Hedera helix*
Honeysuckle	*Lonicera periclymenum.*

In the south of the Atlantic region additional shrubs include the box, *Buxus sempervirens; Rubus ulmifolius;* and the herbaceous climber black bryony, *Tamus communis.* Hedges are common in the more oceanic part of the Atlantic region, where rich pastures are grazed by livestock, which have to be kept within bounds. Further east and south, on less productive soils, livestock have to cover wider grazing areas to obtain their food, and hedges (and walls and fences) become less frequent. Modern agriculture has also eliminated many hedges in the Atlantic region. Hedges are often rich in attractive herbaceous plants, again much reduced by modern weed-killers and artificial manuring.

Atlantic heaths

These form very distinct communities in the Atlantic region, from southwestern Scandinavia to northern Spain, and inland to northwestern Germany (Map 7). In all probability, heaths are nowhere the natural climax vegetation, except possibly in the extreme Atlantic coastal region. They are largely maintained as heaths by such man-controlled activities as recurrent fires, livestock grazing, tree clearance, maintenance of grouse moors, etc. During the last 50–100 years,

there has been a large reduction in the area covered by heaths in the Atlantic region as a result of afforestation, changing agricultural practices, and reduced grazing pressures.

Heaths are found in moist temperate climates, with mild winters and with at least four months with a mean temperature above 10°C. They cannot tolerate drought. They are formed on acid siliceous soils and they develop a typical soil-profile—the *podsol* (Fig. 14). Podsols have a distinctive layered structure. At the surface a thin layer of peat accumulates. Below this lies a pale, often white layer of sand, out of which the mineral salts and peat have been washed downwards (leached) by percolating rainwater; these accumulate in the lower soil layers. These two upper layers are known as the *A horizon.* The *B horizon* is the layer of soil into which the peat and mineral salts have been washed and these usually accumulate in two distinct layers; an upper dark-brown layer, rich in peat, and a lower rust-coloured layer, in which the iron and other minerals accumulate. The iron tends to cement the sand grains together and thus often forms an impervious iron-pan. Below this is the usually reddish sandy subsoil—the *C horizon.* Similar podsols are developed on boulder clays but are not so distinctly layered.

Heaths usually form a closed canopy of dwarf ericaceous shrubs, 30–40 cm in height in the case of the heather, *Calluna vulgaris,* but considerably higher and up to 1 m or more in some of the southern heaths. There is often a second discontinuous layer 10–20 cm high, of such plants as the bell

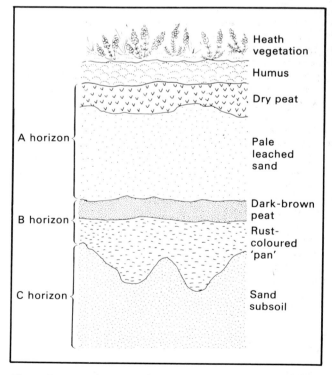

Fig. 14. Podsol soil profile. This kind of soil is characteristic of Atlantic heaths (and also of boreal taiga) and develops in areas of high rainfall.

heather, *Erica cinerea*; cowberry, *Vaccinium vitis-idaea*; crowberry, *Empetrum nigrum*; and bearberry, *Arctostaphylos uva-ursi,* which may also be a creeping shrublet (Fig. 15). Grasses and sedges commonly occur in this layer. Below this is a layer up to about 10 cm high of large mosses and fruticose lichens as well as the basal rosettes of rushes, sedges, and grasses. Over the surface of the soil grow mat-forming mosses, liverworts, and lichens.

Atlantic heaths may be subdivided into northern heaths, oceanic heaths, and southern heaths.

Northern heaths

These are found in southern Norway, southwestern Sweden, and northern Britain, and have the following typical ericaceous shrubby species

Heather	*Calluna vulgaris*
Bilberry	*Vaccinium myrtillus*
Cowberry	*Vaccinium vitis-idaea*
Crowberry	*Empetrum nigrum*

with mosses such as

Hylocomium splendens
Pleurozium schreberi
Hypnum cupressiforme.

Another northern type of heath has the bell heather, *Erica cinerea,* dominant, with lady's bedstraw, *Galium verum,* and with grasses such as marram grass, *Ammophila arenaria* (on dune-heaths near the sea); red fescue, *Festuca rubra*; and the moss, *Rhytidiadelphus triquetrus.* Lichens present include

Cladonia arbuscula
Cladonia portentosa
Cetraria aculeata.

More oceanic northern heaths commonly contain the tormentil, *Potentilla erecta*, and the common bent, *Agrostis capillaris*, along with the hard fern, *Blechnum spicant.* Typical mosses are *Hypnum cupressiforme* and *Dicranum scoparium.*

Oceanic heaths (Plate 85)

These are distinguished largely by the presence of gorse species. The most typical heath species are the dwarf gorse, *Ulex minor*, and the western gorse, *Ulex gallii* (Fig. 16). The much taller, more widely distributed common gorse, *Ulex europaeus*, is more characteristic of abandoned dry grasslands, commons, and along footpaths. Heaths of southern England and northwestern France have the following widespread species

Common gorse	*Ulex europaeus*
Dwarf gorse	*Ulex minor*
Heather	*Calluna vulgaris*
Bilberry	*Vaccinium myrtillus*
Wavy hair-grass	*Deschampsia flexuosa*
Purple moor-grass	*Molinia caerulea*
Bracken	*Pteridium aquilinum.*

Further west and south, the western gorse, *Ulex gallii*, becomes commoner, and St. Dabeoc's heath, *Daboecia cantabrica*, may occur. In Spain and Portugal, the heath, *Erica australis*, and the broom-like *Chamaespartium tridentatum* are

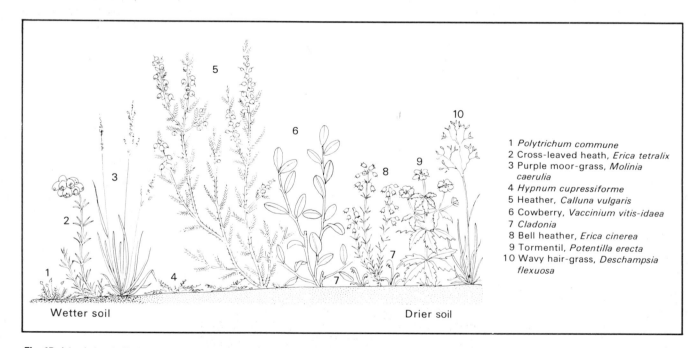

1 *Polytrichum commune*
2 Cross-leaved heath, *Erica tetralix*
3 Purple moor-grass, *Molinia caerulia*
4 *Hypnum cupressiforme*
5 Heather, *Calluna vulgaris*
6 Cowberry, *Vaccinium vitis-idaea*
7 *Cladonia*
8 Bell heather, *Erica cinerea*
9 Tormentil, *Potentilla erecta*
10 Wavy hair-grass, *Deschampsia flexuosa*

Wetter soil   Drier soil

Fig. 15. Atlantic heath. Typical layered structure of an northern Atlantic heath.

widespread. Other heather species form heathlands in the more oceanic region, such as, for example, the Cornish heath, *Erica vagans*, which in France (Belle Isle) iş associated with

Gorse	*Ulex europaeus*
Bell heather	*Erica cinerea*
	Cirsium filipendulum
Tor-grass	*Brachypodium pinnatum.*

On the Lizard peninsula, Cornwall, the dominant Cornish heath, *Erica vagans*, typically occurs with

Gorse	*Ulex europaeus*
Western gorse	*Ulex gallii*
Bell heather	*Erica cinerea*
Heather	*Calluna vulgaris*
Wild thyme	*Thymus praecox*
Glaucous sedge	*Carex flacca*

and, in wetter areas, *Erica vagans* occurs with

Tormentil	*Potentilla erecta*
Purple moor-grass	*Molinia caerulea*
Black bog-rush	*Schoenus nigricans*
Flea sedge	*Carex pulicaris.*

Another heather, Mackay's heath, *Erica mackaiana*, forms heaths in northwestern Spain and western Ireland, often in a mosaic with mat-grass, *Nardus stricta*, grassland.

Continental heathlands extend to northern Germany and further east into Central Europe. They typically have such distinctive leguminous shrubs as the petty whin, *Genista anglica*; hairy greenweed, *Genista pilosa*; and the broom, *Sarothamnus scoparius*.

Southern heaths (Plate 86)

These are in general richer in ericaceous species. They include the green heather, *Erica scoparia*; the more Mediterranean tree heath, *Erica arborea*; and *Erica erigena*, all taller shrubs which may reach heights of 1–5 m and form distinct thickets.

In montane regions of northern Portugal and Spain, as a result of the destruction of woods, the tall green heather, *Erica scoparia*, and the much smaller Dorset heath, *Erica ciliaris*, may be co-dominant, with such associated Iberian species as

Echinospartum lusitanicum
Genista falcata
Chamaespartium tridentatum
Polygala microphylla
Tuberaria globularifolia
Erica australis
Erica umbellata
Luzula lactea.

Mires

These are a characteristic type of vegetation where peat accumulates in the surface soil. Mires occur in the high rainfall areas of the western Atlantic region, both in the lowlands and montane zone, particularly in western Ireland and northern Scotland. On acid, mineral-deficient soils, where drainage is poor, and peat accumulates year by year, bogs are formed which cover large areas of level and sloping ground, like a blanket, and these are known as *blanket bogs*. *Raised bogs*, by contrast, are gradually built up over relatively mineral-rich fen peat. As peat accumulates and rises above the level of the ground-water, the upper layers of peat are fed largely by rainwater and become acid and deficient in minerals, and they thus support a bog vegetation. *Valley bogs* may, by contrast,

Fig. 16. Oceanic heath dominated by western gorse *Ulex gallii*. S. Devon, UK.

Rosemary Wise.

Plate 16. Atlantic oceanic heaths (s=southern; c=continental)
1. Dwarf gorse, *Ulex minor* (inset×1); **2**. Western gorse, *Ulex gallii* (inset×1); **3**. St. Dabeoc's heath, *Daboecia cantabrica*; **4**. Mackay's heath, *Erica mackaiana*; **5**. Spanish heath, *Erica australis* (s); **6**. Green heath, *Erica scoparia* (s); **7**. Irish heath, *Erica erigena* (s); **8**. Dorset heath, *Erica ciliaris*; **9**. *Erica umbellata* (s); **10**. Broom, *Cytisus scoparius* (inset×1); **11**. *Genista florida* (s); **12**. Petty whin, *Genista anglica* (c) (inset×1); **13**. Hairy greenweed, *Genista pilosa* (c) (inset×1); **14**. Cornish heath, *Erica vagans* (inset×6); **15**. *Luzula lactea* (s).

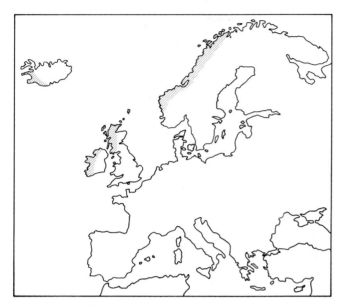

Map 8. Region of blanket-bogs.

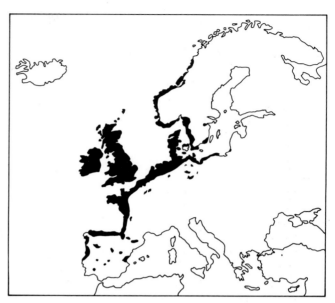

Map 9. Cross-leaved heather *Erica tetralix*, an important species of heath and bog with a strongly atlantic distribution.

develop independently of rainwater. They occur in valleys where drainage comes from acid rocks or subsoils, and where the acid water tends to accumulate in hollows in the valley.

Fens, on the other hand, develop over peat layers which are usually neutral or alkaline, being supplied with water draining from rocks or subsoils rich in minerals, and thus in consequence have quite different vegetation communities.

Blanket-bogs (Map 8; Plate 87)

These are usually characterized by the dominance of the purple moor-grass, *Molinia caerulea*; and the deergrass, *Scirpus cespitosus*, and in western Ireland, by the black bog-rush, *Schoenus nigricans*. In montane areas on the other hand, the hare's-tail cottongrass, *Eriophorum vaginatum*, is very common, and the heather, *Calluna vulgaris*, is widespread, except in the wettest areas.

Blanket-bogs in the hills of western Scotland commonly contain the species

Round-leaved sundew	*Drosera rotundifolia*
Tormentil	*Potentilla erecta*
Cross-leaved heather	*Erica tetralix*
Heather	*Calluna vulgaris*
Bog asphodel	*Narthecium ossifragum*
Deergrass	*Scirpus cespitosus.*

Bogmosses and mosses include

Sphagnum capillifolium
Sphagnum magellanicum
Sphagnum papillosum
Sphagnum subnitens
Hypnum cupressiforme
Racomitrium lanuginosum

with the characteristic lichen

Cladonia portentosa.

The pools which occur frequently in blanket-bogs commonly contain

Bogbean	*Menyanthes trifoliata*
Lesser bladderwort	*Utricularia minor*
Common cottongrass	*Eriophorum angustifolium*
Hare's-tail cottongrass	*Eriophorum vaginatum*
White beak-sedge	*Rhynchospora alba*
Bog-sedge	*Carex limosa*

with the bogmosses, *Sphagnum cuspidatum* and *Sphagnum subsecundum*.

In western Ireland the pools in the blanket bogs usually have the marsh St. John's wort, *Hypericum elodes*; and the bog pondweed, *Potamogeton polygonifolius*, together with the bogbean, *Menyanthes trifoliata*. In the bogs of northwest Spain the following are typical

Bog myrtle	*Myrica gale*
Grass-of-Parnassus	*Parnassia palustris*
Marsh violet	*Viola palustris*
Cross-leaved heather	*Erica tetralix*
Purple moor-grass	*Molinia caerulea*
Hare's-tail cottongrass	*Eriophorum vaginatum*
Black bog-rush	*Schoenus nigricans*
Star sedge	*Carex echinata.*

Raised bogs

These are often formed over either fen or bog peat and are developed in valleys where drainage is impeded. New layers of peat are built up year by year by colonies of bogmosses which

Plate 17. Atlantic mires; bogs.

1. Brown beak-sedge, *Rhynchospora fusca*;   **2**. Star sedge, *Carex echinata*;   **3**. Marsh St. John's-wort, *Hypericum elodes*;   **4**. Bog pondweed, *Potamogeton polygonifolius*;   **5**. Deergrass, *Scirpus cespitosus*;   **6**. Sharp-flowered rush, *Juncus acutiflorus*;   **7**. Black bog-rush, *Schoenus nigricans*;   **8**. Marsh clubmoss, *Lepidotis inundata*;   **9**. Marsh pennywort, *Hydrocotyle vulgaris* (inset×6);   **10**. Cross-leaved heather, *Erica tetralix*;   **11**. Bog myrtle, *Myrica gale*;   **12**. Many-stalked spike-rush, *Eleocharis multicaulis*;   **13**. Common butterwort, *Pinguicula vulgaris*;   **14**. Pale butterwort, *Pinguicula lusitanica* (s);   **15**. Hare's-tail cottongrass, *Eriophorum vaginatum*.

grow together above the general water-level, and gradually they form a convex raised bog overlying the original fen or bog. The surface drainage water runs off in channels, or *rills*, and accumulates in the depressions or *laggs* usually round the margins of the raised bogs. The outer peripheral part or *rand* of the raised bog is usually static and no longer growing and slopes down to the lagg, while the central portion is actively rising and is covered with small watery hollows and intervening hummocks. In these hollows, the dominant bogmoss is *Sphagnum cuspidatum* which is a very active peat-former. Round the bases of the hummocks usually grows the white beak-sedge, *Rhyncosphora alba.* Higher up on the hummocks the bog asphodel, *Narthecium ossifragum;* cross-leaved heather, *Erica tetralix;* and the common cottongrass, *Eriophorum angustifolium,* are dominant, often with the cranberry, *Vaccinium oxycoccos;* and the bog rosemary, *Andromeda polifolia,* while the most important bogmoss at this level is *Sphagnum papillosum.* Higher still on the hummocks grow the hare's-tail cottongrass, *Eriophorum vaginatum;* and the deergrass, *Scirpus cespitosus;* and on the drier summit the reddish bogmoss, *Sphagnum capillifolium,* with the heather, *Calluna vulgaris;* and the lichens, *Cladonia arbuscula* and *Cladonia portentosa.*

Raised bogs have in general an assemblage of species similar to that of blanket bogs, but examples in northwestern Spain often contain the following additional species

Oblong-leaved sundew	*Drosera intermedia*
Marsh pennywort	*Hydrocotyle vulgaris*
Pale butterwort	*Pinguicula lusitanica*
Bog asphodel	*Narthecium ossifragum*
Sharp-flowered rush	*Juncus acutiflorus*
Bulbous rush	*Juncus bulbosus*
Few-flowered sedge	*Carex pauciflora*
Many-stalked spike-rush	*Eleocharis multicaulis*
Marsh clubmoss	*Lepidotis inundata.*

Fen communities

Poor fens occur on neutral or slightly acid soils. They are floristically mid-way between bogs and true fens. They are often associated with bogs but occur where there is local mineral seepage. Such mires contain fewer species than true fens. Examples in Scotland have the dominant sharp-flowered rush, *Juncus acutiflorus,* with

Meadow buttercup	*Ranunculus acris*
Marsh willowherb	*Epilobium palustre*
Self-heal	*Prunella vulgaris*
Sneezewort	*Achillea ptarmica*
Marsh hawk's-beard	*Crepis paludosa*
Yorkshire-fog	*Holcus lanatus*
Carnation sedge	*Carex panicea*

and the mosses, *Calliergon cuspidatum* and *Bryum pseudo-triquetrum.* In Ireland similar poor fens may have the following characteristic plants

| Lesser spearwort | *Ranunculus flammula* |
| Silverweed | *Potentilla anserina* |

Marsh pennywort	*Hydrocotyle vulgaris*
Water mint	*Mentha aquatica*
Creeping bent	*Agrostis stolonifera*
Jointed rush	*Juncus articulatus*
Common spike-rush	*Eleocharis palustris*
Common sedge	*Carex nigra.*

Elsewhere other poor fen species may include the shrubby bog myrtle, *Myrica gale;* and the marsh lousewort, *Pedicularis palustris;* the bulbous rush, *Juncus bulbosus;* and the star sedge, *Carex echinata.*

Rich fens are developed on peat which is alkaline or neutral, as a result of the water draining largely from mineral-rich and especially calcareous rocks. When the peat is built up above the general water level it is commonly colonized by woody species with the formation of alder and willow carrs, which may ultimately develop into damp oak woods or ash woods (Fig. 17). Many rich fens are encountered at different stages in the succession to woodland. They are often maintained in a sub-climax state by man's activities, such as regular or intermittent cutting for litter, thatching, or kindling, or by the control of water levels, etc.

In cases where the water level is above the peat or mud layers, then reed-swamp communities are dominant, and such communities will be described in the wetlands section of this book. Alder carr is described under the wet woodland sections of the Atlantic, Boreal, and Central European regions.

Fens are characteristically dominated by sedges and rushes, with grasses much less in evidence. There are also many distinctive herbaceous species which render fen communities readily identifiable. The commoner sedges and rushes include

Black bog-rush	*Schoenus nigricans*
Great fen-sedge	*Cladium mariscus*
Jointed rush	*Juncus articulatus*
Blunt-flowered rush	*Juncus subnodulosus*
Common yellow-sedge	*Carex demissa*
Star sedge	*Carex echinata*
Glaucous sedge	*Carex flacca*
Slender sedge	*Carex lasiocarpa*
Common sedge	*Carex nigra*
Carnation sedge	*Carex panicea*
Flea sedge	*Carex pulicaris.*

The following distinctive flowering plants, many of which occur in alder carr, are widely distributed in rich fens

Ragged-robin	*Lychnis flos-cuculi*
Marsh marigold	*Caltha palustris*
Lesser spearwort	*Ranunculus flammula*
Common meadow-rue	*Thalictrum flavum*
Meadowsweet	*Filipendula ulmaria*
Marsh cinquefoil	*Potentilla palustris*
Marsh pea	*Lathyrus palustris*
Purple loosestrife	*Lythrum salicaria*
Marsh pennywort	*Hydrocotyle vulgaris*
Wild angelica	*Angelica sylvestris*
Milk-parsley	*Peucedanum palustre*
Yellow loosestrife	*Lysimachia vulgaris*
Marsh valerian	*Valeriana dioica*

Plate 18. Atlantic mires—sedges and mosses

1. *Sphagnum imbricatum*; **2**. *Sphagnum cuspidatum*; **3**. *Sphagnum fuscum*; **4**. *Cladonia portentosa*; **5**. *Hypnum cupressiforme*; **6**. *Racomitrium lanuginosum*; **7**. *Bryum pseudotriquetrum*; **8**. *Cratoneuron commutatum*; **9**. *Sphagnum papillosum*; **10**. Common sedge, *Carex nigra*; **11**. Carnation sedge, *Carex panicea*; **12**. Common yellow-sedge, *Carex demissa*; **13**. *Calliergon cuspidatum*; **14**. Flea sedge, *Carex pulicaris*; **15**. Few-flowered spike-rush, *Eleocharis quinqueflora*; **16**. Glaucous sedge, *Carex flacca*.

Plate 19. Atlantic mires; fen communities

1. Lesser spearwort, *Ranunculus flammula*; **2**. Meadow thistle, *Cirsium dissectum*; **3**. Yellow Iris, *Iris pseudacorus*; **4**. Meadow buttercup, *Ranunculus acris*; **5**. Flea sedge, *Carex pulicaris* (inset×3); **6**. Silverweed, *Potentilla anserina*; **7**. Jointed rush, *Juncus articulatus*; **8**. Blunt-flowered rush, *Juncus subnodulosus*; **9**. Ragged-robin, *Lychnis flos-cuculi*; **10**. Marsh valerian, *Valeriana dioica*; **11**. Common meadow-rue, *Thalictrum flavum*; **12**. Devil's-bit scabious, *Succisa pratensis*; **13**. Early marsh-orchid, *Dactylorhiza incarnate*; **14**. Marsh pea, *Lathyrus palustris*; **15**. Wild angelica, *Angelica sylvestris*.

1

2

3

4

5

6

7

8

9

10

11

12

13

14

15

R. WISE.

Devil's-bit scabious	*Succisa pratensis*
Meadow thistle	*Cirsium dissectum*
Yellow iris	*Iris pseudacorus*
Marsh helleborine	*Epipactis palustris*
Early marsh-orchid	*Dactylorhiza incarnata*

and the marsh fern, *Thelypteris palustris.*

Atlantic grasslands

Grasslands are a characteristic, important, and widespread type of vegetation of the Atlantic region, though in almost every case they are semi-natural and are maintained in this state by man's activities and those of his livestock. The moist climate, with comparatively mild winters, results in a rich growth of vegetation which can provide food and pasture for livestock throughout the year, either directly as grazing, or from hay crops, ensilage, etc. In the northern parts of the Atlantic region grasslands occur from sea level to montane areas while, further south, grasslands are more restricted to montane levels where they have often replaced forests.

Neutral grasslands (Plate 90)

These are the richest grasslands. They are found on neutral or alluvial soils and are rich in mineral salts; they are well drained but remain moderately moist throughout the year. Typical examples are the enclosed hedged, fenced, or walled meadows of the Atlantic west, while woodland and roadside verges, and commons and greens have similar grasslands. They yield rich grazing and very good hay or ensilage crops. The composition of these grasslands depends largely on the types of management, involving grazing, cropping, fertilizing, etc. Common and widely distributed grasses are

Perennial rye-grass	*Lolium perenne*
Cock's-foot	*Dactylis glomerata*
Crested dog's-tail	*Cynosurus cristatus*
False oat-grass	*Arrhenatherum elatius*
Sweet vernal-grass	*Anthoxanthum odoratum*
Yorkshire-fog	*Holcus lanatus*
Common bent	*Agrostis capillaris*
Timothy	*Phleum pratense.*

Other common and characteristic herbaceous plants include

Meadow buttercup	*Ranunculus acris*
Bulbous buttercup	*Ranunculus bulbosus*
Creeping buttercup	*Ranunculus repens*
Narrow-leaved vetch	*Vicia angustifolia*
Common vetch	*Vicia sativa*
Lesser trefoil	*Trifolium dubium*
Red clover	*Trifolium pratense*
White clover	*Trifolium repens*
Common bird's-foot-trefoil	*Lotus corniculatus*
Meadow vetchling	*Lathyrus pratensis*
Self-heal	*Prunella vulgaris*
Ribwort plantain	*Plantago lanceolata*
Yarrow	*Achillea millefolium*
Common ragwort	*Senecio jacobaea*
Common knapweed	*Centaurea nigra*
Meadow saffron	*Colchicum autumnale.*

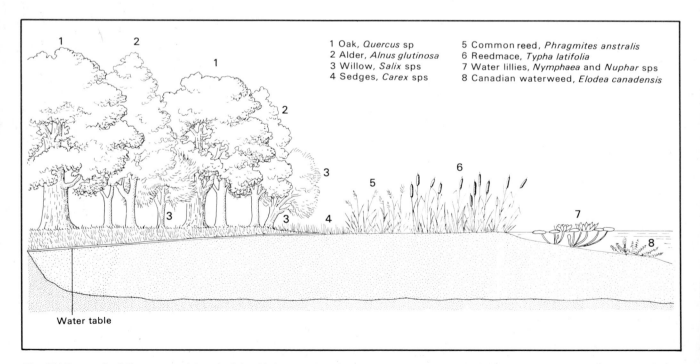

1 Oak, *Quercus* sp
2 Alder, *Alnus glutinosa*
3 Willow, *Salix* sps
4 Sedges, *Carex* sps
5 Common reed, *Phragmites anstralis*
6 Reedmace, *Typha latifolia*
7 Water lillies, *Nymphaea* and *Nuphar* sps
8 Canadian waterweed, *Elodea canadensis*

Water table

Fig. 17. Wet woodland. Here we see the seral relationship between Atlantic wet woodland, reed swamp, and open water. (Redrawn from Moore 1982.)

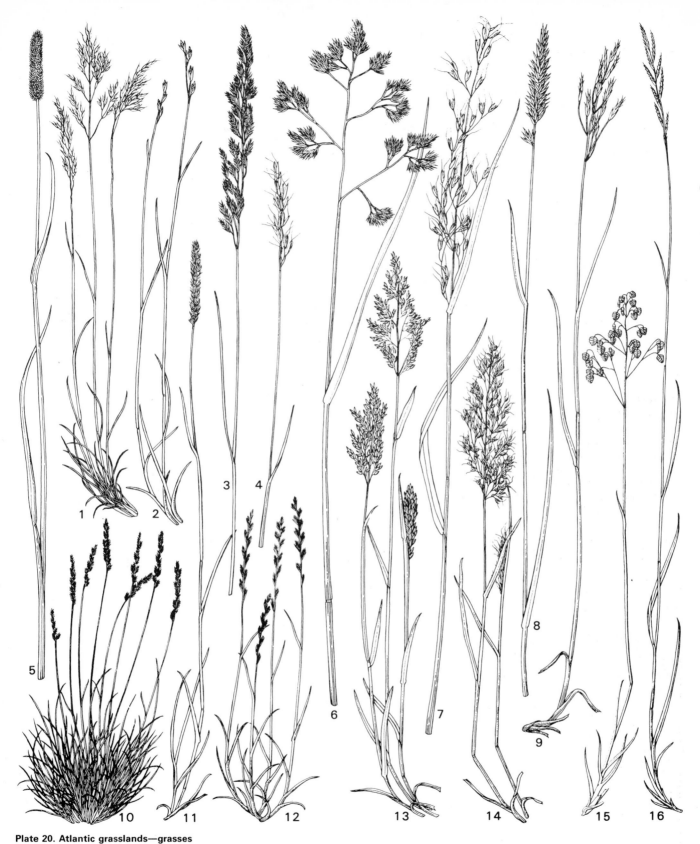

Plate 20. Atlantic grasslands—grasses

1. Wavy hair-grass, *Deschampsia flexuosa*; **2**. *Danthonia decumbens*; **3**. Red fescue, *Festuca rubra*; **4**. Meadow oat-grass, *Avenula pratensis*; **5**. Timothy, *Phleum pratense*; **6**. Cock's-foot, *Dactylis glomerata*; **7**. False oat-grass, *Arrhenatherum elatius*; **8**. Sweet vernal-grass, *Anthoxanthum odoratum*; **9**. Upright brome, *Bromus erectus*; **10**. Sheep's-fescue, *Festuca ovina*; **11**. Crested dog's-tail, *Cynosurus cristatus*; **12**. Perennial rye-grass, *Lolium perenne*; **13**. Yorkshire-fog, *Holcus lanatus*; **14**. Yellow oat-grass, *Trisetum flavescens*; **15**. Quaking-grass, *Briza media*; **16**. Tor-grass, *Brachypodium pinnatum*.

Plate 21. Atlantic grasslands—herbaceous species (excluding grasses)
1. Small scabious, *Scabiosa columbaria*;   **2**. Salad burnet, *Sanguisorba minor*;   **3**. Common rockrose, *Helianthemum nummularium*;   **4**. Kidney vetch, *Anthyllis vulneraria*;   **5**. Bulbous buttercup, *Ranunculus bulbosus* (inset×5);   **6**. Self-heal, *Prunella vulgaris*;   **7**. Fragrant orchid, *Gymnadenia conopsea*;   **8**. Carline thistle, *Carlina vulgaris*;   **9**. Horseshoe vetch, *Hippocrepis comosa*;   **10**. Red clover, *Trifolium pratense*;   **11**. Pyramidal orchid, *Anacamptis pyramidalis*;   **12**. Common knapweed, *Centaurea nigra* (inset×1½);   **13**. Wild thyme, *Thymus praecox*;   **14**. Autumn gentian, *Gentianella amarella*;   **15**. Bee orchid, *Ophrys apifera*.

Calcareous grasslands (Plate 89)

Grasslands on chalk and limestone are found usually on slopes or valley-sides, where the soil is thin and drainage is strong and cultivation is not possible. Again these grasslands are semi-natural and are maintained by grazing. Otherwise they would be quickly colonized by shrubs and later trees; many examples of this colonization by woody species can be seen on the open steep-sided chalk downlands. Sheep and cattle grazing is important in maintaining these grasslands, and so was grazing by rabbits, but recently the effects of myxomatosis have tended to reduce this pressure.

One type of calcareous grassland is dominated by the fescue grasses. These include the sheep's-fescue, *Festuca ovina* and the red fescue, *Festuca rubra,* commonly along with the downy oat-grass, *Avenula pubescens;* the meadow oat-grass, *Avenula pratensis;* and the common bent, *Agrostis capillaris.* Another type of grassland is dominated by the upright brome, *Bromus erectus,* often with the tor-grass, *Brachypodium pinnatum.* The non-grass species are numerous and in many cases very distinctive. The following list includes the most typical species

Bulbous buttercup	*Ranunculus bulbosus*
Dropwort	*Filipendula vulgaris*
Salad burnet	*Sanguisorba minor*
Red clover	*Trifolium pratense*
Common bird's-foot-trefoil	*Lotus corniculatus*
Kidney vetch	*Anthyllis vulneraria*
Horsehoe vetch	*Hippocrepis comosa*
Fairy flax	*Linum catharticum*
Hairy violet	*Viola hirta*
Common rockrose	*Helianthemum nummularium*
Burnet-saxifrage	*Pimpinella saxifraga*
Wild carrot	*Daucus carota*
Cowslip	*Primula veris*
Autumn gentian	*Gentianella amarella*
Self-heal	*Prunella vulgaris*
Wild thyme	*Thymus praecox*
Squinancywort	*Asperula cynanchica*
Hedge bedstraw	*Galium mollugo*
Lady's bedstraw	*Galium verum*
Small scabious	*Scabiosa columbaria*
Round-headed rampion	*Phyteuma orbiculare*
Harebell	*Campanula rotundifolia*
Yarrow	*Achillea millefolium*
Carline thistle	*Carlina vulgaris*
Dwarf thistle	*Cirsium acaule*
Common knapweed	*Centaurea nigra*
Rough hawkbit	*Leontodon hispidus.*

Orchid species include

Fragrant orchid	*Gymnadenia conopsea*
Common spotted-orchid	*Dactylorhiza fuchsii*
Burnt orchid	*Orchis ustulata*
Man orchid	*Aceras anthropophorum*
Pyramidal orchid	*Anacamptis pyramidalis*
Bee orchid	*Ophrys apifera.*

Acid grasslands (Plate 88) are more frequent in montane areas, but they may occur on dry sandy lowland soils where for some reason or other heath or woodland has not developed. The main dominant grasses are the bents, *Agrostis* species, and on the grass 'moors' at higher altitudes the mat-grass, *Nardus stricta* is dominant, with the wavy hair-grass, *Deschampsia flexuosa* in drier areas. On damper sites the purple moor-grass, *Molinia caerulea,* and the heath rush, *Juncus squarrosus,* are dominant. Other plants commonly associated with the montane grasslands are

Tormentil	*Potentilla erecta*
Heath bedstraw	*Galium saxatile*
Sheep's-fescue	*Festuca ovina*
Sweet vernal-grass	*Anthoxanthum odoratum*
Brown bent	*Agrostis canina*
Creeping bent	*Agrostis stolonifera*
Common bent	*Agrostis capillaris*
Bracken	*Pteridium aquilinum.*

Occasionally the gorse, *Ulex europaeus,* is present. In the higher montane moor grasslands the mat-grass, *Nardus stricta,* is commonly dominant and associated with

Lousewort	*Pedicularis sylvatica*
Devil's-bit scabious	*Succisa pratensis*
Heath wood-rush	*Luzula multiflora*
Common bent	*Agrostis capillaris*
Heath-grass	*Danthonia decumbens.*

Lowland grass-heaths on sandy soils are rich in such annuals as

Sheep's sorrel	*Rumex acetosella*
Common whitlowgrass	*Erophila verna*
Biting stonecrop	*Sedum acre*
Rue-leaved saxifrage	*Saxifraga tridactylites*

with the sand sedge, *Carex arenaria,* and the lichen, *Cladonia arbuscula.*

9 Central European vegetation (Plates 22–29; Plates 91–103)

In the plains and lowlands of Central Europe (Map 10), the original natural vegetation consisted largely of forests of broad-leaved deciduous trees with beech, oak, and hornbeam largely dominating the climax communities (Fig. 18). Once again, most of the forests have been cleared for agricultural purposes by the dense populations, so that little of the semi-natural woodland remains. However, in the montane regions, considerable forests dominated by both deciduous and ever-green coniferous trees do remain largely in a semi-natural state. In the eastern part of Central Europe where the climate is more continental, different species of oak become dominant, but again in the lowland and submontane regions, few relicts of natural forests still persist.

In the montane regions coniferous forests are widespread (Fig. 19), though beech woods are also common. Coniferous forests of spruce, silver fir, pine, and larch cover wide areas in a semi-natural state, but in many cases, particularly at lower altitudes, they have been extended by planting for timber, etc.

Wet woodlands are restricted to river valleys and flood plains and are largely dominated by alder, willow, poplar, and ash.

Mires, such as raised bogs and fens, occur in the damper areas of Central Europe, but they have been much reduced by drainage and changes in water-level. They are more frequent in valleys of the submontane and montane regions. Further east, in the drier climates, intermediate mires and fens are widely distributed by rivers and lake verges and in montane valleys.

Heaths are largely found in the western part of Central Europe and are closely related to those of the Atlantic region. They are largely sub-climax communities resulting from forest clearance and grazing.

The grasslands of Central Europe are almost entirely man-made, except locally in some alpine regions and in the steppe-like grassland regions of the drier parts of Eastern Europe. Here the feather-grasses, *Stipa* species, come into their own.

Central European beech woods

Beech is probably the natural climax forest tree over much of Europe, from the lowlands to the mountainous regions (Map 11), on relatively rich soils with average water content. It does not occur on poor dry sandy soils or on the deeper river-valley soils of Central Europe. Much of the land on which beech formerly grew has been cleared for agricultural purposes, and the lowland beech woods that remain have been heavily affected by man's selective felling, for beech is a valuable timber tree used for many purposes.

More natural forests occur in most of the mountain ranges of Europe, including the Pyrenees, Alps, Carpathians, and central Balkans, and penetrate via the Apennines as far south as Sicily where beech grows to 2100 m on Mount Etna. Beech forests are particularly well developed in western Bulgaria, and in Czechoslovakia, where in Slovakia they form 32 per cent of the forest cover. In general, as one travels further south, the beech woods occupy a higher and higher montane zone. Thus in southern Sweden they are entirely lowland; in southern England they have their upper limit at 300 m; in the northern Alps at 1200 m; in the Pyrenees at 1500 m; in the Tyrol at 1550 m; and in the Apennines at 1850 m. These montane beech woods have a lower limit as well, ranging from 700 m in the Rhodope mountains of Bulgaria to 1000 m on Mount Olympus, Greece.

For the optimum development of beech woods (Fig. 20), the climate must be relatively mild—from sub-oceanic to mildly continental—with an average annual temperature of about 10°C and an annual rainfall of about 1000 mm, though in Central Europe in cooler climates, the rainfall can be as low as 500 mm. Beech woods are perhaps the most uniform and distinctive of deciduous forest communities in Europe. The experience of visiting a montane beech wood in the Pyrenees and one in the Rhodope mountains of Bulgaria is unnerving; the same species are present, and only after careful searching for western or eastern plants in the field layer can one be sure of one's whereabouts. The effect of man's management is of course partly responsible for this surprising uniformity, though it has relatively little effect on the floristic composition of the distinctive field layer.

Usually beech trees are grown as standards of uniform age, forming a closed canopy of foliage above, which during the summer months allows as little as 2 per cent of the light to penetrate the field layer. This has a profound and selective effect on both the shrub and field layers. Beech trees mature in about 60–100 years and grow to an average height of 30 m; after about 160 years they begin to show signs of decay. Selective felling, or clear felling, results in the rapid regeneration of young trees, as well as a rich and rapid growth of the shrub and field layers, with certain species becoming locally dominant during this 'light' phase.

The shrub layer of beech woods is usually poorly developed, and the woods are often surprisingly open, with long vistas

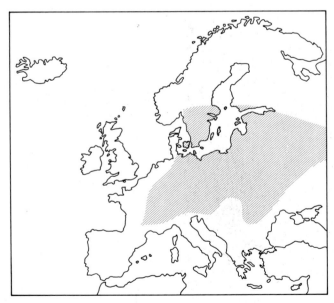

Map 10. Central European region.

through the straight grey trunks. The field layer likewise is often sparse and does not always cover the ground, leaving bare patches covered with dead beech leaves. Here certain saprophytic species are characteristic, living on the rich humus layers. Many of the field layer species are *pre-vernal* or *vernal* in their flowering time, that is to say they flower before, or as, the foliage of the beech canopy develops.

The following are the most important types of beech wood described by plant sociologists. Each type has one or more distinctive species in the shrub or field layers by which it can be identified. Space will not allow more than a brief mention of most of these beech wood types.

Beech woods with orchids and sedges are found in Switzerland, Austria, Italy, Greece and northwards to northern France and Britain. The characteristic orchids are (Fig. 21)

White helleborine	*Cephalanthera damasonium*
Narrow-leaved helleborine	*Cephalanthera longifolia*
Red helleborine	*Cephalanthera rubra*
Bird's-nest orchid	*Neottia nidus-avis*

with the sedges

White sedge	*Carex alba*
Fingered sedge	*Carex digitata*
Soft-leaved sedge	*Carex montana.*

Beech woods with blue moor-grass are found in the montane and sub-alpine regions of southern Germany, Poland, Czechoslovakia, and Hungary, on dry limestone slopes. Dominant in the field layer is the blue moor-grass, *Sesleria caerulea,* commonly with the false daisy, *Aster bellidiastrum,* and marjoram, *Origanum vulgare.*

Beech woods with yew, although rare in Central Europe, are easily recognized by the heavy evergreen foliage of the yew, *Taxus baccata.* This beech wood with yew is more characteristic of the Atlantic region.

Beech woods with wood melick are widely distributed from southern Sweden and the Baltic region, and from the Pyrenees to the lower montane zone of the northern Alps, through Germany, Czechoslovakia, the Carpathians, the Dinaric Alps of Yugoslavia and central Italy. Characteristic is wood melick, *Melica uniflora,* with Ramsons, *Allium ursinum.*

Beech woods with fir occur widely scattered in the montane regions and tend to occur in a sub-oceanic climate at altitudes of 900–1250 m, forming a zone between the lower beech woods on limestone and the sub-alpine beech–sycamore zone. They prefer richer soils, and develop in steep valleys with northern or northwestern exposures. The silver fir, *Abies alba,* and the Norway spruce, *Picea abies,* are often co-dominant. Distinctive herbaceous species in the field layer are

	Ranunculus aconitifolius
	Cardamine enneaphyllos
Trifoliate bitter-cress	*Cardamine trifolia*
	Veronica urticifolia
White butterbur	*Petasites albus*
	Adenostyles alliariae
	Prenanthes purpurea.

Beech woods with sycamore have developed in the lower sub-alpine zone, at altitudes from 600 to 1700 m, largely in the southwest of Central Europe. On mountains which receive much snow, but have relatively mild winters, beech with sycamore forms the upper tree line in place of the conifers.

Fig. 18. The Bialowieza National Park in Poland is one of the best examples of a virtually untouched natural mixed forest.

Central European vegetation

Typical shrubs include alpine rose, *Rosa pendulina*; alpine honeysuckle, *Lonicera alpigena*; and black-berried honeysuckle, *Lonicera nigra*. Like the previous beech-with-fir type of forest, much of it has been cleared to provide sub-alpine pastures for grazing and hay-making.

Beech woods with wood-rush are found on acidic soils in the lowlands and hills, and are widespread in the Central European region, from southern Sweden to Switzerland, and from the Ardennes to the Carpathians. They also occur in sub-montane and montane zones. Characteristic is the white wood-rush, *Luzula luzuloides,* with the great wood-rush, *Luzula sylvatica,* in more Atlantic or moister climates (Fig. 22).

Three contrasting examples of typical beech woods in Central Europe are given in more detail below. They are beech woods on limestone, beech woods on acid soils, and beech woods in the sub-alpine zone.

Beech woods on limestone have many local variants, with different species dominating the field layer; the species listed below may be present in many such beech woods. The beech, *Fagus sylvatica,* is dominant in the tree layer with

Silver fir	*Abies alba*
Wych elm	*Ulmus glabra*
Norway maple	*Acer platanoides*
Sycamore	*Acer pseudoplatanus*
Ash	*Fraxinus excelsior*

Shrub layer

Spurge-laurel	*Daphne laureola*
Mezereon	*Daphne mezereum*
Ivy	*Hedera helix*
Alpine honeysuckle	*Lonicera alpigena*
Black-berried honeysuckle	*Lonicera nigra*
Fly honeysuckle	*Lonicera xylosteum*

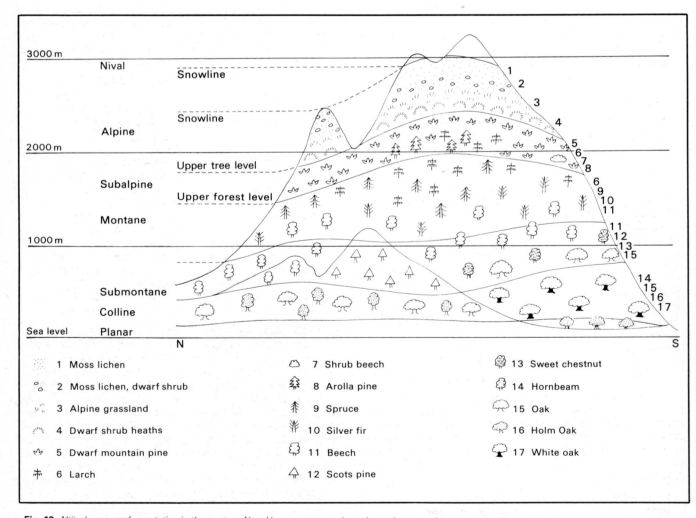

Fig. 19. Altitude zones of vegetation in the eastern Alps. Here we can see how the main vegetation types are influenced by the altitude and climatic changes across a mountain chain; in this case the Alps.

Natural range of Beech (*Fagus sylvatica*)

Map 11. Beech. Natural range of Beech *Fagus sylvatica* and major areas of Beech woodland (dark).

Field layer

Wood anemone	*Anemone nemorosa*
Yellow wood-anemone	*Anemone ranunculoides*
Bulbous corydalis	*Corydalis bulbosa*
Coralroot	*Cardamine bulbifera*
Wood-sorrel	*Oxalis acetosella*
Dog's mercury	*Mercurialis perennis*
Early dog-violet	*Viola reichenbachiana*
Enchanter's nightshade	*Circaea lutetiana*
Woodruff	*Galium odoratum*
Yellow archangel	*Lamiastrum galeobdolon*
	Veronica urticifolia
Ramsons	*Allium ursinum*
Lords-and-ladies	*Arum maculatum*
Wood fescue	*Festuca altissima*
Wood melick	*Melica uniflora*
Wood barley	*Hordelymus europaeus*
Wood millet	*Milium effusum*
Fingered sedge	*Carex digitata*
Wood-sedge	*Carex sylvatica*
Lady-fern	*Athyrium filix-femina.*

Beech woods on acid soils are distinguished by the presence of wood-rushes which dominate the field layer. Beech is dominant in the tree layer, with the sessile oak, *Quercus petraea;* the sycamore, *Acer pseudoplatanus;* and the rowan, *Sorbus aucuparia,* while the shrub layer may be dominated by the raspberry, *Rubus idaeus,* and numerous tree seedlings. The field layer contains

Wood-sorrel	*Oxalis acetosella*
Bilberry	*Vaccinium myrtillus*
Chickweed wintergreen	*Trientalis europaea*

Heath speedwell	*Veronica officinalis*
Common cow-wheat	*Melampyrum pratense*
White wood-rush	*Luzula luzuloides*
Great wood-rush	*Luzula sylvatica*
Wavy hair-grass	*Deschampsia flexuosa*
Bracken	*Pteridium aquilinum*
Oak fern	*Gymnocarpium dryopteris*

with the mosses

Polytrichum attenuatum
Leucobryum glaucum
Dicranella heteromalla.

Sub-alpine beech woods (Plate 91) may have well-developed trees up to about 28 m, but at higher altitudes the trees are often dwarfed and twisted by the action of wind and snow. They are found in the western Alps, the Jura (Fig. 23), and the Pyrenees, and they sometimes form the upper tree limit.

The dominant beech is commonly accompanied by sycamore, *Acer pseudoplatanus,* while the shrub layer has the alpine rose, *Rosa pendulina,* the alpine honeysuckle, *Lonicera alpigena;* and the black-berried honeysuckle, *Lonicera nigra.* The field layer typically contains the following species

The sorrel	*Rumex arifolius*
Aconite-leaved buttercup	*Ranunculus aconitifolius*
Large white-buttercup	*Ranunculus platanifolius*
Round-leaved saxifrage	*Saxifraga rotundifolia*
Wood-sorrel	*Oxalis acetosella*
Sanicle	*Sanicula europaea*
Yellow pimpernel	*Lysimachia nemorum*
Yellow archangel	*Lamiastrum galeobdolon*
Wood scabious	*Knautia dipsacifolia*
White butterbur	*Petasites albus*
Adenostyles	*Adenostyles alliariae*
Wood ragwort	*Senecio nemorensis*
Alpine sow-thistle	*Cicerbita alpina*
Whorled Solomon's-seal	*Polygonatum verticillatum*
Beech fern	*Thelypteris phegopteris*
Lady-fern	*Athyrium filix-femina*
Male-fern	*Dryopteris filix-mas*
Oak fern	*Gymnocarpium dryopteris*

and the moss, *Ctenidium molluscum.*

In the extreme south-east of the Central European region, the oriental beech, *Fagus orientalis* (Plate 115), forms extensive woods. These are found in Greece, eastern Bulgaria, Romania, and Turkey-in-Europe. They spread eastwards in the Pontus mountains of northern Turkey. Further west the oriental beech hybridizes with the common beech and forms woods in Bulgaria and northern Greece. Oriental beech woods have much the same appearance as the Central European beech woods and have similar requirements of both soil and climate. However they are rich in distinctive Eastern European species. The oriental beech, *Fagus orientalis,* is dominant with other southern or eastern species in the tree layer, such as

Walnut	*Juglans regia*
Turkey oak	*Quercus cerris*

Plate 22. Central Europe beech woods

1. Mezereon, *Daphne mezereum*; **2**. Fly honeysuckle, *Lonicera xylosteum*; **3**. Alpine honeysuckle, *Lonicera alpigena*; **4**. Blue moor-grass, *Sesleria caerulea*; **5**. Bulbous corydalis, *Corydalis bulbosa*; **6**. Purple lettuce, *Prenanthes purpurea*; **7**. Ramsons, *Allium ursinum*; **8**. Coralroot, *Cardamine bulbifera*; **9**. White wood-rush, *Luzula luzuloides*; **10**. Wood melick, *Melica uniflora*; **11**. Raspberry, *Rubus idaeus*; **12**. Fingered sedge, *Carex digitata*; **13**. Red helleborine, *Cephalanthera rubra*; **14**. Woodruff, *Galium odoratum*; **15**. Yellow anemone, *Anemone ranunculoides*.

Hungarian Oak	*Quercus frainetto*
Caucasian Ash	*Fraxinus angustifolia*
	spp. *oxycarpa*
Manna ash	*Fraxinus ornus.*

The shrub layer has the following more eastern species

	Rubus hirtus
Medlar	*Mespilus germanica*
Black-berried hawthorn	*Crataegus pentagyna*
Wild service tree	*Sorbus torminalis*
Cherry-laurel	*Prunus laurocerasus*
	Daphne pontica
Cornelian cherry	*Cornus mas*
Rhododendron	*Rhododendron ponticum*
	Smilax excelsa.

The field layer includes central and eastern European species such as

Ranunculus constantinopolitanus
Epimedium pubigerum
Cardamine bulbifera
Potentilla micrantha
Lathyrus laxiflorus
Lathyrus niger
Hypericum calycinum
Viola suavis
Primula vulgaris ssp. *sibthorpii*
Periploca graeca
Pulmonaria officinalis
Symphytum tauricum

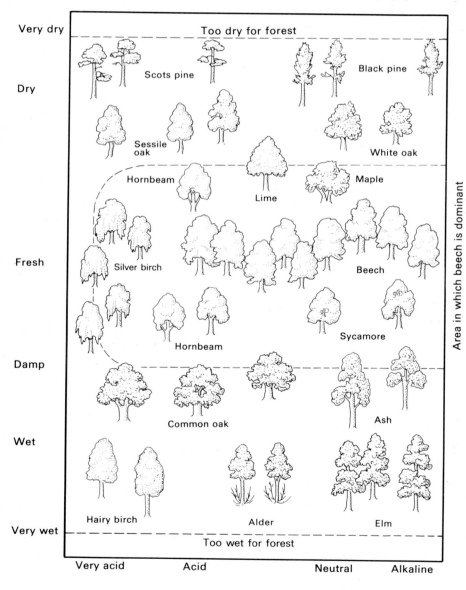

Fig. 20. The ecological 'position' of beech. The approximate ecological requirements of the major European trees in terms of soil, water content, and acidity. Notice the wide tolerance of beech, the tendency for Scots pine and birch to favour acid soil (although Scots pine is also found on limestone when conditions are dry), and the restriction of alder to wet soil. (Redrawn from Ellenberg 1982.)

Trachystemon orientalis
Ajuga genevensis
Lamium maculatum
Salvia forskaohlei
Salvia glutinosa
Lathraea squamaria
Nectaroscordum siculum
Fritillaria pontica
Scilla bifolia
Scilla bithynica
Ruscus hypoglossum.

Oak–hornbeam woods (Plates 92 and 93)

These woodlands are characteristic, like the beech woods, of Central Europe (Fig. 24, Map 12). They are most widespread between the latitudes of 45°N and 55°N, and are found at altitudes of between 200 m and 500 m above sea level. They require an annual rainfall of 500–600 mm, and July temperatures of 17–19°C and a mean annual temperature of about 9°C, which can be said to be the average climate of the tem-

Fig. 21. The red helleborine *Cephalanthera rubra* is characteristic of certain types of calcareous beech woods.

perate zone. In the western areas, oak–hornbeam woods occur in relatively drier habitats while, further east, they require moister conditions.

The relationship between oak–hornbeam woods and beech woods is complex; in general the former are found on damper soils than the latter, but they also occur on drier soils than beech. Also oak–hornbeam woods favour more acid soils than beech, while the activities of man, with selective felling and livestock-grazing may encourage oak–hornbeam woods to replace beech woods.

Oak–hornbeam woods have the hornbeam, *Carpinus betulus,* co-dominant with either the pedunculate oak, *Quercus robur,* or the sessile oak, *Quercus petraea,* in various combinations. With them are mixed the following broad-leaved trees: ash, *Fraxinus excelsior;* small-leaved lime, *Tilia cordata;* and sycamore, *Acer pseudoplatanus;* with, less frequently, the silver fir, *Abies alba;* and spruce, *Picea abies.*

The pedunculate oak is more common in the Atlantic areas, as in northern Germany and Poland, while on light sandy soils and gravels in the drier Alpine regions, the most important oak is the sessile oak. This latter oak occurs on the upper Rhine lowlands, and on the lower slopes of the mountain ranges of Central Europe including the lower Danube and the Carpathians, where it occurs below the zone of beech woods. The shrub layer of oak–hornbeam woods is not so well developed, but the field layer is rich in herbaceous plants, particularly in the more western woods where it covers approximately 60–80 per cent of the ground; consequently the moss layer is poor or absent. The following are the main types of oak–hornbeam woods in central Europe:

Northwestern type with honeysuckle, *Lonicera periclymenum,* with high constancy, and greater stitchwort, *Stellaria holostea,* characteristic.

Sub-Atlantic Central-European type characterized by barren strawberry, *Potentilla sterilis;* field rose, *Rosa arvensis;* and wood bedstraw, *Galium sylvaticum,* and with the less constant *Carex umbrosa.*

Sub-continental type with typically hepatica, *Hepatica nobilis;* wood bedstraw, *Galium sylvaticum;* and various-leaved fescue, *Festuca heterophylla.*

Eastern type with a high constancy of *Viola mirabilis* and associated species including wild service-tree, *Sorbus torminalis;* bastard balm, *Melittis melissophyllum;* and various-leaved fescue, *Festuca heterophylla.*

Carpathian type with constant *Carex pilosa,* and with bastard balm, *Melittis melissophyllum;* and tuberous comfrey, *Symphytum tuberosum.*

Eastern type with hepatica, *Hepatica nobilis,* constant, but with the absence of western and southern species, and the presence of Central European, Boreal, and continental species.

Southern type includes the northern Appenine woods with *Anemone trifolia* and bladderseed, *Physospermum cornubiense.*

Illyric type of the lower uplands of the northwestern Balkans, with *Helleborus dumetorum;* barrenwort, *Epimedium alpinum;* *Knautia drymeia,* and white crocus, *Crocus albiflorus.*

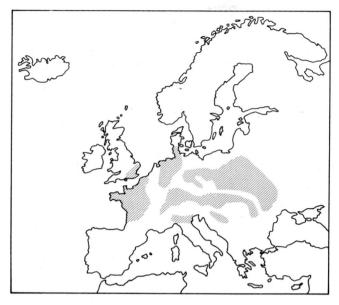

Map 12. Oak–hornbeam woodland—the major areas in which this community may be found.

The shrub layer has hawthorns, *Crataegus* species, and the field layer commonly includes

Wild strawberry	*Fragaria vesca*
Ivy	*Hedera helix*
Wood bedstraw	*Galium sylvaticum*
Lily-of-the-valley	*Convallaria majalis*
Various-leaved fescue	*Festuca heterophylla*
Wood meadow-grass	*Poa nemoralis*
Wood-sedge	*Carex sylvatica.*

Typical lowland examples of oak–hornbeam woods on fertile loam and marl, spreading from France to Poland, have the pedunculate oak and ash mixed with hornbeam in the tree layer. Much of this type of forest has been destroyed for agricultural purposes. Hazel is predominant in the shrub layer and other herbaceous species typical of the field layer are

Wood anemone	*Anemone nemorosa*
Early dog-violet	*Viola reichenbachiana*
Solomon's-seal	*Polygonatum multiflorum*
Wood millet	*Milium effusum.*

Examples of southern oak–hornbeam woods have the following composition (only the more widespread species are listed here)

Tree layer

Pedunculate oak	*Quercus robur*
Hornbeam	*Carpinus betulus*
Small-leaved elm	*Ulmus minor*
Wild cherry	*Prunus avium*
Field maple	*Acer campestre*
Tartarian maple	*Acer tataricum*

Sub-Atlantic types of oak–hornbeam woods are usually found on mineral-rich soils in the warmer lowlands, ranging in altitude from 200 to 450 m. Both the pedunculate and sessile oaks are present with the hornbeam. Other trees include

Beech	*Fagus sylvatica*
Field maple	*Acer campestre*
Small-leaved lime	*Tilia cordata*
Ash	*Fraxinus excelsior.*

Fig. 22. An acid beech wood with the field layer dominated by the great woodrush *Luzula sylvatica.* Katzenstein, W. Germany.

Plate 23. Central Europe beech woods—montane and sub-alpine

1. Alpine rose, *Rosa pendulina*;   **2**. Round-leaved saxifrage, *Saxifraga rotundifolia*;   **3**. Drooping bittercress, *Cardamine enneaphyllos*;   **4**. Trifoliate bittercress, *Cardamine trifolia*;   **5**. White butterbur, *Petasites albus*;   **6**. Mountain dock, *Rumex arifolius*;   **7**. Wood ragwort, *Senecio nemorensis* (inset×1);   **8**. Yellow pimpernel, *Lysimachia nemorum* (inset×3);   **9**. *Veronica urticifolia*;   **10**. Aconite-leaved buttercup, *Ranunculus aconitifolius*;   **11**. Adenostyles, *Adenostyles alliariae*;   **12**. Wood scabious, *Knautia dipsacifolia*;   **13**. Oak fern, *Gymnocarpium dryopteris*;   **14**. Beech fern, *Thelypteris phegopteris*;   **15**. *Ranunculus platanifolius* (leaf).

Plate 24. Central Europe Oriental beech woods

1. *Salvia forskaohlei*;   **2**. Large butcher's-broom, *Ruscus hypoglossum*;   **3**. *Fritillaria pontica*;   **4**. Toothwort, *Lathraea squamaria*;   **5**. Eastern borage, *Trachystemon orientalis*;   **6**. *Epimedium pubigerum*;   **7**. *Lathyrus laxiflorus*;   **8**. Spotted deadnettle, *Lamium maculatum*;   **9**. *Nectaroscordum siculum*;   **10**. Rose of Sharon, *Hypericum calycinum*;   **11**. Silk-vine, *Periploca graeca*;   **12**. *Symphytum tauricum*;   **13**. Pink barren-strawberry, *Potentilla micrantha*;   **14**. *Ranunculus constantinopolitanus*;   **15**. Oriental beech, *Fagus orientalis*.

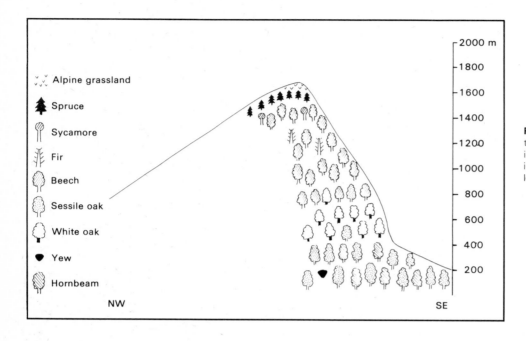

Fig. 23. Montane beech forest. Here in the Swiss Jura, beech forest grades into natural spruce forest higher up and into white oak and oak–hornbeam at lower altitudes.

Shrub layer (common species)

Dewberry	*Rubus caesius*
Hawthorn	*Crataegus monogyna*
Spindle	*Euonymus europaeus*
Dogwood	*Cornus sanguinea*
Privet	*Ligustrum vulgare*

Field layer (common species)

Wood spurge	*Euphorbia amygdaloides*
Early dog-violet	*Viola reichenbachiana*
Lungwort	*Pulmonaria officinalis*
Alpine squill	*Scilla bifolia*
Lords-and-ladies	*Arum maculatum*

Fig. 24. Oak–hornbeam wood with lime in the Kottenforst Nature Reserve, near Bonn, W. Germany.

R. Wise

Plate 25. Central Europe oak–hornbeam woods

1. Barren strawberry, *Potentilla sterilis*;   **2**. Bastard balm, *Melittis melissophyllum*;   **3**. Small-leaved lime, *Tilia cordata*;   **4**. White crocus, *Crocus vernus* ssp. *albiflorus*;   **5**. *Helleborus dumetorum*;   **6**. *Anemone trifolia*;   **7**. Hornbeam, *Carpinus betulus*;   **8**. Barrenwort, *Epimedium alpinum*;   **9**. *Physospermum cornubiense* (inset×5);   **10**. *Viola mirabilis*;   **11**. Wood bedstraw, *Galium sylvaticum*;   **12**. Tuberous comfrey, *Symphytum tuberosum*;   **13**. *Knautia drymeia*;   **14**. *Carex pilosa*;   **15**. *Carex umbrosa*;   **16**. *Festuca heterophylla* (inset×3).

Central European vegetation

False brome	*Brachypodium sylvaticum*
Wood millet	*Milium effusum*
Wood-sedge	*Carex sylvatica.*

Oak woods

On acid, usually siliceous soils, the sessile oak, *Quercus petraea,* is commonly dominant, though in some areas such as in southern Sweden, the pedunculate oak, *Quercus robur,* may dominate on some acid soils. Typical sessile oak woods occur in the foothills of the main mountain ranges of Central Europe, up to about 550 m, on podsolized brown-earth soils, or on sands or loams. The tree layer is dominated by *Quercus petraea,* and the shrub layer by the bilberry, *Vaccinium myrtillus.* The field layer characteristically includes

Common cow-wheat	*Melampyrum pratense*
The hawkweed	*Hieracium umbellatum*
White wood-rush	*Luzula luzuloides*
Sheep's-fescue	*Festuca ovina*
Wavy hair-grass	*Deschampsia flexuosa.*

The moss layer includes

Dicranum scoparium
Hypnum cupressiforme
Leucobryum glaucum
Polytrichum attenuatum

with the lichen, *Cladonia fimbriata.*

In the northern part of Central Europe, around the Baltic, the pedunculate oak, *Quercus robur,* is often dominant, with the aspen, *Populus tremula,* and the rowan, *Sorbus aucuparia.* In some cases the sessile oak may be dominant in this region. The shrub layer has the juniper, *Juniperus communis,* and the bilberry, *Vaccinium myrtillus.* The field layer typically contains the golden-rod, *Solidago virgaurea;* wavy hair-grass, *Deschampsia flexuosa;* and the polypody fern, *Polypodium vulgare.*

On sandy loam, or degraded loess, in the lowlands in more continental areas such as Poland, the Scots pine, *Pinus sylvestris,* is often co-dominant with the sessile oak, *Quercus petraea,* with, in addition, silver birch, *Betula pendula;* aspen, *Populus tremula;* small-leaved lime, *Tilia cordata;* and, less commonly, pedunculate oak. The shrub layer includes

Hazel	*Corylus avellana*
Hawthorn species	*Crataegus* spp.
Rowan	*Sorbus aucuparia*
Guelder-rose	*Viburnum opulus*

with a field layer of

Bilberry	*Vaccinium myrtillus*
Common cow-wheat	*Melampyrum pratense*
Sheep's-fescue	*Festuca ovina*
Wavy hair-grass	*Deschampsia flexuosa*
Pill sedge	*Carex pilulifera*
Bracken	*Pteridium aquilinum.*

In the southern part of Central Europe, in the warm rainy-valleys of the southern Alps, sweet chestnut, *Castanea sativa,* may be co-dominant with the sessile oak. It has been introduced by man into this region for over 2000 years, and it has spread under his influence as far north as the northern Alps. These oak–chestnut woods are usually associated with the ash, *Fraxinus excelsior;* the black poplar, *Populus nigra;* lime and elm.

In the south-east of the Central European region, in Yugoslavia, Romania, Bulgaria, and northern Greece, oak woods dominated by the Hungarian oak, *Quercus frainetto,* with the Turkey oak, *Quercus cerris,* are widespread in both the lowland and sub-montane regions. Other trees commonly present in these woods include

Oriental hornbeam	*Carpinus orientalis*
White oak	*Quercus pubescens*
Service tree	*Sorbus domestica*
Wild service tree	*Sorbus torminalis*
Wild pear	*Pyrus pyraster*
Silver lime	*Tilia tomentosa*
Manna ash	*Fraxinus ornus.*

Shrubs and small trees include

Almond-leaved pear	*Pyrus amygdaliformis*
Hairy broom	*Chamaecytisus hirsutus*
Tartarian maple	*Acer tataricum*
Field maple	*Acer campestre*
Cornelian cherry	*Cornus mas*
Privet	*Ligustrum vulgare*
Wayfaring tree	*Viburnum lantana.*

The field layer is rich in species.

Coniferous forests

Pines, silver fir, spruce, and larch form extensive forests in the montane regions of Central Europe. In the central Alps, the larch, *Larix decidua* (Plate 98), Arolla pine, *Pinus cembra,* and dwarf mountain pine, *Pinus mugo,* commonly form the upper limit of coniferous forests and scrub. Below this limit belts of spruce, *Picea abies,* and Scots Pine, *Pinus sylvestris,* usually occur.

Pine woods

The Scots pine, *Pinus sylvestris,* is the most widespread pine in Central Europe (Map 13). It occurs on acid soils, spreading from the lowlands right up to the tree line in some mountain regions. In the lowlands it has been extensively planted by man and increased by forestry selection. The black pine, *Pinus nigra,* occurs in the south east of the Central European region, while the five-needled Arolla pine, *Pinus cembra,* is restricted to the sub-alpine zones of the Alps and Carpathians. The dwarf mountain pine, *Pinus mugo* (Plate 94), is an important bush-former in the alpine zone, and the closely related mountain pine, *Pinus uncinata,* with which it frequently hybridizes, is found in mountains in the west of Central Europe and occasionally in bogs in the foothills.

Scots pine woods are typical of sandy soils, particularly in the east of the region, and they are most widely distributed in Poland, with many variant types from lichen-rich woodland

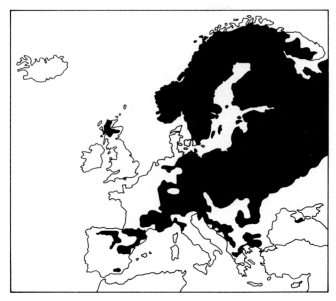

Map 13. Scots pine *Pinus sylvestris*—natural range.

with *Cladonia* and *Cetraria* species, developed on sands, to Baltic coastal types with such characteristic species as crowberry, *Empetrum nigrum;* twinflower, *Linnaea borealis;* and lesser twayblade, *Listera cordata.* In western and central Poland and northern Germany, Scots pine woods commonly include the Pedunculate oak, *Quercus robur,* in the tree layer.

The shrub layer includes

Heather	*Calluna vulgaris*
Bilberry	*Vaccinium myrtillus*
Cowberry	*Vaccinium vitis-idaea*

while the field layer includes

Common cow-wheat	*Melampyrum pratense*
Sheep's-fescue	*Festuca ovina*
Wavy hair-grass	*Deschampsia flexuosa.*

with the common mosses, *Pleurozium schreberi, Dicranum undulatum,* and *Leucobryum glaucum.*

In the mountain valleys up to sub-alpine levels in the Alps, Scots pines form woods on dry calcareous gravels (Fig. 25). Commonly associated with these pine woods are the spring heath, *Erica herbacea,* with the cowberry, *Vaccinium vitis-idaea;* shrubby milkwort, *Polygala chamaebuxus;* the juniper, *Juniperus communis;* and the grass, *Calamagrostis varia.*

The five-needled Arolla pine, *Pinus cembra* (Map 14) (Plate 96) forms sparse woods on poor acid soils at altitudes of 1400–2500 m in the Alps and Carpathians. An example, with the larch, *Larix decidua* (Map 15), in the tree layer, has the following typical species in the shrub layer

Alpenrose	*Rhododendron ferrugineum*
Bilberry	*Vaccinium myrtillus*
Cowberry	*Vaccinium vitis-idaea.*

The field layer commonly includes

Wood-sorrel	*Oxalis acetosella*
Small cow-wheat	*Melampyrum sylvaticum*
Twinflower	*Linnaea borealis*

Fig. 25. Scots pine wood on river gravel in the Pupplinger Au Nature Reserve, Bavaria, W. Germany.

Central European vegetation

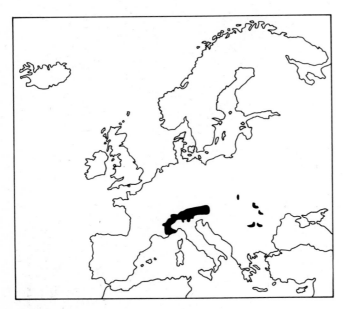

Map 14. Arolla pine *Pinus cembra*.

Alpine coltsfoot   *Homogyne alpina*
Great wood-rush   *Luzula sylvatica*
White wood-rush   *Luzula luzuloides*
The grass   *Calamagrostis villosa*
Wavy hair-grass   *Deschampsia flexuosa*.

Black pine woods (Plate 95) occur on the dry dolomitic soils of Austria and are dominated by the Austrian black pine, the subspecies *nigra*. Further south-east the larger-leaved and larger-coned Balkan black pine, subspecies *pallasiana*, may take its place. Austrian black pine woods typically have a shrub layer of

Snowy mespilus   *Amelanchier ovalis*
Common whitebeam   *Sorbus aria*
Wild cotoneaster   *Cotoneaster integerrimus*
                                        Cotoneaster nebrodensis.

The field layer includes

Buckler mustard   *Biscutella laevigata*
Shrubby milkwort   *Polygala chamaebuxus*
Garland flower   *Daphne cneorum*
Spring heath   *Erica herbacea*
Matted globularia   *Globularia cordifolia*
Blue sesleria   *Sesleria albicans*

and the Austrian endemic *Callianthemum anemonoides*. Growing on the pines is the parasitic misletoe, *Viscum album* subspecies *austriacum*, with usually yellow berries.

The dwarf mountain pine, *Pinus mugo* (Map 16, Plate 94), is really a bush-forming shrub which grows to 1 to 3 m in height. It often forms the highest coniferous zone in the sub-alpine region, from about 1400 to 2000 m in altitude, and is especially characteristic of limestone mountains. The closely related mountain pine, *Pinus uncinata*, which is a small tree, is found in the western Alps, and the Pyrenees, and central Spain. However, heavy grazing often results in the disappearance of these pines, and grasslands characterized by the moor-grass, *Sesleria albicans*, occur in their place. The following shrubs are commonly associated with the dwarf mountain pine

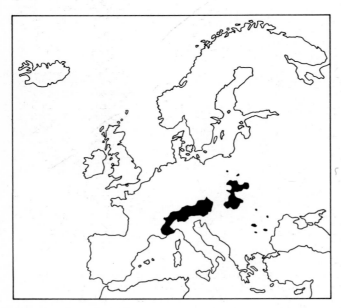

Map 15. Larch *Larix decidua*.

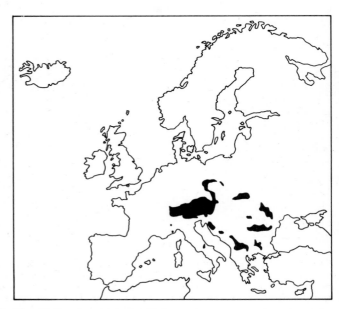

Map 16. Dwarf mountain pine *Pinus mugo*.

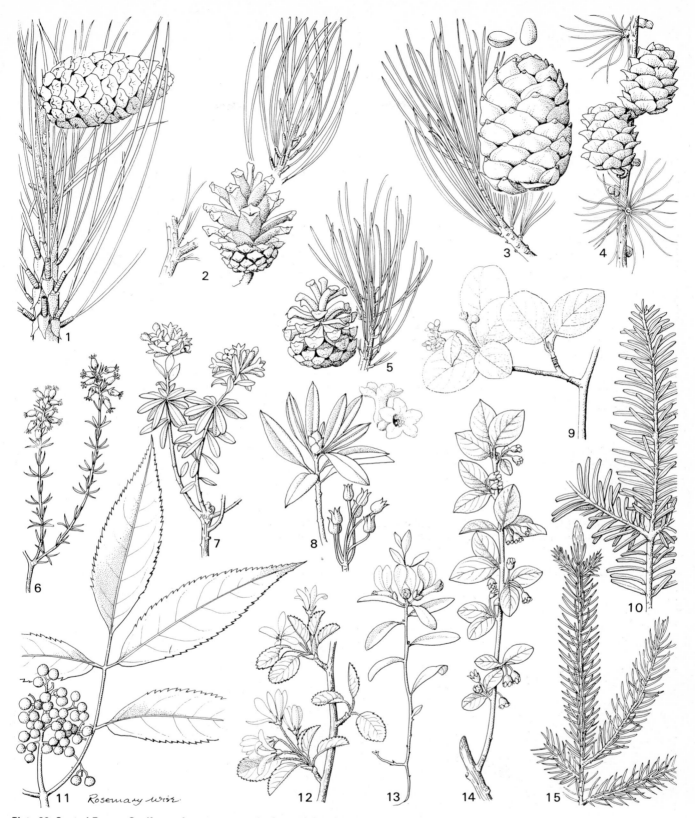

Plate 26. Central Europe Coniferous forests—trees, shrubs, and dwarf shrubs

1. Black pine, *Pinus nigra*;　　**2**. Scots pine, *Pinus sylvestris*;　　**3**. Arolla pine, *Pinus cembra*;　　**4**. Larch, *Larix decidua*;　　**5**. Dwarf mountain pine, *Pinus mugo*;　　**6**. Spring heath, *Erica herbacea*;　　**7**. Garland flower, *Daphne cneorum*;　　**8**. Alpenrose, *Rhododendron ferrugineum*;　　**9**. *Cotoneaster nebrodensis*;　　**10**. Silver fir, *Abies alba*;　　**11**. Red-berried elder, *Sambucus racemosa*;　　**12**. Snowy mespilus, *Amelanchier ovalis*;　　**13**. Shrubby milkwort, *Polygala chamaebuxus*;　　**14**. Wild cotoneaster, *Cotoneaster integerrimus*;　　**15**. Spruce, *Picea abies*.

Dwarf juniper	*Juniperus communis* spp. *nana*
False medlar	*Sorbus chamaemespilus*
	Daphne striata
Hairy alpenrose	*Rhododendron hirsutum*
Spring heath	*Erica herbacea*
Bilberry	*Vaccinium myrtillus*
Cowberry	*Vaccinium vitis-idaea*

with, less frequently,

| Mountain avens | *Dryas octopetala* |
| Purple coltsfoot | *Homogyne alpina* |

and the blue sesleria, *Sesleria albicans*.

In southeastern Europe, dwarf mountain pine thickets which spread as far south as Albania and Greece, may contain additional species such as stone bramble, *Rubus saxatilis*; bearberry, *Arctostaphylos uva-ursi*; and great wood-rush, *Luzula sylvatica*. In the eastern Pyrenees the mountain pine, *Pinus uncinata*, grows to 12 m, together with the Scots pine. Typical of the shrub layer are

Rowan	*Sorbus aucuparia*
	Cytisus purgans
Heather	*Calluna vulgaris*
Bearberry	*Arctostaphylos uva-ursi*
Bilberry	*Vaccinium myrtillus*.

The mountain pine is also occasionally found in bogs in the Alps and Black Forest, at altitudes of 600 to 1100 m.

Spruce woods

Norway spruce, *Picea abies*, is an important forest-forming tree in the northeast of the Central European region, occurring in the lowlands of southern Sweden and south-western Finland, and also much further north in the Boreal

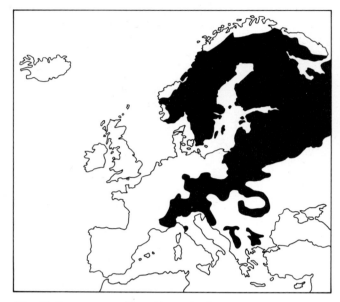

Map 17. Norway spruce *Picea abies*—natural range.

region (Map 17). It also forms important forests in the Alps and Carpathians, in the sub-alpine zone (Fig. 26). However, perhaps more than any other forest species, it has been widely planted by man in the lowlands and lower montane regions, and pure stands of even-aged Norway spruce are usually an indication that they have been planted. Natural and semi-natural forests are more or less restricted to the montane and sub-alpine zones. In the northern forests, Scots pine is often mixed with Norway spruce, while, further south, beech is fre-

Fig. 26. Naturally regenerating spruce wood *Picea abies* in the Tatra Mountains, Czechoslovakia.

quent in these forests. The phenomenon of climatic inversion, where colder heavier air accumulates in the valleys in winter, may result in the spruce forming forests in the bottom of the valleys, while higher up beech woods occur (Fig. 27).

Norway spruce trees may grow very tall, up to 42 m in Poland, and pure stands cast heavy shade, so that both the shrub and field layers are often poorly developed. Sub-alpine spruce forests may have the following composition

Shrub layer

Rowan	*Sorbus aucuparia*
Bilberry	*Vaccinium myrtillus*
Cowberry	*Vaccinium vitis-idaea*

Field layer

Wood-sorrel	*Oxalis acetosella*
Wood cow-wheat	*Melampyrum sylvaticum*
Nodding wintergreen	*Orthilia secunda*
One-flowered wintergreen	*Moneses uniflora*
Purple coltsfoot	*Homogyne alpina*
Goldenrod	*Solidago virgaurea*
The hawkweed	*Hieracium sylvaticum* group
	Calamagrostis villosa
Wavy hair-grass	*Deschampsia flexuosa.*

In the Carpathians, Norway spruce is occasionally mixed with silver fir, and the shrub layer is dominated by the bilberry. The field layer commonly includes the interrupted clubmosses, *Lycopodium annotinum*, and the fir clubmoss, *Huperzia selago*, with such ferns as the hard fern, *Blechnum spicant;* the broad buckler-fern, *Dryopteris dilatata;* and the lemon-scented fern, *Thelypteris limbosperma.* Characteristic flowering plants of the field layer include purple coltsfoot, *Homogyne alpina;* lesser tway-blade, *Listera cordata;* and the grass, *Calamagrostis villosa.*

The silver fir, *Abies alba* (Map 18 Plate 97), unlike the Norway spruce, usually forms mixed forests, largely in the southern and eastern parts of Central Europe. Its climatic and soil requirements are rather similar to those of beech, and it is often mixed with beech. It is more sensitive to frost and requires a higher moisture level than spruce. Silver fir forests are easily damaged by grazing, and the extension of spruce due to planting and selective felling has resulted in their being steadily reduced in Central Europe. Silver fir forests found on richer soils are usually mixed with beech with some spruce and sycamore in the tree layer, and have a very poorly developed shrub layer. The field layer includes

Wood-sorrel	*Oxalis acetosella*
Dog's mercury	*Mercurialis perennis*
Early dog-violet	*Viola reichenbachiana*
Woodruff	*Galium odoratum*
Yellow archangel	*Lamiastrum galeobdolon*
Wall lettuce	*Mycelis muralis*
Purple lettuce	*Prenanthes purpurea*
Wood fescue	*Festuca altissima*

and the ferns: lady fern, *Athyrium filix-femina;* and oak fern, *Gymnocarpium dryopteris.*

Silver fir forests mixed with Norway spruce, occurring in the Black Forest, Germany, have a sparse shrub layer of rowan, *Sorbus aucuparia,* and red-berried elder, *Sambucus racemosa.* The field layer includes

Alpine enchanter's nightshade	*Circaea alpina*
Nodding wintergreen	*Orthilia secunda*
Small wintergreen	*Pyrola minor*
One-flowered wintergreen	*Moneses uniflora*
Round-leaved bedstraw	*Galium rotundifolium*

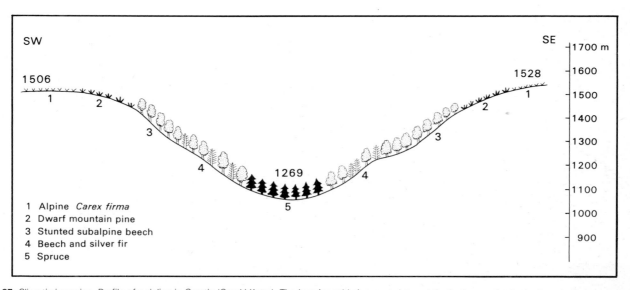

Fig. 27. Climatic inversion. Profile of a doline in Croatia (Gorski Kotar). The heavier cold air accumulates at the bottom and only the frost-resistant spruce forest survives. (Redrawn from Horvat *et al.* 1974.)

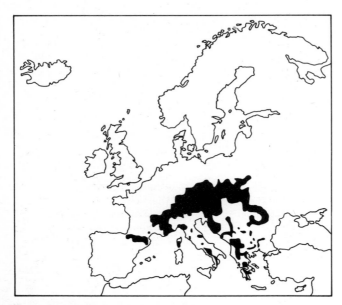

Map 18. Silver fir *Abies alba* and Greek fir *A. cephalonica*.

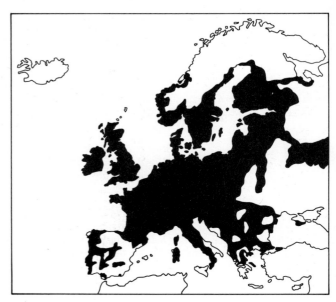

Map 19. Alder *Alnus glutinosa*.

Bilberry	*Vaccinium myrtillus*
Cowberry	*Vaccinium vitis-idaea*
Wavy hair-grass	*Deschampsia flexuosa*

and the hard fern, *Blechnum spicant*, with the broad buckler-fern, *Dryopteris dilatata*.

Wet woodlands (Plate 99)

The most important deciduous woods on wet soils are alder-woods, which occur commonly in river valleys and on peaty soils. The two widespread species are the common alder, *Alnus glutinosa* (Map 19), and the grey alder, *Alnus incana* (Map 20). The former is characteristic of lowland communities in Central Europe and the latter of montane and sub-alpine zones. Alders may be associated with softwood trees such as willows and poplars, or with hardwood trees such as ash, elm, or oak.

River-valley alder woods are composed of tall well-grown trees (if they have not been coppiced), and they often form drier hummocks around the bases of the trunks above the water level, while the field layer species occupy the hollows which remain wet most of the year round. These river-valley alder woods are dominated by the common alder, *Alnus glutinosa*, and usually have the following composition (Fig. 28)

Tree layer

Downy birch	*Betula pubescens*
Ash (occasional)	*Fraxinus excelsior*

Shrub layer

Dewberry	*Rubus caesius*
Hop	*Humulus lupulus*
Black currant	*Ribes nigrum*
Alder buckthorn	*Frangula alnus*

Field layer

Nettle	*Urtica dioica*
Marsh violet	*Viola palustris*
Bittersweet	*Solanum dulcamara*
Bog arum	*Calla palustris*
Elongated sedge	*Carex elongata*
Royal fern	*Osmunda regalis*
Narrow buckler-fern	*Dryopteris carthusiana*.

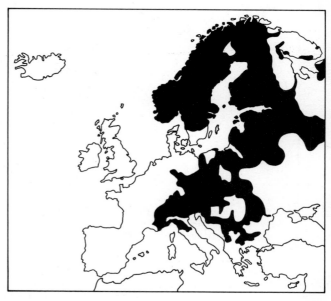

Map 20. Grey alder *Alnus incana*.

Plate 27. Central Europe wet woodlands

1. Dark-leaved willow, *Salix nigricans*; **2**. Purple willow, *Salix purpurea* (leaf); **3**. Mountain currant, *Ribes alpinum*; **4**. Hairy chervil, *Chaerophyllum hirsutum*; **5**. Marsh thistle, *Cirsium palustre*; **6**. Great meadow-rue, *Thalictrum aquileqifolium*; **7**. Cabbage thistle, *Cirsium oleraceum*; **8**. Touch-me-not balsam, *Impatiens noli-tangere*; **9**. Myricaria, *Myricaria germanica*; **10**. Large-leaved lime, *Tilia platyphyllos*; **11**. Hedge woundwort, *Stachys sylvatica*; **12**. Wood dock, *Rumex sanguineus* (inset×6); **13**. Dewberry, *Rubus caesius* (inset×1); **14**. Perennial honesty, *Lunaria rediviva*; **15**. Black poplar, *Populus nigra*; **16**. Elongated sedge, *Carex elongata*.

Central European vegetation

Grey alder woods dominated by *Alnus incana* are common along many streams and rivers, particularly in the inner alpine areas of Central Europe. The soil is usually composed of well-aerated gravels. These grey alder woods are usually the climax of the willow–myricaria communities which first colonize shifting and flooded river gravels. An example in southern Germany has the composition

Shrub layer

Dark-leaved willow	*Salix nigricans*
Dewberry	*Rubus caesius*
Bird cherry	*Prunus padus*

Field layer

Great meadow-rue	*Thalictrum aquilegifolium*
Meadowsweet	*Filipendula ulmaria*
Touch-me-not balsam	*Impatiens noli-tangere*
Hairy chervil	*Chaerophyllum hirsutum*
Ground elder	*Aegopodium podagraria*
Wild angelica	*Angelica sylvestris*
Hedge woundwort	*Stachys sylvatica*
Fly honeysuckle	*Lonicera xylosteum*
Wood scabious	*Knautia dipsacifolia*
Cabbage thistle	*Cirsium oleraceum*
False brome	*Brachypodium sylvaticum*
Tufted hair-grass	*Deschampsia cespitosa.*

Willow woods or thickets occur commonly along river valleys, and the dominant tree species are usually the crack willow, *Salix fragilis*, and the white willow, *Salix alba*. In the southeast and south of the Central European region in particular, poplars often grow with, and overtop the willows (Plate 138). The white poplar, *Populus alba*, and the black poplar, *Populus nigra*, are the commonest species, but hybrid poplars are often planted for timber in these river valleys. Ash and alder may also be present. The shrub layer commonly has the following composition

Purple willow	*Salix purpurea*
Almond willow	*Salix triandra*
Dewberry	*Rubus caesius*
Dogwood	*Cornus sanguinea*

Field layer

Nettle	*Urtica dioica*
Common chickweed	*Stellaria media*
Hop	*Humulus lupulus*
Hedge bindweed	*Calystegia sepium*
Bittersweet	*Solanum dulcamara*
Spotted deadnettle	*Lamium maculatum*
Reed canary-grass	*Phalaris arundinacea.*

Such willow–poplar woods are common along the larger river valleys of Central Europe such as those of the Rhine, Elbe, Danube, and Vistula. Willow thickets also occur along alpine river valleys, with such dominant willows as *Salix daphnoides* and *Salix elaeagnos*, often with the German tamarisk, *Myricaria germanica*, and the sea-buckthorn, *Hippophae rhamnoides*.

Mixed woods are often formed on wet, very fertile river-valley soils, whether weakly acid, neutral, or alkaline, with such hardwood trees as ash, elm, alder, and oak. They can

Fig. 28. Wet alder wood with tussocks of elongated sedge, *Carex elongata*, and bittersweet, *Solanum dulcamara*, and yellow iris, *Iris pseudacorus*, 'Heiliges Meer', Nr. Ibbenbüren, W. Germany.

either be permanently waterlogged or subject to periodic flooding. On drier soils, oak–hornbeam woods develop. On rich heavy loams in the major valleys of Central Europe, which are subject to occasional flooding, mixed woods with the pedunculate oak, *Quercus robur;* ash, *Fraxinus excelsior;* small-leaved elm, *Ulmus minor;* fluttering elm, *Ulmus laevis;* and field maple, *Acer campestre,* comprise the tree layer. The shrub layer includes

Hazel	*Corylus avellana*
Dewberry	*Rubus caesius*
Hawthorn	*Crataegus monogyna*
Spindle	*Euonymus europaeus*
Dogwood	*Cornus sanguinea.*

Common in the field layer are

Nettle	*Urtica dioica*
Wood anemone	*Anemone nemorosa*
Celandine	*Ranunculus ficaria*
Wood avens	*Geum urbanum*
Enchanter's nightshade	*Circaea lutetiana*
Ground-elder	*Aegopodium podagraria*
Cleavers	*Galium aparine*
Ground-ivy	*Glechoma hederacea*
Hedge woundwort	*Stachys sylvatica*
Yellow star-of-Bethlehem	*Gagea lutea*
Giant fescue	*Festuca gigantea*
Tufted hair-grass	*Deschampsia cespitosa*
Wood-sedge	*Carex sylvatica.*

On loam or sandy soils, which are fed by calcareous water, such additional species as the wood dock, *Rumex sanguineus;* alternate-leaved golden saxifrage, *Chrysosplenium alternifolium;* touch-me-not balsam, *Impatiens noli-tangere;* and remote sedge, *Carex remota,* may occur with many of the above species. The great horsetail, *Equisetum telmateia,* is common in wet mixed woods in the Polish Carpathians.

On rocky lime-rich soils on steep slopes in the northern Alps, where there is both high humidity and much sunshine, mixed woods occur with a tree layer of

Wych elm	*Ulmus glabra*
Sycamore	*Acer pseudoplatanus*
Norway maple	*Acer platanoides*
Ash	*Fraxinus excelsior*
Small-leaved lime	*Tilia cordata*
Large-leaved lime	*Tilia platyphyllos.*

The shrub layer, in addition to hazel, includes

Mountain currant	*Ribes alpinum*
Gooseberry	*Ribes uva-crispa*

The field layer includes

Perennial honesty	*Lunaria rediviva*
Touch-me-not balsam	*Impatiens noli-tangere*
Goat's-beard	*Aruncus dioicus*
Broad-leaved willowherb	*Epilobium montanum*
Hart's-tongue fern	*Phyllitis scolopendrium*
Hard shield-fern	*Polystichum aculeatum*

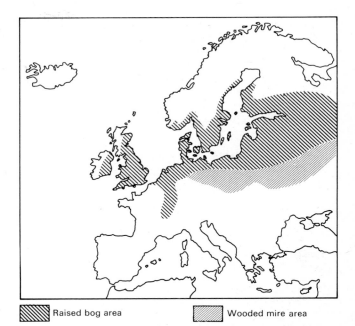

▨ Raised bog area	▧ Wooded mire area

Map 21. Regions of raised bogs and wooded mires.

Male fern	*Dryopteris filix-mas*
Lady-fern	*Athyrium filix-femina.*

Mires

The most important mires, or peat-generating vegetation, occurring in the Central European region are the raised bogs (Map 21, Plate 100), the formation of which has already been described in the Atlantic mire section. Raised bogs depend on a delicate balance between rainfall, water run-off, and evaporation, in places where there is restricted drainage. Mires embrace all stages between acid bogs and alkaline fens. While the vegetation of these two extremes is quite different, there are distinctive intermediate stages. *Intermediate mires* occur where some of the water supply is from water containing mineral salts, while *transitional mires* are those in which bogs and fens are closely associated, either as a mosaic, or where acid bogs and fens succeed each other due to changing water sources, resulting in changing mineral supplies.

Raised bogs have in most cases the characteristic hummock-and-hollows structure, each with their own distinctive vegetation. The hollows, though wet in winter, may in the summer dry out more than the hummocks. In some cases, in montane areas, the hollows may not be present; only the hummock vegetation occurs. The *rands*, or outer sloping margins of the bogs, often dry out and are colonized by Scots pine, *Pinus sylvestris,* and heather, *Calluna vulgaris,* in many Central European mires (Fig. 29). At the present day, only small fragments of the original mires remain intact in Central Europe; they are constantly threatened by water drainage or the lowering of the water table on adjacent land as a result of farming activities. The intermediate and transitional mires are particularly vulnerable as there is such a delicate balance between bog and fen.

Examples of raised bogs in Poland have the following composition. The hollows are dominated by

Oblong-leaved sundew	*Drosera intermedia*
Rannoch-rush	*Scheuchzeria palustris*
Brown beak-sedge	*Rhynchospora fusca* (rare)
Bog-sedge	*Carex limosa*

with the bogmosses, *Sphagnum cuspidatum* and *Sphagnum recurvum*. The hummocks by contrast have

Round-leaved sundew	*Drosera rotundifolia*
Heather	*Calluna vulgaris*
Bog rosemary	*Andromeda polifolia*
Cranberry	*Vaccinium oxycoccos*
Deergrass	*Scirpus cespitosus*
Hare's-tail cottongrass	*Eriophorum vaginatum*

and some species of lichen, *Cladonia.*

In southwestern Sweden, raised bogs often have the following species in the hollows

Oblong-leaved sundew	*Drosera intermedia*
Rannoch-rush	*Scheuchzeria palustris*
White beak-sedge	*Rhynchospora alba*
Bog-sedge	*Carex limosa*

with the bogmoss, *Sphagnum cuspidatum*. The hummocks contain

Cloudberry	*Rubus chamaemorus*
Heather	*Calluna vulgaris*
Crowberry	*Empetrum nigrum*
Hare's-tail cottongrass	*Eriophorum vaginatum*

with the bogmoss, *Sphagnum fuscum.*

In the west of the Central European region, the bogs are similar to those described in the Atlantic region, with such additional species as bog asphodel, *Narthecium ossifragum*, and common cottongrass, *Eriophorum angustifolium*, in the hollows, and with the bell heather, *Erica cinerea;* the cranberry, *Vaccinium oxycoccos;* and the purple moor-grass, *Molinia caerulea*, on the hummocks.

In somewhat drier continental climates, such as in eastern Poland, the raised bogs may be wooded with Scots pine over much of their surface; only the laggs may be treeless. The whole raised bog may dry out in summer—even the laggs may be dry—but in spring they may become very wet with the snow-melt water. Where however the water level is unusually high, the whole bog may remain very wet throughout the year, with an abundance of the bogmosses, *Sphagnum cuspidatum* and *Sphagnum recurvum*. Under these conditions the Scots pine may die out.

Well developed raised bogs may have a central lake, or a series of small lakelets, in which floating vegetation may develop. Due possibly to more active water movement these

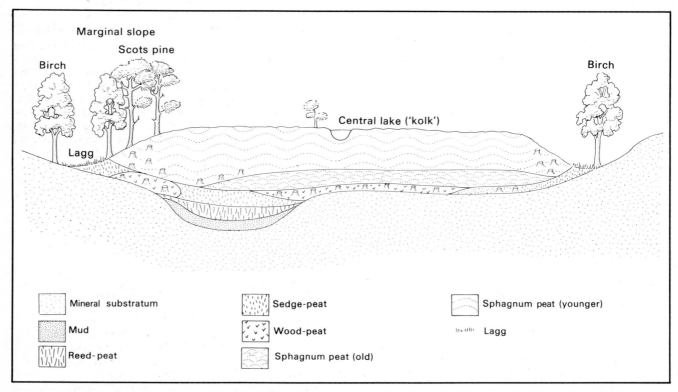

Fig. 29. Raised bog. Structure of classic raised bog, developed partly over previous woodland (note old tree stumps), partly over an infilled lake. Birch and Scots pine are colonizing the edges. (Redrawn from Strasburger.)

central lakes or small lakelets are somewhat richer in minerals than the surrounding bog and the following plants may be present

Round-leaved sundew	*Drosera rotundifolia*
Bog rosemary	*Andromeda polifolia*
Cranberry	*Vaccinium oxycoccos*
Rannoch-rush	*Scheuchzeria palustris*
White beak-sedge	*Rhynchospora alba*
Common cottongrass	*Eriophorum angustifolium*

with the bogmosses, *Sphagnum cuspidatum* and *Sphagnum recurvum*.

Intermediate mires are more in evidence in Eastern Europe, as for example in eastern Poland, where evaporation of moisture is strong in relation to rainfall and thus the mire may still come under the influence of some mineral-containing soil-water. In consequence the transition from fen to bog, which takes place in wetter climates, is halted at an intermediate stage. The bog species which are most likely to survive longest in these intermediate stages are the bogbean, *Menyanthes trifoliata*; the common cottongrass, *Eriophorum angustifolium*; often with the rannoch-rush, *Scheuchzeria palustris*; and the bog-sedge, *Carex limosa*.

Fens. These are not so dependent on the local climate, for they develop where there is ground-water or drainage-water which is relatively rich in mineral salts. River valleys, lake verges, valley and montane slopes with water seepage, all may support fen communities and these occur locally throughout Europe. Many of the Central European fen communities are semi-natural and are maintained in this state by regular cutting; in many cases they would be colonized by willow, alder, or birch if left unattended.

The wet fens, which may be transitional to swamps (described in the Wetlands section), commonly occur by the margins of rivers or lakes where they may be partly submerged for periods during the year. Examples of wet fens in Central Europe may be dominated by the slender tufted-sedge, *Carex acuta*, with

Marsh-marigold	*Caltha palustris*
Lesser spearwort	*Ranunculus flammula*
Purple loosestrife	*Lythrum salicaria*
Common marsh-bedstraw	*Galium palustre*
Water horsetail	*Equisetum fluviatile*.

A wet fen, dominated by the great fen-sedge, *Cladium mariscus*, rarely occurs in Central Europe.

Drier fens, which are not usually submerged at any time of the year, may be divided into poor fens, with a pH of usually 5–6.5 and rich, calcareous fens (pH 6.5–8.2). Poor fens commonly occur at montane levels and on the margins of raised bogs, or in parts of transitional mires. The most widespread type of poor fen is dominated by the white sedge, *Carex curta*, and the common sedge, *Carex nigra*. Less common dominants include the string sedge, *Carex chordorrhiza*, and the slender sedge, *Carex lasiocarpa*, which tends to occur more in transitional mires.

Examples of the more widespread type developed around the margin of bogs in Sweden, have in addition to the two sedges, *Carex curta* and *Carex nigra*,

Tormentil	*Potentilla erecta*
Marsh cinquefoil	*Potentilla palustris*
Marsh violet	*Viola palustris*
Milk-parsley	*Peucedanum palustre*
Common marsh bedstraw	*Galium palustre*
Bulbous rush	*Juncus bulbosus*
Brown bent	*Agrostis canina*
Star sedge	*Carex echinata*.

Examples of *poor fens* on mildly acid soils, at montane levels, which may occur above the tree-line, have the dominant slender sedge, *Carex lasiocarpa*, with

Marsh pennywort	*Hydrocotyle vulgaris*
Milk-parsley	*Peucedanum palustre*
Bogbean	*Menyanthes trifoliata*
Common cottongrass	*Eriophorum angustifolium*

and the moss, *Scorpidium scorpioides*. Other examples of poor fens, particularly in montane regions, may be dominated by the sharp-flowered rush, *Juncus acutiflorus*, with

Eared willow	*Salix aurita*
Downy birch	*Betula pubescens*
Tormentil	*Potentilla erecta*
Marsh cinquefoil	*Potentilla palustris*
Heather	*Calluna vulgaris*
Devil's-bit scabious	*Succisa pratensis*
Purple moor-grass	*Molinia caerulea*
Common cottongrass	*Eriophorum angustifolium*

and the bogmoss, *Sphagnum recurvum,* and the common moss, *Polytrichum commune.*

Rich calcareous fens, with a pH ranging from 6.5 to 8.2, are rather rare in Central Europe, especially in the lowlands where they have been destroyed. They are now largely restricted to montane and sub-alpine levels, particularly in the Alps and the Carpathians. They have a rich and interesting flora. In the forest and sub-alpine regions of Germany, in the Alps and the Black Forest, calcareous fens are often related to late snow patches. They may have *Carex frigida* abundant, with other sedges such as the star sedge, *Carex echinata*, and the small-fruited yellow-sedge, *Carex serotina*. Characteristic of such fens are

Yellow saxifrage	*Saxifraga aizoides*
Grass-of-Parnassus	*Parnassia palustris*
Tormentil	*Potentilla erecta*
Lady's-mantle	*Alchemilla vulgaris*
Alpine lovage	*Ligusticum mutellina*
Alpine snowbell	*Soldanella alpina*
Alpine bartsia	*Bartsia alpina*
Common butterwort	*Pinguicula vulgaris*
	Aster bellidiastrum
	Calycocorsus stipitatus
	Tofieldia calyculata
Alpine rush	*Juncus alpinus*
Tufted hair-grass	*Deschampsia cespitosa*
Lesser clubmoss	*Selaginella selaginoides*

Central European vegetation

Rich calcareous fens in southern Sweden, occurring mainly in the lowlands, commonly include the following members of the sedge family: broad-leaved cottongrass, *Eriophorum latifolium*; the brown bog-rush, *Schoenus ferrugineus*; and black bog-rush, *S. nigricans* (e.g. in Scane); long-stalked yellow-sedge, *Carex lepidocarpa*; and carnation sedge, *Carex panicea*. Other species include

Grass-of-Parnassus	*Parnassia palustris*
Tormentil	*Potentilla erecta*
Fairy flax	*Linum catharticum*
Bird's-eye primrose	*Primula farinosa*
Common butterwort	*Pinguicula vulgaris*
Devil's-bit scabious	*Succisa pratensis*
Marsh arrowgrass	*Triglochin palustris*
Alpine rush	*Juncus alpinus*
Quaking-grass	*Briza media*
Blue moor-grass	*Sesleria caerulea*
Purple moor-grass	*Molinia caerulea*
Marsh helleborine	*Epipactis palustris*
Orchid species	*Dactylorhiza* sps.

In the lower montane zone of the western Carpathians in Poland, fen communities with the following sedges occur: *Carex davalliana*; dioecious sedge, *Carex dioica*; large yellow sedge, *Carex flava*; carnation sedge, *Carex panicea*. Additional species include *Alchemilla glabra*; eastern marsh valerian, *Valeriana dioica* ssp. *simplicifolia*; broad-leaved cottongrass, *Eriophorum latifolium*; and marsh horsetail, *Equisetum palustre*.

Heaths

Heath communities are widespread in Central Europe, particularly in the more oceanic regions, as for example in southern Sweden, northwestern Germany, and in western Poland. They are however most characteristic of the Atlantic region where the main types of heaths are described. In Central Europe additional types of heath will be briefly described. A distinctive and truly Central European community is the heather–broom heath. This is relatively common on sandy soils, at altitudes of above about 150 m. It usually occurs as a result of the destruction of such climax woods as acid beech woods or beech–oak woods. Should these heaths be protected and left ungrazed they will often be recolonized by trees and shrubs and eventually develop into climax woodlands. Typical species of heather–broom heaths are

Juniper	*Juniperus communis*
Sessile oak	*Quercus petraea*
Broom	*Cytisus scoparius*
Hairy greenweed	*Genista pilosa*
Heather	*Calluna vulgaris*
Bilberry	*Vaccinium myrtillus*
Sheep's-fescue	*Festuca ovina*
Wavy hair-grass	*Deschampsia flexuosa*
Common bent	*Agrostis capillaris*

and the moss, *Pleurozium schreberi*.

A montane type of heath, typified by the frequency of heather, bilberry, and cowberry is found in the Central European montane zone above 650 m, on dry peat. It includes

Tormentil	*Potentilla erecta*
Rowan	*Sorbus aucuparia*
Heather	*Calluna vulgaris*
Bilberry	*Vaccinium myrtillus*
Cowberry	*Vaccinium vitis-idaea*
Heath bedstraw	*Galium saxatile*
Wavy hair-grass	*Deschampsia flexuosa*

and the moss, *Pleurozium schreberi*, with lichens (*Cladonia*).

Widespread on acid soils in the montane region are grass-heaths, in which both grasses and rushes are abundant; these are mixed with heather. An example in Germany, with both the mat-grass, *Nardus stricta*, and the heath rush, *Juncus squarrosus*, abundant, includes

Tormentil	*Potentilla erecta*
Heather	*Calluna vulgaris*
Cowberry	*Vaccinium vitis-idaea*
Lousewort	*Pedicularis sylvatica*
Arnica	*Arnica montana*.

Another German montane example characteristically includes the winged broom, *Chamaespartium sagittale*, with heather and such grasses as

Sheep's-fescue	*Festuca ovina*
Red fescue	*Festuca rubra*
Sweet vernal-grass	*Anthoxanthum odoratum*
Mat-grass	*Nardus stricta*
Heath-grass	*Danthonia decumbens*.

Grasslands

In the Central European region, grasslands are important and widespread, but very few are natural; the majority are maintained as grasslands by man and his animals. The only natural grasslands are possibly those found in alpine habitats (described in the Alpine section), and grasslands on very dry sites forming steppe-like communities, as well as grasslands on saline soils, and possibly some fen grasslands on flooded alluvial soils. All widespread lowland grasslands are almost entirely artificial, though they may contain many natural species, depending on the cutting and grazing regimes.

Dry grasslands (Plate 101)

Steppe-like grasslands are more characteristic of the drier Pannonic region where they are described. In Central Europe, however, they occur on dry, sunny, south-facing slopes, on well-drained neutral or alkaline soils. Here such tussocky, drought-tolerant grasses as the following tend to dominate

Festuca pallens
Festuca rupicola
Festuca valesiaca
Koeleria macrantha
Stipa capillata
Stipa pennata

while on less extreme dry sites the Hungarian brome, *Bromus inermis,* and tor-grass, *Brachypodium pinnatum,* are dominant.

In many of these dry grasslands there is a typical cycle of growth and development. The peak of growth is reached in late spring or early summer, when many species come into flower, such as the yellow adonis, *Adonis vernalis;* the grey cinquefoil, *Potentilla cinerea;* and the dwarf sedge, *Carex humilis.* Many small annuals also flower before the advent of the dry summer; these include the thyme-leaved sandwort, *Arenaria serpyllifolia;* the annual rock-cress, *Arabis recta;* and species of *Cerastium* and *Veronica,* etc. There follows a period of rest through the dry summer, when no species flower. In autumn there is renewed growth, with the flowering of such species as the yellow odontites, *Odontites lutea;* the European michaelmas daisy, *Aster amellus;* and others. It is doubtful whether these dry grasslands are the climax communities of these dry soils; in many cases they would in all probability be re-colonized by dry woodlands. However as far north as southern Sweden, a possibly natural community of *Stipa pennata* occurs on south-facing esker slopes. Feather-grass-grasslands, dominated by *Stipa capillata* in many cases, occur in southeastern Germany, Austria, Czechoslovakia, and Poland. They contain the grasses

	Festuca pallens
	Poa badensis
	Melica transsilvanica
	Koeleria pyramidata
Purple-stem cat's-tail	*Phleum phleoides*
	Stipa joannis.

Other frequently occurring herbaceous species in these feather-grass grasslands are

Spanish catchfly	*Silene otites*
Fastigiate gypsophilia	*Gypsophila fastigiata*
Carthusian pink	*Dianthus carthusianorum*
	Erysimum crepidifolium
Grey cinquefoil	*Potentilla cinerea*
Purple milk-vetch	*Astragalus danicus*
	Astragalus exscapus
Cypress spurge	*Euphorbia cyperissias*
Teesdale violet	*Viola rupestris*
Common rockrose	*Helianthemum nummularium*
	Seseli hippomarathrum
Golden-drop	*Onosma arenaria*
Wall germander	*Teucrium chamaedrys*
Thyme species	*Thymus* sps.
	Orobanche amethystea
Squinancywort	*Asperula cynanchica*
	Scabiosa canescens
	Aster linosyris
Field wormwood	*Artemisia campestris*
Mouse-ear hawkweed	*Hieracium pilosella*
	Carex supina.

Another type of grassland, without feather-grasses, *Stipa* species, develops on marl or loam, after the destruction of xerothermic woods. Here tor-grass, *Brachypodium pinnatum,*

is dominant. An example from the Garchinger Heide, near Munich, has such characteristic species as

	Thesium linophyllon
Yellow adonis	*Adonis vernalis*
Snowdrop windflower	*Anemone sylvestris*
Woolly milk-vetch	*Oxytropis pilosa*
Mountain clover	*Trifolium montanum*
Large self-heal	*Prunella grandiflora*
	Scorzonera purpurea
	Anthericum ramosum.

On warm rocky calcareous sites, the blue fescue, *Festuca glauca,* may be dominant often with *Sesleria albicans.* Interesting species may be present in this grassland, such as

Cheddar pink	*Dianthus gratianopolitanus*
White stonecrop	*Sedum album*
Cypress spurge	*Euphorbia cyparissias*
	Allium senescens.

Dry grasslands in the southwestern part of Central Europe are commonly dominated by the upright brome, *Bromus erectus,* and they have many similarities with the sub-Mediterranean grasslands. These upright brome grasslands are rich in attractive species, many of which may not necessarily occur together in a single community. They include

Bulbous buttercup	*Ranunculus bulbosus*
Kidney vetch	*Anthyllis vulneraria*
Sainfoin	*Onobrychis viciifolia*
Cypress spurge	*Euphorbia cyparissias*
Cross gentian	*Gentiana cruciata*
	Gentianella ciliata
German gentian	*Gentianella germanica*
Large thyme	*Thymus pulegioides*
Meadow clary	*Salvia pratensis*
Small scabious	*Scabiosa columbaria*
Round-headed rampion	*Phyteuma orbiculare*
Hoary ragwort	*Senecio erucifolius*
Stemless carline-thistle	*Carlina acaulis*
Dwarf thistle	*Cirsium acaule*
Greater knapweed	*Centaurea scabiosa*
Autumn lady's-tresses	*Spiranthes spiralis*
Military orchid	*Orchis militaris*
Green-winged orchid	*Orchis morio*
Monkey orchid	*Orchis simia*
Toothed orchid	*Orchis tridentata*
Burnt orchid	*Orchis ustulata*
Man orchid	*Aceras anthropophorum*
Lizard orchid	*Himantoglossum hircinum*
Pyramidal orchid	*Anacamptis pyramidalis*
Bee orchid	*Ophrys apifera*
Late spider-orchid	*Ophrys fuciflora*
Fly orchid	*Ophrys insectifera*
Early spider-orchid	*Ophrys sphegodes.*

Drier upright brome grasslands, occurring on south-facing slopes in the south, have such attractive species as

| Carthusian pink | *Dianthus carthusianorum* |
| Small pasqueflower | *Pulsatilla pratensis* |

Rosemary Win

Plate 28. Central Europe grasslands—dry
1. Grey cinquefoil, *Potentilla cinerea*;   **2**. Sainfoin, *Onobrychis viciifolia*;   **3**. Meadow clary, *Salvia pratensis*;   **4**. Field wormwood, *Artemisia campestris*;   **5**. Round-headed rampion, *Phyteuma orbiculare*;   **6**. Small pasque-flower, *Pulsatilla pratensis*;   **7**. Hairy melick, *Melica ciliata*;   **8**. Feather-grass, *Stipa pennata*;   **9**. German gentian, *Gentianella germanica*;   **10**. Large self-heal, *Prunella grandiflora*;   **11**. Spanish catchfly, *Silene otites*;   **12**. Yellow adonis, *Adonis vernalis*;   **13**. Burnt orchid, *Orchis ustulata*;   **14**. Military orchid, *Orchis militaris*;   **15**. Carthusian pink, *Dianthus carthusianorum*.

Pasqueflower	*Pulsatilla vulgaris*
Horseshoe vetch	*Hippocrepis comosa*
Wall germander	*Teucrium chamaedrys*
Mountain germander	*Teucrium montanum*
Common globularia	*Globularia punctata.*

Steppe-like feather-grass grasslands are scattered throughout the southwestern part of Central Europe. They are dominated by such grasses as *Stipa pennata;* purple-stem cat's-tail, *Phleum phleoides;* tor-grass, *Brachypodium pinnatum;* and the dwarf sedge, *Carex humilis.*

Meadows and pastures

These are all artificial, and are either grazed, or mown regularly once, twice, or even three times a year to produce rich crops of hay or ensilage. Certain species growing under these conditions have become adapted to these regimes. Such genera as the eyebrights, *Euphrasia;* yellow rattles, *Rhinanthus;* and some *Gentianella* species have evolved as distinct races, which flower either late or early in the year, in relation to the time of mowing or grazing. Where crops are taken either directly, or indirectly through grazing, regular manuring must take place to maintain the grassland's productivity as well as the structure of the community. If these grasslands are abandoned, they are quickly colonized by shrubs, and in time develop into woodland.

Such artificial grasslands are common in the western part of Central Europe where the climate and soil are sufficiently moist. Further south and east they are restricted to river valleys and flood plains, or to montane or sub-alpine areas. Despite their artificial origin and relatively temporary nature, these grasslands do show distinct types such as wet meadows and pastures, fresh meadows and pastures, and poor grasslands. Wet meadows and pastures occur on soils with a high water level, and they may be flooded periodically in winter or early spring. Some are transitional to fen vegetation. The purple moor-grass, *Molinia caerulea,* is dominant in these wet meadows, which, in their unimproved state, are usually mown once a year. These develop a rich and beautiful association of species, some of them rather rare. Improvement of such grasslands, by manuring or drainage, may increase their productivity, but results in the loss of many interesting species.

Such unimproved *wet meadows* commonly include the species

Superb pink	*Dianthus superbus*
Meadow buttercup	*Ranunculus acris*
Great burnet	*Sanguisorba officinalis*
Tormentil	*Potentilla erecta*
Winged pea	*Tetragonolobus maritimus*
Cambridge milk-parsley	*Selinum carvifolia*
	Laserpitium prutenicum
Willow gentian	*Gentiana asclepiadea*
Marsh gentian	*Gentiana pneumonanthe*
Northern bedstraw	*Galium boreale*
Betony	*Stachys officinalis*
Devil's-bit scabious	*Succisa pratensis*
Irish fleabane	*Inula salicina*

Sneezewort	*Achillea ptarmica*
Tuberous thistle	*Cirsium tuberosum*
Saw-wort	*Serratula tinctoria*
Brown knapweed	*Centaurea jacea*
	Allium angulosum
Fragrant onion	*Allium suaveolens*
	Iris sibirica
Marsh gladiolus	*Gladiolus palustris*
	Carex tomentosa
Adder's-tongue	*Ophioglossum vulgatum.*

Another frequent type of *Molinia* grassland, which is regularly flooded in lowland areas, has such rush species co-dominant as the compact rush, *Juncus conglomeratus;* the sharp-flowerd rush, *Juncus acutiflorus;* and the soft rush, *Juncus effusus.*

The tufted hair-grass, *Deschampsia cespitosa,* is another common grass which may dominate in poor wet pastures on clay or sandy-loam soils. It commonly has the associated species

Common sorrel	*Rumex acetosa*
Ragged robin	*Lychnis flos-cuculi*
Creeping buttercup	*Ranunculus repens*
Great burnet	*Sanguisorba officinalis*
Fen violet	*Viola persicifolia*
Pepper-saxifrage	*Silaum silaus*
	Cnidium dubium
Sneezewort	*Achillea ptarmica*
Oxeye daisy	*Leucanthemum vulgare.*

Similar wet grasslands in more hilly and mountainous regions may have such species as the tall pale yellow cabbage thistle, *Cirsium oleraceum;* the marsh-marigold *Caltha palustris;* the common bistort, *Polygonum bistorta;* the meadowsweet, *Filipendula ulmaria;* and wild angelica, *Angelica sylvestris*—all distinctive plants.

Fresh meadows and pastures are the most familiar of the managed grasslands of the Central European region. They occur on fresh, that is to say moderately moist, soils and though they involve more maintainance, they yield a much better crop than the wet grasslands already described. They are manured regularly, unless they are situated on flood plains where they are to some extent fertilized naturally.

The commonest and most productive type of such fresh meadows is the hay-meadow, dominated by the false oat-grass, *Arrhenatherum elatius.* These communities are most widespread in the valleys of the hilly and montane areas up to about 600 m. Growth is rapid and continues over a long period and consequently two or three hay crops may be taken each year. Other grasses present include

Meadow fescue	*Festuca pratensis*
Smooth meadow-grass	*Poa pratensis*
Soft-brome	*Bromus hordeaceus*
Sweet vernal-grass	*Anthoxanthum odoratum*
Yorkshire-fog	*Holcus lanatus*
Meadow foxtail	*Alopecurus pratensis.*

Plate 29. Central Europe grasslands—wet and fresh

1. Irish fleabane, *Inula salicina*; **2**. Saw-wort, *Serratula tinctoria*; **3**. Sneezewort, *Achillea ptarmica*; **4**. Brown knapweed, *Centaurea jacea* (inset×2); **5**. Meadow crane's-bill, *Geranium pratense*; **6**. Oxeye daisy, *Leucanthemum vulgare*; **7**. Cambridge milk-parsley, *Selinum carvifolia* (inset×6); **8**. Common bistort, *Polygonum bistorta*; **9**. Rough hawkbit, *Leontodon hispidus*; **10**. Betony, *Stachys officinalis*; **11**. Great burnet, *Sanguisorba officinalis*; **12**. Willow gentian, *Gentiana asclepradea*; **13**. Superb pink, *Dianthus superbus*; **14**. White clover, *Trifolium repens*.

Non-grass species include

Common sorrel	*Rumex acetosa*
	Cerastium fontanum
Meadow buttercup	*Ranunculus acris*
White clover	*Trifolium repens*
Meadow crane's-bill	*Geranium pratense*
Hogweed	*Heracleum sphodylium*
Oxeye daisy	*Leucanthemum vulgare*
Autumn hawkbit	*Leontodon autumnalis*
Dandelion	*Taraxacum officinale.*

Above about 400 m, a type of *montane hay meadow* (Plates 102 and 103) is found which is dominated by the yellow oat-grass, *Trisetum flavescens,* with the sweet vernal-grass, *Anthoxanthum odoratum.* These meadows contain such typical species as

Lady's-mantle	*Alchemilla vulgaris*
Red clover	*Trifolium pratense*
Hogweed	*Heracleum sphondylium*
Greater burnet-saxifrage	*Pimpinella major*
Cow parsley	*Anthriscus sylvestris*
Caraway	*Carum carvi*
Rough hawkbit	*Leontodon hispidus.*

The usually white-flowered crocus *Crocus vernus* ssp. *albiflorus,* is often seen in these grasslands, while in the Salzkammergut, Austria, in rainy areas, the pheasant's-eye narcissus, *Narcissus poeticus* ssp. *radiiflorus,* grows.

Another very common type of grassland occurring in the lowlands and lower hills and developing particularly on sandy loams, is dominated by the perennial rye-grass, *Lolium perenne,* with other grasses including

Smooth meadow-grass	*Poa pratensis*
Crested dog's-tail	*Cynosurus cristatus*
Sweet vernal-grass	*Anthoxanthum odoratum*
Yorkshire-fog	*Holcus lanatus*
Timothy	*Phleum pratense.*

This forms a fine grazing grassland and contains many of the common grassland species.

This latter grassland is often replaced above about 300 m, by the densely tufted *Festuca nigrescens,* with the crested dog's-tail, *Cynosurus cristatus.*

Poor grasslands are found on acid soils and are commonly dominated by the mat-grass, *Nardus stricta.* Such grasslands show all gradations with grass-heaths, which are described in the previous section, and these grasslands also occur in the alpine regions. Where fresh rich meadows cease to be fertilized, and where cropping and grazing still persist, these meadows may revert to poor mat-grasslands, in which the upper layers of the soil become acid and podsolized. Such acid depauperate grasslands commonly have such indicative species as the heath dog-violet, *Viola canina;* common milkwort, *Polygala vulgaris;* and imperforate St. John's wort, *Hypericum maculatum.*

10 Mediterranean vegetation (Plates 30–42; Plates 104–131)

The vegetation of the Mediterranean region (Map 22) is quite distinct from that of any other region in Europe. It is dominated by evergreen trees, shrubs, and shrublets which can survive the long hot summers without rain. Most of the herbaceous plants die right down and remain inactive in the summer with dormant buds in the soil while the annuals complete their life cycle by the summer. Such a contrasting and unique type of climate has favoured types of vegetation that occur nowhere else in Europe. At the same time man's early colonization of the Mediterranean shores has resulted in his prolonged and intensive influence on the vegetation of the region. Consequently today little remains of the natural plant communities. Plant communities persisting to the present day are either scattered woodlands which survive in localities that have not been destroyed by man or his animals, or more commonly dense evergreen scrub known as *maquis* or, more widespread still, dwarf, scattered, mostly evergreen shrublets, the *garigue (phrygana)*. Many of the woody species have small thick leathery leaves which reduce transpiration during the dry summer; many are aromatic, releasing ethereal oils, thus possibly reducing water-loss and deterring grazing. Active growth and flowering takes place in the autumn and often throughout the winter, and reaches its peak in the spring.

The dominating trees in the Mediterranean zone are evergreen oaks and pines, with many evergreen shrubs such as juniper, heathers, cistus, spiny broom, strawberry trees, lentisc, and others, while the olive (Map 23) and carob are characteristic. However, due to the long and intensive occupation of the Mediterranean coasts by man, almost nothing remains of the true natural woodland communities, and shrub communities maintained by cutting, firing, grazing, and the resultant erosion of the soil have taken the place of the evergreen woods. In places the soil has become so eroded that the exposed bed-rock supports only a sparse steppe-like community of scattered herbaceous plants which is nevertheless composed of an interesting and rich assortment of species which flower in spring. Elsewhere olive trees have been widely planted, often on terraced hill slopes (Plate 120).

Inland in the hill region a sub-Mediterranean zone occurs which is less affected by the long drought and hot summers and has a higher rainfall. Here deciduous trees and shrubs largely replace the evergreens; these bush communities are known as *šibljak* in the Balkan region. Several species of deciduous oaks, with other trees such as hop-hornbeam, maples, manna ash, beech etc., are dominant, and coniferous woods, largely of black pine, occur. Other much more local deciduous woods in the east are dominated by sweet chestnut, horse chestnut, walnut, and plane. There are also some very local and degraded juniper woods in the hills in both the east and west.

The higher mountain regions bordering the Mediterranean have distinctive coniferous forests composed of several species of pine and fir, both in the Iberian and Balkan peninsulas. Hedgehog-heath communities of dwarf cushion-forming shrubs often occur above these forests.

Grasslands in the Mediterranean are largely steppe-like, with feather-grasses, esparto grasses, etc. They are largely maintained as grasslands by the grazing of goats and sheep, or as a result of continued soil erosion due to the short but heavy local rainstorms. *Montane grasslands* are also largely man-made following forest and shrub clearance and they are maintained in this condition by the regular grazing of flocks, which are brought up from the lower hills and are found grazing over the highest mountains in the summer months.

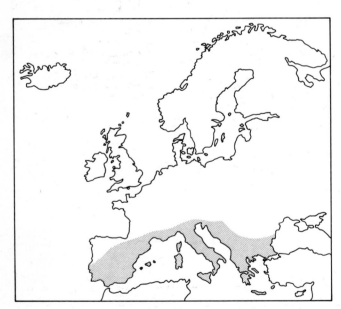

Map 22. Mediterranean region.

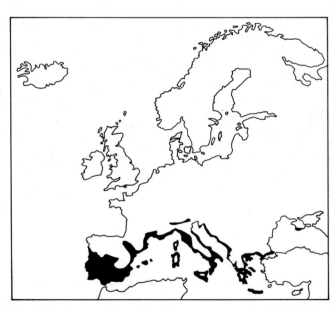

Map 23. The cultivated Olive *Olea europaea* var. *europaea* has a truly Mediterranean distribution.

Evergreen and semi-evergreen oak woods

Four examples of *holm oak* woods (Maps 24 and 25; Plate 104) have been chosen to show the range of variation in these woods, but there are many intermediate stages to be seen in the Mediterranean region and likewise many stages of degradation to *maquis* and *garigue*. In southern France and southern Spain, typical examples of holm oak woods (Fig. 30) have the following structure and composition. The dominant tree, the holm oak, *Quercus ilex*, grows to a height of 15–18 m, and usually because of selective felling, forms an open wood with widely spaced trees with well developed shrub and field layers. Less commonly it is found growing in closed canopy with a sparse undergrowth.

The shrub layer is usually 3–5 m high, rarely up to 12 m. Characteristic shrubs include

Fragrant clematis	*Clematis flammula*
	Rosa sempervirens
Mastic tree	*Pistacia lentiscus*
Turpentine tree	*Pistacia terebinthus*
Box	*Buxus sempervirens*
Mediterranean buckthorn	*Rhamnus alaternus*
Ivy	*Hedera helix*
Strawberry tree	*Arbutus unedo*
	Phillyrea angustifolia
	Phillyrea latifolia
Laurustinus	*Viburnum tinus*
	Lonicera etrusca
Minorca honeysuckle	*Lonicera implexa.*

The field layer includes

Large Mediterranean-spurge	*Euphorbia characias*
	Viola alba
Wall germander	*Teucrium chamaedrys*
Betony	*Stachys officinalis*
Wild madder	*Rubia peregrina*
	Asparagus acutifolius
Butcher's-broom	*Ruscus aculeatus*
	Carex distachya
Black spleenwort	*Asplenium adiantum-nigrum.*

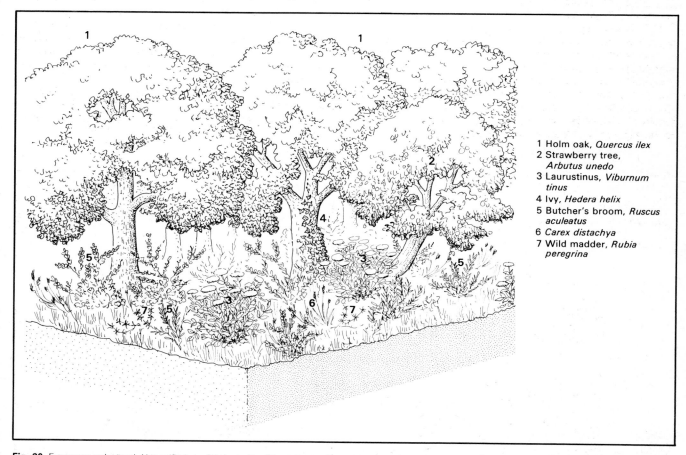

1 Holm oak, *Quercus ilex*
2 Strawberry tree, *Arbutus unedo*
3 Laurustinus, *Viburnum tinus*
4 Ivy, *Hedera helix*
5 Butcher's broom, *Ruscus aculeatus*
6 *Carex distachya*
7 Wild madder, *Rubia peregrina*

Fig. 30. Evergreen oak wood. Natural holm oak woods like this are rare today. When well grown they may be shady and have a dense shrub-layer of species familiar from the maquis. Species shown here: Holm oak *Quercus ilex*; Strawberry tree *Arbutus unedo*; Laurustinus *Viburnum tinus*; Ivy *Hedera helix*; Butcher's Broom *Ruscus aculeatus*; *Carex distachya*; Wild Madder *Rubia peregrina*.

Mediterranean vegetation

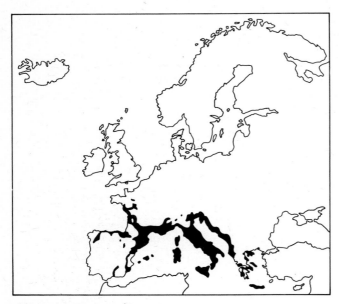

Map 24. Holm oak *Quercus ilex*.

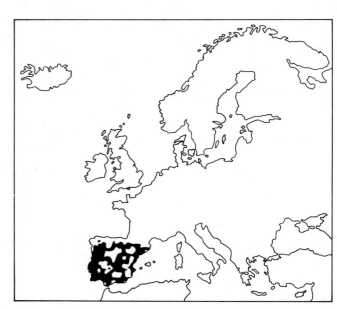

Map 25. *Quercus rotundifolia*.

An example of a coppiced holm oak wood found in Monte Argentario, Italy, has both the holm oak and the white oak, *Quercus pubescens*, dominant, with the wild service tree, *Sorbus torminalis*, and Montpellier maple, *Acer monspessulanum*. The shrub layer contains several *Cistus* species as well as

Mediterranean mezereon	*Daphne gnidium*
Myrtle	*Myrtus communis*
	Phillyrea latifolia
Strawberry tree	*Arbutus unedo*
Tree heath	*Erica arborea*
Laurustinus	*Viburnum tinus*.

The field layer includes

The parasite	*Cytinus hypocistis*
Repand cyclamen	*Cyclamen repandum*
Tuberous comfrey	*Symphytum tuberosum*
Dense-flowered orchid	*Neotinea moculata*

as well as several bee orchids, *Ophrys* species.

Holm oak woods on the Adriatic coast of Yugoslavia have in addition in the tree layer, the manna ash, *Fraxinus ornus*; the laurel, *Laurus nobilis*; and the Aleppo pine, *Pinus halepensis*. The shrub layer comprises

Prickly juniper	*Juniperus oxycedrus*
	Rosa sempervirens
Scorpion senna	*Coronilla emerus*
Turpentine tree	*Pistacia terebinthus*
Mediterranean buckthorn	*Rhamnus alaternus*
Laurustinus	*Viburnum tinus*
Minorca honeysuckle	*Lonicera implexa*.

The field layer includes

Early dog-violet	*Viola reichenbachiana*
Repand cyclamen	*Cylamen repandum*
	Sesleria autumnalis
	Piptatherum virescens
	Carex distachya
	Asplenium onopteris.

Holm oak woods in Greece usually have the white oak, *Quercus pubescens*, in the tree layer, and either the Aleppo pine, *Pinus halepensis*, or the closely related *Pinus brutia*, depending on the locality of the wood. The shrub layer commonly includes

Prickly juniper	*Juniperus oxycedrus*
Kermes oak	*Quercus coccifera*
Turpentine tree	*Pistacia terebinthus*
	Cistus incanus
	Cistus salvifolius
Eastern strawberry tree	*Arbutus andrachne*
Strawberry tree	*Arbutus unedo*
	Phillyrea latifolia
Olive	*Olea europaea*
Minorca honeysuckle	*Lonicera implexa*
Butcher's-broom	*Ruscus aculeatus*.

The cork oak, *Quercus suber* (Plate 106), is primarily a western Mediterranean tree (Map 26), with its main concentration in Portugal and western Spain, where it is found both native and in extensive plantations. It spreads eastwards in a narrow coastal belt as far east as Italy and Sardinia, and into the adjacent hills to an altitude of about 400–600 m. It requires a relatively moist climate with rainfall of about 1000

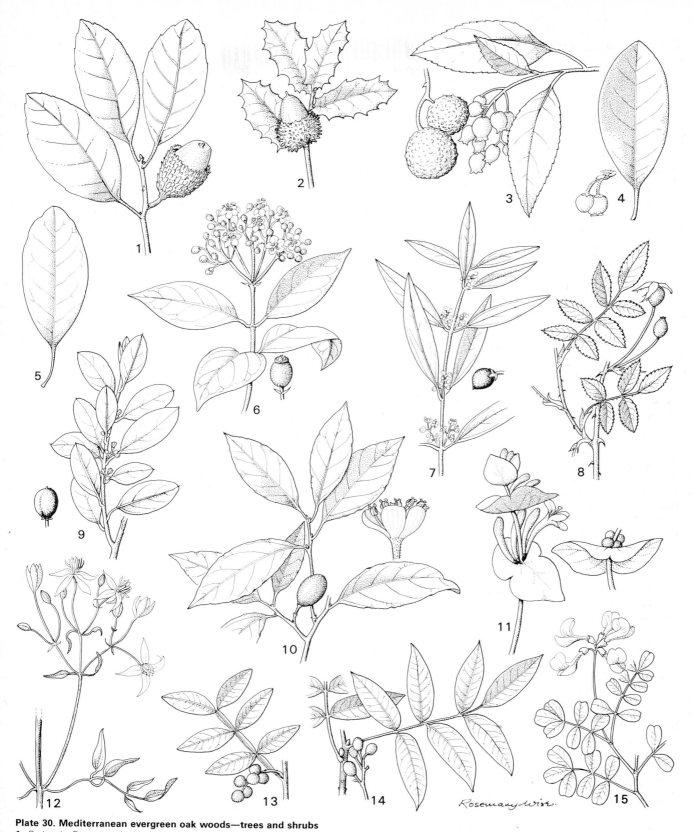

Plate 30. Mediterranean evergreen oak woods—trees and shrubs

1. Cork oak, *Quercus suber*; **2**. Kermes oak, *Quercus coccifera*; **3**. Strawberry tree, *Arbutus unedo*; **4**. Eastern strawberry tree, *Arbutus andrachne* (leaf and flower); **5**. Holm oak, *Quercus ilex* (leaf); **6**. Laurustinus, *Viburnum tinus* (inset×2); **7**. *Phillyrea angustifolia* (inset×2); **8**. *Rosa sempervirens*; **9**. Mediterranean buckthorn, *Rhamnus alaternus*; **10**. Laurel, *Laurus nobilis* (inset×3); **11**. Minorca honeysuckle, *Lonicera implexa*; **12**. Fragrant clematis, *Clematis flammula*; **13**. Mastic tree, *Pistacia lentiscus*; **14**. Turpentine tree, *Pistacia terebinthus*; **15**. Scorpion senna, *Coronilla emerus*.

Mediterranean vegetation

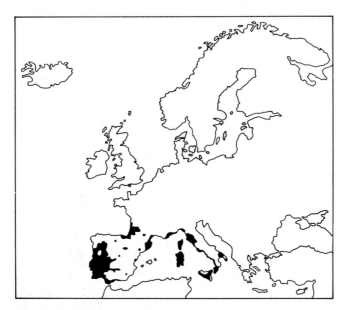

Map 26. Cork oak *Quercus suber*.

Olive — *Olea europaea*
French lavender — *Lavandula stoechas*
Dwarf fan palm — *Chamaerops humilis*.

The field layer includes *Lithodora diffusa* and tree germander, *Teucrium fruticans*.

In Italy, in Montepescali, the shrub layer of cork oak woods has the following composition

Holm oak	*Quercus ilex*
White oak	*Quercus pubescens*
Christ's thorn	*Paliurus spina-christi*
Mediterranean mezereon	*Daphne gnidium*
	Cistus albidus
Narrow-leaved cistus	*Cistus monspeliensis*
Sage-leaved cistus	*Cistus salvifolius*
Myrtle	*Myrtus communis*
Strawberry tree	*Arbutus unedo*
Tree heath	*Erica arborea*
Green heath	*Erica scoparia*
	Phillyrea angustifolia
	Smilax aspera.

The field layer includes

	Dorycnium hirsutum
	Ornithopus compressus
Honeywort	*Cerinthe major*
Yellow centaury	*Centaurium maritimum*
Great quaking-grass	*Briza maxima*
Quaking-grass	*Briza minor*
	Orchis morio ssp. *picta*
Limodore	*Limodorum abortivum*.

mm a year, the minimum being about 500 mm. It tends to replace the holm oak on poor acid stony soils, where slow growth gives the best quality cork. The outer cork layer is removed every 8–12 years from the trunk and the main branches; the acorns are important as pig fodder. Extensive plantations of cork oak are particularly important in the River Tajo region in Portugal where they are found up to an altitude of 1600 m in the surrounding hills.

Cork oaks grow to a height of 15 m. They have stout trunks and spreading branches and are usually found in open canopy, often widely spaced, and with a well developed shrub growth below. Alternatively, the intervening ground flora may be grazed, or cultivated with crops. A typical Spanish cork oak wood of Andalusia may have the following rich shrub layer

Kermes oak	*Quercus coccifera*
	Stauracanthus boivinii
	Calicotome spinosa
	Chamaespartium tridentatum
	Adenocarpus telonensis
Mastic Tree	*Pistacia lentiscus*
	Thymelaea lanuginosa
Mediterranean mezereon	*Daphne gnidium*
	Cistus crispus
Gum cistus	*Cistus ladanifer*
Narrow-leaved cistus	*Cistus monspeliensis*
Sage-leaved cistus	*Cistus salvifolius*
	Halimium halimifolium
Myrtle	*Myrtus communis*
Strawberry tree	*Arbutus unedo*
Spanish heath	*Erica australis*
Green heath	*Erica scoparia*
	Phillyrea angustifolia

The kermes or holly oak, *Quercus coccifera* (Plate 105), is also a widespread evergreen Mediterranean tree (Map 27), but because of grazing, cutting, and fires, it is rarely found forming a wood. Probably the best extant examples are the woods on the rocky slopes of Crete, at altitudes of 350–1000 m, on marly rocks or shales. More commonly it occurs as a shrub in the maquis. However in the Aegean region it often replaces the holm oak, and probably in the past formed climax woods in this region.

Examples of kermes oak woods in Crete fall into two types. One is characterized by the presence of *Pyrus amygdaliformis*, with *Phillyrea latifolia*, and the turpentine tree, *Pistacia terebinthus*. The shrub layer commonly includes

Thorny burnet	*Sarcopoterium spinosum*
Spiny broom	*Calicotome villosa*
Large Mediterranean spurge	*Euphorbia characias*
	Rhamnus prunifolius
	Hypericum empetrifolium
	Cistus incanus
Olive	*Olea europaea*
Jerusalem sage	*Phlomis fruticosa*
	Teucrium microphyllum
	Asparagus stipularis.

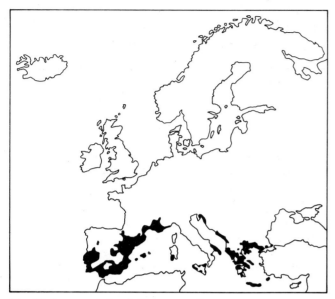

Map 27. Kermes or holly oak *Quercus coccifera*.

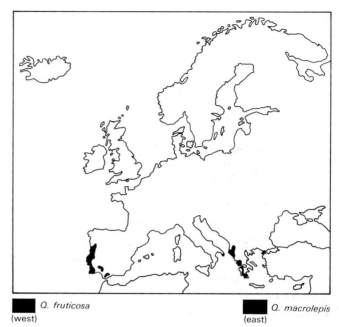

■ *Q. fruticosa*
(west)

■ *Q. macrolepis*
(east)

Map 28. *Quercus fruticosa* (west) and *Quercus macrolepis* (east).

The other variant is characterized by the storax, *Styrax officinalis*, with *Pyrus amygdaliformis* and *Prunus webbii*.

Other Mediterranean oaks include the semi-evergreen Valonia oak, *Quercus macrolepis* (Plate 109, Map 28), which is a stout-trunked tree growing to a height of 15 m. Its main centre of distribution is in Greece, where it has been selectively cultivated for its large acorn-cups used in tanning, dyeing, and ink-making. Open woods of Valonia oak occur on stony sites on mainland Greece, and on some Aegean islands and Crete but, due to its long history of cultivation, its native range is obscure.

In Crete, the Valonia oak is often associated with the funeral cypress, *Cupressus sempervirens;* the kermes oak, *Quercus coccifera;* and the carob, *Ceratonia siliqua*. Plants occurring in the lower layers include

	Osyris alba
	Hypericum empetrifolium
Jerusalem sage	*Phlomis fruticosa*
	Origanum heracleoticum
Bracken	*Pteridium aquilinum.*

The dwarf semi-evergreen oak, *Quercus fruticosa* (Map 28), grows to a height of about 2 m. It is an important constituent of the maquis on sandy sites in Portugal and southwestern Spain.

Sub-Mediterranean and montane–Mediterranean deciduous and semi-evergreen oakwoods

These are the characteristic climax communities occurring inland and at higher altitudes than the true Mediterranean evergreen oak and coniferous woods of the south. However in the north of the Mediterranean region, where lower average temperatures and a shorter growing season prevail, these deciduous woods replace the evergreen woods, and their upper altitude range extends to about 900 m. The climate still shows a marked summer reduction in rainfall, but the total annual rainfall is higher than in the true mediterranean, rising to 700–1000 mm per year. The average July temperatures are 20–24°C, and the January mean is 0–5°C.

The deciduous white oak, *Quercus pubescens* (Plate 108), is the most widespread and typically dominant tree (Map 29). It usually grows to 4–7 m in height, but occasionally grows up to 15 m. Other oaks are often associated with or replace the white oak (Fig. 31). In Spain the Pyrenean oak, *Quercus pyrenaica*, takes its place and in the Balkan peninsula the Turkey oak, *Quercus cerris*, and the Hungarian oak, *Quercus frainetto*, form woods often in place of the white oak, or in association with it. Typical trees associated with these deciduous oakwoods include

Oriental hornbeam	*Carpinus orientalis*
Hop-hornbeam	*Ostrya carpinifolia*
Sweet chestnut	*Castanea sativa*
Nettle tree	*Celtis australis*
	Pyrus amygdaliformis
	Pyrus elaeagrifolia
Service tree	*Sorbus domestica*
Wild service tree	*Sorbus torminalis*
St. Lucie's cherry	*Prunus mahaleb*
Judas tree	*Cercis siliquastrum*
Montpellier maple	*Acer monspessulanum*
	Acer opalus
	Acer obtusatum
Manna ash	*Fraxinus ornus.*

Characteristic of the shrub layers are

Snowy mespilus	*Amelanchier ovalis*

Mediterranean vegetation

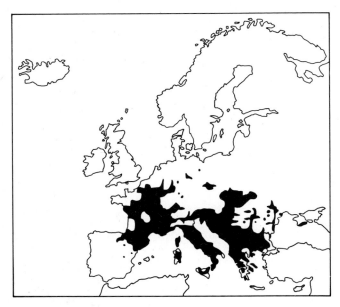

Map 29. White or downy oak *Quercus pubescens*.

Wig tree   *Cotinus coggygria*
Bladder-nut   *Staphylea pinnata*
Box   *Buxus sempervirens*
Cornelian cherry *Cornus mas*
Privet   *Ligustrum vulgare*
Wayfaring tree   *Viburnum lantana.*

Much of this woodland has been destroyed, or degraded into a low coppice or bush community, known as *šibljak* in the Balkans. Charcoal-burning and grazing are largely responsible for this and, if the latter is intensive, a dry grassland may result.

An example of white oak woods on Monte Argano, Italy at an altitude of 450 m, has the dominant white oak with the field maple, *Acer campestre*, in the tree layer.

Shrub layer

                         Rosa sempervirens
Spanish broom   *Spartium junceum*
Mastic tree   *Pistacia lentiscus*
Mediterranean buckthorn *Rhamnus alaternus*
                         Asparagus acutifolius
Butcher's-broom   *Ruscus aculeatus*

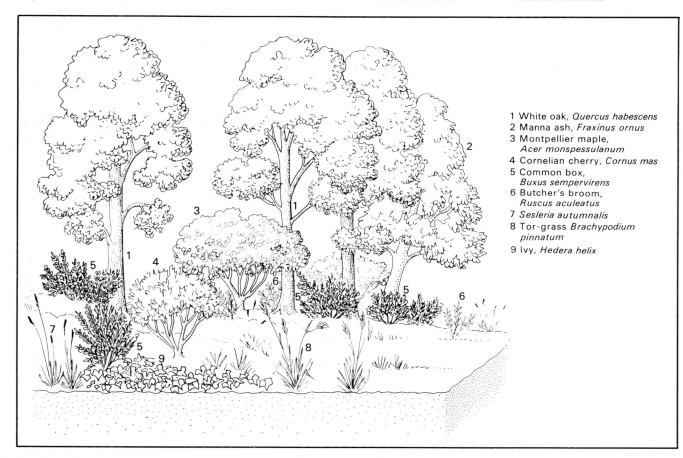

1 White oak, *Quercus habescens*
2 Manna ash, *Fraxinus ornus*
3 Montpellier maple, *Acer monspessulanum*
4 Cornelian cherry, *Cornus mas*
5 Common box, *Buxus sempervirens*
6 Butcher's broom, *Ruscus aculeatus*
7 *Sesleria autumnalis*
8 Tor-grass *Brachypodium pinnatum*
9 Ivy, *Hedera helix*

Fig. 31. Deciduous oak wood showing some of the characteristic species.

Plate 31. Mediterranean deciduous oak woods—trees

1. Portuguese oak, *Quercus faginea*;   **2**. Pyrenean oak, *Quercus pyrenaica*;   **3**. Turkey oak, *Quercus cerris*;   **4**. Hungarian oak, *Quercus frainetto*;   **5**. Macedonian oak, *Quercus trojana*;   **6**. White oak, *Quercus pubescens* (leaf);   **7**. Italian maple, *Acer opalus*;   **8**. Montpellier maple, *Acer monspessulanum*;   **9**. Sweet chestnut, *Castanea sativa*;   **10**. Manna ash, *Fraxinus ornus*;   **11**. *Acer obtusatum*;   **12**. *Acer granatense*;   **13**. Service tree, *Sorbus domestica* (inset×1);   **14**. Wild service tree, *Sorbus torminalis*.

Plate 32. Mediterranean deciduous oak woods—trees and shrubs
1. St. Lucie's cherry, *Prunus mahaleb*; **2**. *Pyrus elaeagrifolia*; **3**. *Pyrus amygdaliformis*; **4**. Judas tree, *Cercis siliquastrum*; **5**. Oriental hornbeam, *Carpinus orientalis*; **6**. Hop-hornbeam, *Ostrya carpinifolia*; **7**. Portugal laurel, *Prunus lusitanica* (inset×1); **8**. *Genista scorpius*; **9**. Christ's thorn, *Paliurus spina-christi*; **10**. Cornelian cherry, *Cornus mas*; **11**. Wig tree, *Cotinus coggygria*; **12**. Prickly juniper, *Juniperus oxycedrus*; **13**. Nettle tree, *Celtis australis*; **14**. Bladder-nut, *Staphylea pinnata*.

Plate 33. Mediterranean deciduous woods—herbaceous species

1. Bloody crane's-bill, *Geranium sanguineum* (inset×1); **2**. *Anemone hortensis*; **3**. Repand cyclamen, *Cyclamen repandum*; **4**. Scorpion vetch, *Coronilla coronata*; **5**. Bastard agrimony, *Aremonia agrimonoides*; **6**. Snake's-head iris, *Hermodactylus tuberosus*; **7**. Western peony, *Paeonia broteroi*; **8**. Hyssop, *Hyssopus officinalis*; **9**. Italian catchfly, *Silene italica*; **10**. Plantain-leaved leopard's-bane, *Doronicum plantagineum*; **11**. Italian lords-and-ladies, *Arum italicum*; **12**. Narrow-leaved inula, *Inula ensifolia*; **13**. Angular Solomon's-seal, *Polygonatum odoratum*; **14**. Spiked speedwell, *Veronica spicata*.

Mediterranean vegetation

Field layer

	Anemone hortensis
Bastard agrimony	*Aremonia agrimonoides*
Repand cyclamen	*Cyclamen repandum*
	Asphodelus ramosus
Yellow asphodel	*Asphodelus lutea*
	Ornithogalum collinum
Black bryony	*Tamus communis*
Snake's-head iris	*Hermodactylus tuberosus*
Italian arum	*Arum italicum*
Bumble-bee orchid	*Ophrys bombyliflora*
	Orchis italica
Provence orchid	*Orchis provincialis*
Greater butterfly-orchid	*Plantanthera chlorantha*
Bracken	*Pteridium aquilinum.*

In France, white oak woods may have the box, *Buxus semper-virens*, common in the shrub layer, while in the tree layer the St. Lucie's cherry, *Prunus mahaleb*, and whitebeam, *Sorbus aria*, are often present. Distinctive field layer species may include: stinking hellebore, *Helleborus foetidus*; bastard balm, *Melittis melissophyllum*; *Tanacetum corybosum*; and angular Solomon's-seal, *Polygonatum odoratum*.

On the coast of Croatia, Yugoslavia, white oak woods commonly have the following composition

Tree layer

White oak	*Quercus pubescens*
Turkey oak	*Quercus cerris*
Oriental hornbeam	*Carpinus orientalis*
Hop–hornbeam	*Ostrya carpinifolia*
Montpellier maple	*Acer monspessulanum*
Field maple	*Acer campestre*
Manna ash	*Fraxinus ornus*

Shrub layer

	Rubus ulmifolius
	Rosa sempervirens
St. Lucie's cherry	*Prunus mahaleb*
Scorpion senna	*Coronilla emerus*
Wig tree	*Cotinus coggygria*
Rock buckthorn	*Rhamnus saxatilis*
Buckthorn	*Rhamnus catharticus*
Christ's-thorn	*Paliurus spina-christi*

Field layer

	Cnidium silaifolium
Winter savory	*Satureja montana*
Spiked speedwell	*Veronica spicata*
	Asparagus acutifolius
Butcher's-broom	*Ruscus aculeatus*
The grass	*Sesleria autumnalis.*

Mixed deciduous oak woods with the Turkey oak, *Quercus cerris* (Map 30), and the Hungarian oak, *Quercus frainetto* (Map 31, Plate 107), are characteristic of the Balkan region, but they also occur in the eastern Central European region and in the Pannonic region. They usually occupy a zone between the true sub-Mediterranean zone and the more con-

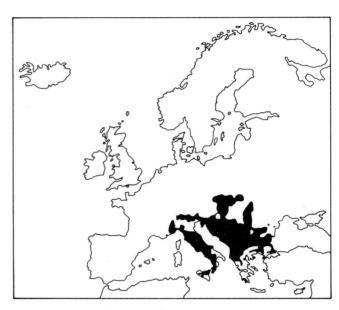

Map 30. Turkey oak *Quercus cerris*.

tinental oak–hornbeam woods. These mixed oak woods require a more continental climate, with an average annual temperature of 10–11.5°C, and with an annual rainfall of 500–650 mm. In the karstlands of Yugoslavia they occur at altitudes of between 200 and 700 m, usually below the zone of beech woods; further south mixed deciduous oak woods may be found as high as 1000 m. At the lower altitudes the Turkey and Hungarian oaks may be exclusive, and the associated vegetation is largely composed of sub-Mediterranean species. At the higher altitudes, at montane level, the sessile oak, *Quercus petraea*, may be common, or become dominant.

Typical examples of mixed deciduous oak woods in northern Greece have the composition

Tree layer

Turkey oak	*Quercus cerris*
Hungarian oak	*Quercus frainetto*
White oak	*Quercus pubescens*
Sessile oak	*Quercus petraea*
Service tree	*Sorbus domestica*

Shrub layer

	Pyrus amygdaliformis
Blackthorn	*Prunus spinosa*
Cornelian cherry	*Cornus mas*

Field layer

Pink barren-strawberry	*Potentilla micrantha*
Wood spurge	*Euphorbia amygdaloides*
Sweet violet	*Viola odorata*
Southern wood-rush	*Luzula forsteri*
Slender false-brome	*Brachypodium sylvaticum*
Cock's-foot	*Dactylis glomerata*
Wood meadow-grass	*Poa nemoralis.*

Sub-Mediterranean and montane–Mediterranean deciduous and semi-evergreen oak woods

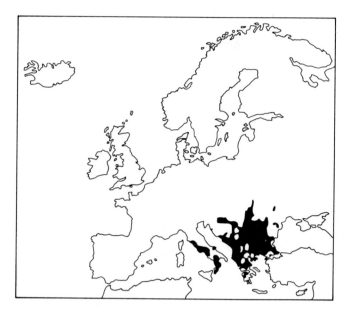

Map 31. Hungarian oak *Quercus frainetto.*

The *karstwoods* of the Dalmation coast are mixed oak woods with the trees

Turkey oak	*Quercus cerris*
Hungarian oak	*Quercus frainetto*
White oak	*Quercus pubescens*
Sessile oak	*Quercus petraea*
Pedunculate oak	*Quercus robur*
Holm oak	*Quercus ilex*
Sweet chestnut	*Castanea sativa*
Oriental hornbeam	*Carpinus orientalis*
Hop-hornbeam	*Ostrya carpinifolia*
Montpellier maple	*Acer monspessulanum*
Manna ash	*Fraxinus ornus.*

The shrub layer commonly includes

St. Lucie's cherry	*Prunus mahaleb*
Bladder senna	*Colutea arborescens*
Wig tree	*Cotinus coggygria*
Christ's-thorn	*Paliurus spina-christi.*

The Macedonian oak, *Quercus trojana*, has a restricted range in southern Yugoslavia, Albania, and Macedonia. It forms woods on marble or dolomite, on warm dry slopes out of the influence of cold winds. An example in Macedonia has the composition

Tree layer

Macedonian oak	*Quercus trojana*
White oak	*Quercus pubescens*
Turkey oak	*Quercus cerris*
Oriental hornbeam	*Carpinus orientalis*
Hop-hornbeam	*Ostrya carpinifolia*
Whitebeam	*Sorbus aria*

Wild service-tree	*Sorbus torminalis*
The maple	*Acer obtusatum*
Manna ash	*Fraxinus ornus*

Shrub layer

Scorpion senna	*Coronilla emerus*
Box	*Buxus sempervirens*

Field layer

Italian catchfly	*Silene italica*
Lesser meadow-rue	*Thalictrum minus*
Scorpion vetch	*Coronilla coronata*
Bloody crane's-bill	*Geranium sanguineum*
Wild basil	*Clinopodium vulgare*
	Melampyrum heracleoticum
Narrow-leaved inula	*Inula ensifolia*
	Inula spiraeifolia
Angular Solomon's-seal	*Polygonatum odoratum.*

In the Iberian Peninsula, there are two oaks which form distinctive woods: the semi-evergreen Lusitanian oak, *Quercus faginea* (Map 32), and the deciduous Pyrenean oak, *Quercus pyrenaica*. The Lusitanian oak occurs in a zone between the evergreen oak and the deciduous white oak, *Quercus pubescens*. It favours a Mediterranean climate tempered by cooler rainy winters; it is also found in regions where the winters are milder but the rainfall higher. It grows on both siliceous and limestone soils, with different species dominating the lower layers. On limestone, at montane levels, the woods are often degraded to heath or dry grassland, while on siliceous soils they are degraded to a cistus-heath, or in the sub-oceanic areas to a heather–gorse heath.

A typical example of a Lusitanian oak wood has Montpellier maple, *Acer monspessulanum; Acer granatense;* and nettle tree, *Celtis australis,* in the tree layer with the dominant Lusitanian oak. The shrub layer includes

Blackthorn	*Prunus spinosa*
Buckthorn	*Rhamnus catharticus*
Laurel-leaved cistus	*Cistus laurifolius.*

The field layer commonly includes

	Paeonia broteroi
	Paeonia coriacea
Grass-leaved buttercup	*Ranunculus gramineus*
	Geum sylvaticum
Dropwort	*Filipendula vulgaris*
Cowslip	*Primula veris*
Tuberous valerian	*Valeriana tuberosa*
Plantain-leaved leopard's-bane	*Doronicum plantagineum.*

On limestone soils, for example in Lerida Province, Spain at an altitude of 700 m, in addition to the Lusitanian oak, the Montpellier maple and the black and Aleppo pines may occur with the holm and kermes oaks in the tree layer. The shrub layer includes

Prickly juniper	*Juniperus oxycedrus*
Burnet rose	*Rosa pimpinellifolia*
	Genista scorpius

Mediterranean vegetation

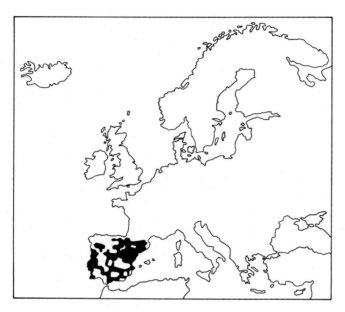

Map 32. Lusitanian oak *Quercus faginea*.

Rock buckthorn — *Rhamnus saxatilis*.

The field layer includes

Spanish gorse	*Genista hispanica*
	Dorycnium pentaphyllum
White flax	*Linum suffruticosum*
	Thymelaea tinctoria
	Euphorbia nicaeensis
	Bupleurum rigidum
	Lavandula latifolia
	Sideritis hirsuta
Thyme	*Thymus vulgaris*
	Brachypodium phoenicoides.

On siliceous soils, the maritime pine, the cork oak, and the sweet chestnut may also occur in the tree layer. The shrub layer includes

Portugal laurel	*Prunus lusitanica*
White spanish-broom	*Cytisus multiflorum*
Broom	*Cytisus scoparius*
	Genista falcata
Poplar-leaved cistus	*Cistus populifolius*
Honeysuckle	*Lonicera periclymenum.*

The field layer includes

Palmate anemone	*Anemone palmata*
Wood sage	*Teucrium scorodonia*
Bracken	*Pteridium aquilinum.*

Other oaks, far less widely distributed and of less importance, form local woods or thickets. They include the semi-evergreen shrub oak, *Quercus infectoria*, of the northern Aegean region, and the semi-evergreen oak, *Quercus canari-*

ensis, of southern Portugal and southwestern Spain. Deciduous oaks forming local woods in southeastern Europe include *Quercus dalechampii*, *Quercus polycarpa*, and *Quercus pedunculiflora*. In the western and central Mediterranean region, *Quercus mas*, a deciduous tree, occurs in northern Spain and southwestern France. Other deciduous oak trees or shrubs include *Quercus sicula* which is endemic to Sicily and *Quercus congesta* restricted to Sicily, southern France, and Sardinia.

Laurel woods

These are rare in the Mediterranean region, but they may form a mixed wood with oaks where laurel is sometimes locally dominant. Laurel woods are usually located above the evergreen oak zone and below the sub-montane deciduous oak woods, usually not far from the coast. Laurel, *Laurus nobilis*, is however a widespread shrub in the maquis, and it is found throughout the Mediterranean region. It is difficult to ascertain the true status of laurel as a dominant Mediterranean woodland tree, for it has been so affected by man's activities.

An example from Algeciras, Spain has laurel in the tree layer mixed with

Black poplar	*Populus nigra*
Alder	*Alnus glutinosa*
Cork oak	*Quercus suber*
Nettle tree	*Celtis australis*
Tree heath	*Erica arborea.*

The shrub layer includes

	Rhamnus ludovici-salvatoris
Ivy	*Hedera helix*
Rhododendron	*Rhododendron ponticum*
	Smilax aspera.

On the Croatian coast of Yugoslavia, mixed laurel woods have in the tree layer

Oriental hornbeam	*Carpinus orientalis*
Hop-hornbeam	*Ostrya carpinifolia*
White oak	*Quercus pubescens*
Turkey oak	*Quercus cerris*
Turpentine tree	*Pistacia terebinthus*
Montpellier maple	*Acer monspessulanum*
Manna ash	*Fraxinus ornus.*

The shrub layer includes

Fragrant clematis	*Clematis flammula*
	Rubus ulmifolius
St. Lucie's cherry	*Prunus mahaleb*
Scorpion senna	*Coronilla emerus*
Wig tree	*Cotinus coggygria*
Christ's-thorn	*Paliurus spina-christi*
Cornelian cherry	*Cornus mas.*

The field layer includes

Ivy	*Hedera helix*
	Asparagus acutifolius

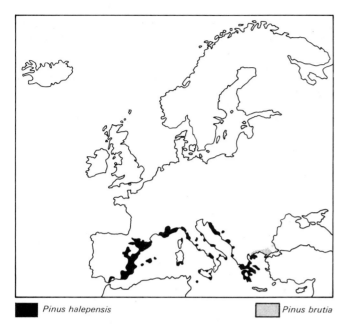

■ *Pinus halepensis*	▨ *Pinus brutia*

Map 33. Aleppo pine *Pinus halepensis* and *Pinus brutia*.

Butcher's-broom	*Ruscus aculeatus*
	Sesleria autumnalis.

Mediterranean pine woods

Four species of pine are native in the true Mediterranean region. They often form distinctive woods, particularly in the coastal regions which are less accessible to the activities of man. They are the Aleppo pine, *Pinus halepensis;* and the closely related *Pinus brutia;* the maritime pine, *Pinus pinaster;* and the stone pine, *Pinus pinea.*

The Aleppo pine, *Pinus halepensis,* (Map 33; Plate 111) is the most widespread species in the Mediterranean coastal region, spreading from southern Spain to Greece and into the western Aegean region. It is less common inland, but forms woods just east of Madrid, and on the southeastern foothills of the Pyrenees. It is common along the western Riviera of France, and along the Adriatic coast in Yugoslavia. It is widely distributed along coastal areas of Greece and the Peloponnese, and spreads as far east as the Khalkidhiki and the Kassandra peninsulas of Greece; further east on the Athos peninsula *Pinus brutia* takes its place (Map 33). The Aleppo pine requires a typical Mediterranean climate with the temperature rarely falling below freezing point, and it can withstand prolonged drought and the absence of summer rainfall. It grows on a wide range of soils but particularly favours dry sands and light calcareous soils. However, on most of the richer soils it has been largely destroyed, often for olive cultivation, etc. Fire is another common cause of destruction. The altitude range of the Aleppo pine is from just above the splash-zone to about 200 m on the north Adriatic coast of Yugoslavia, and to about 1000 m in southern Spain. Individual trees may grow to heights of 25 m, and they are often rather

widely spaced, thus allowing much light to penetrate the shrub layer, with the result that it is often well developed and very dense (Fig. 32). The field layer, by contrast, is absent or very poorly developed. Selective felling also tends to keep these woods more open than they would otherwise be in their natural condition.

Aleppo pine woods characteristically have two types of shrub layer, the first being the taller *maquis* type, which is often dominated by the tree heath, *Erica arborea;* the strawberry tree, *Arbutus unedo;* and the myrtle, *Myrtus communis.* These shrubs may grow to a height of 2 m or more and form very dense impenetrable thickets. The second type of shrub layer is a lower heath-like layer in which *Cistus* species usually dominate, commonly with the heath, *Erica manipuliflora.* This shrub layer grows to a height of about 50 cm or more. This second type of shrub layer often develops after the burning of the original *phrygana* shrub, and the ground is colonized by the Aleppo pine which grows up in close-canopy.

An example of the first type of Aleppo pine wood in Euboea, Greece, has the shrub layer

Kermes oak	*Quercus coccifera*
Mastic tree	*Pistacia lentiscus*
Turpentine tree	*Pistacia terebinthus*
	Cistus incanus
Sage-leaved cistus	*Cistus salvifolius*
Tree heath	*Erica arborea*
	Erica manipuliflora
Strawberry tree	*Arbutus unedo*
	Phillyrea latifolia

with the Greek cyclamen, *Cyclamen graecum,* in the field layer.

In Aleppo pine woods in Italy, additional species may include the white oak, *Quercus pubescens,* and holm oak, *Quercus ilex,* and such orchids as pink butterfly orchid, *Orchis papilionacea;* the tongue orchids, *Serapias lingua* and *Serapias neglecta;* and the bee orchid, *Ophrys apifera.* In southern Spain, additional species may include the dwarf fan palm, *Chamaerops humilis,* and rosemary, *Rosmarinus officinalis.* Other shrubs such as myrtle, *Myrtus communis;* Mediterranean buckthorn, *Rhamnus alaternus;* scorpion senna, *Coronilla emerus;* and the climber, *Smilax aspera,* are widespread in these Aleppo pine woods along the Mediterranean coast.

Aleppo pine woods with the heath-like type of shrub layer, as found in Euboea, Greece, have the sage-leaved cistus, *Cistus salvifolius,* and the heather, *Erica manipuliflora,* dominant with

Thorny burnet	*Sarcopoterium spinosum*
Mastic tree	*Pistacia lentiscus*
Wig tree	*Cotinus coggygria*
	Hypericum empetrifolium
Sage-leaved cistus	*Cistus salvifolius*
	Smilax aspera.

In the field layer, the Greek cyclamen, *Cyclamen graecum,* and the tor-grass, *Brachypodium pinnatum,* are characteristic.

Pinus brutia is very like the Aleppo pine, but it differs in its

113

Mediterranean vegetation

smaller cones, longer needles, and in the branches of mature trees which have less of a tendency to cluster near the apex of the trunk of the tree. It forms woods in the eastern Aegean Islands, in Crete, Thassos, Turkey-in-Europe, and the Athos peninsula. Like the Aleppo pine, *Pinus brutia* may have a maquis-like shrub layer dominated by the shrubby growths of the holm oak, *Quercus ilex*, and the kermes oak, *Quercus coccifera*, with the tree heath, *Erica arborea*, and *Phillyrea latifolia*. Or, alternatively, it may have a lower-growing heath-like shrub layer, but there are many intermediates. The following are characteristically present in the heath-like shrub layer

Prickly juniper	*Juniperus oxycedrus*
	Cistus incanus
Myrtle	*Myrtus communis*
Olive	*Olea europaea*
	Erica manipuliflora
	Asparagus acutifolius

with the herbaceous plants, early-flowered calamint, *Calamintha grandiflora*, and bracken, *Pteridium aquilinum*.

The maritime pine, *Pinus pinaster*, mainly forms woods on the sandy coastal areas of the western Mediterranean, with its main stronghold in the Iberian Peninsula including the Portuguese coast, and along the southwestern coast of France (Map 34). It is native as far east as Italy and Corsica. However it has been extensively planted on coastal dunes and its range has

consequently been extended. It is sensitive to frost, and can only succeed where the average winter temperature does not fall below 6°C. The tree layer may consist of pure maritime pine, which has a pyramidal shape, with the old lower branches devoid of needles remaining attached to the trunk and thus giving the woods a highly characteristic appearance. Alternatively, maritime pine woods may be mixed with the Scots pine and holm oak as for example in Catalonia, or with black pine (the subspecies *salzmannii*) as in Murcia, Spain.

In Italy, examples of maritime pine woods typically have the shrub layer with

Prickly juniper	*Juniperus oxycedrus*
	Calicotome spinosa
Mediterranean mezereon	*Daphne gnidium*
Myrtle	*Myrtus communis*
Heather	*Calluna vulgaris*
Tree heather	*Erica arborea*
Strawberry tree	*Arbutus unedo*

with the bracken, *Pteridium aquilinum*.

Spanish examples have species of broom, *Cytisus;* greenweed, *Genista;* gorse, *Ulex* with *Cistus;* and heather, *Erica* species.

The stone pine, *Pinus pinea* (Plate 110), is another largely coastal tree which, like the maritime pine, has a distribution centred in the Iberian Peninsula, but which spreads further east to Greece and Turkey-in-Europe (Map 35). However it

1 Tree heath, *Erica arborea*
2 Strawberry tree, *Arbutus unedo*
3 Terebinth, *Pistacia terebinthus*
4 Kermes oak, *Quercus coccifera*
5 Tor-grass, *Brachypodium pinnatum*
6 Thorny burnet, *Sarcopoterium spinosum*
7 Aleppo pine, *Pinus halepensis*

Fig. 32. Mediterranean pine wood. This diagram represents an Aleppo pine community from Greece, with some characteristic undergrowth.

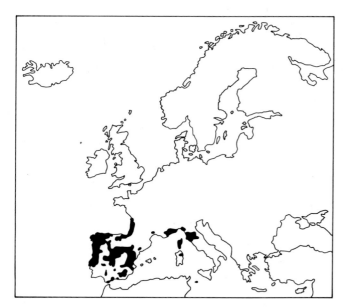

Map 34. Maritime pine *Pinus pinaster*.

Map 35. Stone pine *Pinus pinea*.

has been widely planted, largely for its edible nut-like seeds, and its native distribution is uncertain. Stone pines require deeper and moister soils and a milder climate than the maritime pines. They also occur in the lower montane zone of Spain and Italy, where increased precipitation and the Atlantic influence is felt. The alternative name for the stone pine is the umbrella pine which describes its shape well, for it has a domed flattened crown and a stout trunk devoid of side branches. The woods are usually dominated by pure stone-pine, *Pinus pinea* and because of the richer soils on which they occur, the shrub and field layers are relatively rich in species. Three types of stone pine woods can be distinguished—grass-rich, maquis-type, and those on damper soils, but these depend to some extent on local management and grazing.

Grass-rich stone pine woods have the hare's-tail grass, *Lagurus ovatus*, and the large quaking-grass, *Briza maxima*, characteristic, with such shrubs as

Gorse	*Ulex europaeus*
Grey-leaved cistus	*Cistus albidus*
Sage-leaved cistus	*Cistus salvifolius*
Mediterranean mezereon	*Daphne gnidium*
Wild madder	*Rubia peregrina*
Ivy	*Hedera helix*.

The field layer commonly includes

	Lathyrus clymenum
Perennial flax	*Linum perenne*
	Polygala nicaeensis
Narrow-leaved helleborine	*Cephalanthera longifolia*
Dense-flowered orchid	*Neotinea intacta*
Helleborine species	*Epipactis* sps.

The maquis-type of stone pine wood characteristically

includes the holm oak, *Quercus ilex;* the gorse, *Ulex europaeus;* and the tree heath, *Erica arborea*.

On damper, usually richer soils, stone pine woods have the shrub layer of

Prickly juniper	*Juniperus oxycedrus*
Holm oak	*Quercus ilex*
Fragrant clematis	*Clematis flammula*
	Rosa sempervirens
Blackberry species	*Rubus* sps.
Firethorn	*Pyracantha coccinea*
Gorse	*Ulex europaeus*
Sage-leaved cistus	*Cistus salvifolius*
Myrtle	*Myrtus communis*
Ivy	*Hedera helix*
	Phillyrea latifolia
Wild madder	*Rubia peregrina*.

The field layer includes

Bugle	*Ajuga reptans*
Hemp agrimony	*Eupatorium cannabinum*
Yellow iris	*Iris pseudacorus*
Marsh fern	*Thelypteris palustris*.

Sub-Mediterranean and Mediterranean–montane coniferous woods

In the montane zone of the Mediterranean region, coniferous woods are widespread and well developed on most of the lower and middle slopes of the higher mountain masses. The most widely distributed tree is the black pine, *Pinus nigra*, which has distinctive subspecies, each forming woods in different parts of the Mediterranean region of Europe. In the Balkans and in eastern and southeastern Europe, several other

115

Plate 34. Mediterranean and sub-Mediterranean coniferous woods—trees

1. Italian cypress, *Cupressus sempervirens*; **2**. Spanish juniper, *Juniperus thurifera*; **3**. Syrian juniper. *Juniperus drupacea*: **4**. Bosnian pine, *Pinus leucodermis*; **5**. Grecian juniper, *Juniperus excelsa*; **6**. Macedonian pine, *Pinus peuce*; **7**. Maritime pine, *Pinus pinaster*; **8**. Aleppo pine, *Pinus halepensis*; **9**. Phoenician juniper, *Juniperus phoenicea*; **10**. Alerce, *Tetraclinis articulata*; **11**. Stone pine, *Pinus pinea*; **12**. Crimean pine, *Pinus nigra* ssp. *pallasiana*.

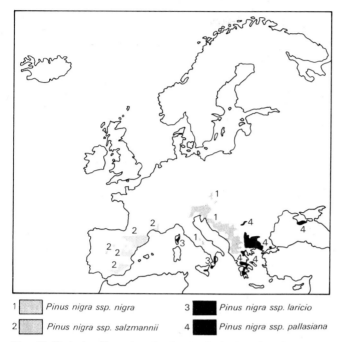

1 ☐ *Pinus nigra ssp. nigra*   3 ■ *Pinus nigra ssp. laricio*

2 ☐ *Pinus nigra ssp. salzmannii*   4 ■ *Pinus nigra ssp. pallasiana*

Map 36. Black pine *Pinus nigra* showing the distribution of the four major European subspecies.

species of pines are important forest-forming trees. They include the Bosnian pine, *Pinus leucodermis;* the closely related white-barked pine, *Pinus heldreichii;* and the five-needled Macedonian pine, *Pinus peuce.*

The black pine, *Pinus nigra,* is an important forest-former in the Mediterranean region, as well as in the Central European region, where an example has already been described (Map 36). It usually forms woods at the montane level, for it can withstand winter frost as well as hot dry summers. It grows on most types of dry soil, as long as they are well drained, but it is most commonly found on 'poor' siliceous soils where other tree species cannot compete. It is typical of the serpentine soils of northern Greece and Albania, as well as of the limestone karst soils of northern Yugoslavia. It also occurs in eastern Spain and southern France, but as a different subspecies. There are five distinct subspecies of black pine: subsp. *nigra,* the Austrian pine, spreads from Austria, Italy, and Yugoslavia to Greece; subsp. *salzmannii,* the Pyrenean pine, occurs in Spain, the Pyrenees, and Cevennes; subsp. *dalmatica,* the Dalmation pine, has a limited distribution in the coastal region and islands of northwestern Yugoslavia; subsp. *pallasiana,* the Crimean pine, forms woods in the Balkan peninsula and the Carpathians; subsp. *laricio,* the Corsican pine, occurs in Corsica, Sicily, and Calabria, Italy.

The Austrian pine, subsp. *nigra,* has a tall, widely branched, somewhat umbrella-like crown, and may grow to 35 m high. On the serpentine mountains of Albania, from 800 to 1500 m in altitude, it is widespread, and also in southwestern Croatia. The montane woods may contain Scots pine, silver fir, and beech in the tree layer.

Lower Croatian woods up to about 850 m altitude, have such sub-Mediterranean trees as the white oak, *Quercus pubescens;* hop-hornbeam, *Ostrya carpinifolia;* Montpellier maple, *Acer monspessulanum;* and manna ash, *Fraxinus ornus.* Hazel, *Corylus avellana;* barberry, *Berberis vulgaris;* snowy mespilus, *Amelanchier ovalis;* wig tree, *Cotinus coggygria;* and cornelian cherry, *Cornus mas* occur in the shrub layer. In the lower shrub layer the spring heath, *Erica herbacea,* often dominates with grasses.

In Macedonia, Austrian pine woods have a more Central European assemblage of species in the tree layer, including sessile oak, *Quercus petraea;* aspen, *Populus tremula;* and rowan, *Sorbus aucuparia.* Common in the shrub layer are bilberry, *Vaccinium myrtillus; Festuca heterophylla;* and bracken, *Pteridium aquilinum.* The Dalmation pine, subspecies *dalmatica,* is similar in many ways to the Aleppo pine in its ecology and woodland composition. It forms woods on the islands of Hvar and Brač on the Dalmation coast, and has the typical maquis shrub layer of that region. In the more montane localities *Juniperus communis* subsp. *nana* may dominate the shrub layer.

The Crimean pine, subspecies *pallasiana,* is the most widespread of the black pines in southeastern Europe, spreading from eastern Yugoslavia and Bulgaria in the Central European region, to Greece in the Mediterranean–montane region, at altitudes of about 1000 to 1500 m. Woods are well developed on the mountains of Olympus, the Pindus, and the eastern Peloponnese. Examples on Mount Olympus have a tree layer comprising

Crimean pine	*Pinus nigra* ssp. *pallasiana*
Hybrid silver fir	*Abies borisii-regis*
Yew	*Taxus baccata*
Beech	*Fagus sylvatica*
Hop-hornbeam	*Ostrya carpinifolia*
Manna ash	*Fraxinus ornus*

and a shrub layer including

Prickly juniper	*Juniperus oxycedrus*
Kermes oak	*Quercus coccifera*
Box	*Buxus sempervirens.*

At lower altitudes, below about 1000 m, there is a characteristic maquis shrub layer, with the holm oak, *Quercus ilex,* dominant. The field layer includes

Wall germander	*Teucrium chamaedrys*
Wild basil	*Clinopodium vulgare*
Marjoram	*Origanum vulgare*
	Staehelina uniflosculosa
Bracken	*Pteridium aquilinum.*

Further north in Macedonia, in the Treska river area of Yugoslavia, at altitudes of 500–1700 m, are light woods of Crimean pines with loosely crowned trees growing to about 20 m. Here white oak, *Quercus pubescens;* silver birch, *Betula pendula;* aspen, *Populus tremula; Acer obtusatum;* hop-hornbeam, *Ostrya carpinifolia;* and manna ash, *Fraxinus ornus,* are also found in the tree layer. The shrub layer includes, in addition to some of the above species of the tree layer, prickly juniper, *Juniperus oxycedrus*; and *Cotoneaster nebrodensis*. The field layer contains

	Pulsatilla sps
Common dog-Violet	*Viola riviniana*
Cowslip	*Primula veris*
The woundwort	*Stachys scardica*
	Cephalaria flava
Tor-grass	*Brachypodium pinnatum*.

The Pyrenean pine, *Pinus nigra* ssp. *salzmannii*, grows on dry slopes in the montane region in the southern Iberian mountains. It is commonly associated with the Montpellier maple, *Acer monspessulanum*, which is co-dominant in some cases, or occasionally with the maritime pine, *Pinus pinaster*, and the Spanish juniper, *Juniperus thurifera*. At lower altitudes it has a typical maquis undergrowth.

The Corsican pine, *Pinus nigra* ssp. *laricio*, forms well developed forests in Corsica, on dry southern and eastern slopes, at altitudes of 900–1200 m. Individual trees may grow to 40 m in height. It is often mixed with silver fir, *Abies alba*, and silver birch. Corsican pine grows at altitudes up to 2000 m on Mount Etna.

The Bosnian pine, *Pinus leucodermis* (Plate 113), forms woods in the montane region of the central part of the Balkans, in southern Yugoslavia, Albania, and northern Greece. These woods are found on rocky sites, at altitudes of 900–1800 m, but individual trees may be found scattered to heights of 2300 m, or as shrubs up to 2600 m on Mount Olympus. Often the trees are widely spaced, on exposed stony or rocky areas where the shrub layer may be lacking. Elsewhere the typical shrubs may include the juniper, *Juniperus communis* ssp. *nana*; the savin, *Juniperus sabina*; or the dwarf mountain pine, *Pinus mugo*, at the higher altitudes. An example in the Prokletije mountains of southern Yugoslavia, includes in the tree layer, in addition to the Bosnian pine, the silver fir, *Abies alba*, and the beech, *Fagus sylvatica*. The shrub layer includes

Juniper	*Juniperus communis*
Bosnian pine	*Pinus leucodermis*
Mezereon	*Daphne mezereum*
Bilberry	*Vaccinium myrtillus*.

Typical of the field layer are

Wood anemone	*Anemone nemorosa*
Wild strawberry	*Fragaria vesca*
Wood spurge	*Euphorbia amygdaloides*
Early dog-violet	*Viola reichenbachiana*
	Verbascum nicolai
Heath speedwell	*Veronica officinalis*.

The white-barked pine, *Pinus heldreichii*, is closely related to the Bosnian pine. It has similar ecological requirements, and grows on limestone mountains and forms forests in the central and western part of the Balkan peninsula and in central Italy. It has a rounded pyramidal crown, and young twigs which turn brown in the second year, while the cone-scales have short straight tips (Bosnian pines have cone-scales with recurved tips at the apex and twigs which remain greyish-white for three years).

The Macedonian pine, *Pinus peuce* (Plate 112), and the Arolla pine, *Pinus cembra*, are the only five-needled pines native in Europe. They are both montane trees.

The Macedonian pine forms pure woods above altitudes of about 1700 m. These are particularly well developed in Bulgaria, where they cover an area of 11 600 hectares (about 3 per cent of Bulgarian coniferous forests). They also form scattered woods in Albania and Macedonia. In Bulgaria the Macedonian pine woods are well developed in the Rila, Pirin, and western Rhodope mountains.

Below about 1700 m, the Macedonian pine is usually mixed with spruce, beech, and silver fir, and sometimes with Scots pine, but between about 1700 and 2000 m the Macedonian pine may form pure woods. An example of a mixed wood in Metohija, Yugoslavia has mixed in the tree layer the Macedonian pine, the silver fir, and the spruce. All three coniferous species also occur in the shrub layer, with the bilberry, *Vaccinium myrtillus*; *Rubus hirtus*; and the juniper, *Juniperus communis*. Common species in the field layer include

Wood anemone	*Anemone nemorosa*
Heath speedwell	*Veronica officinalis*
	Thymus praecox
	Crocus vernus
	Luzula luzulina
Male-fern	*Dryopteris filix-mas*.

Other examples of Macedonian pine woods are characterized by the presence of the bilberry, *Vaccinium myrtillus*, with the great yellow gentian, *Gentiana lutea*. Another type has bracken, with *Digitalis viridiflora*; Greek hellebore, *Helleborus cyclophyllus*; and *Lathyrus laxiflorus*.

Sweet chestnut woods (Plate 114)

The sweet chestnut, *Castanea sativa*, is very characteristic of the sub-Mediterranean region, though it is probably only native in southeastern Europe (Map 37). During Roman times it was cultivated in Italy and Central Europe, where it has now become naturalized and has spread to northern France and southeastern Britain, and into the Iberian peninsula. Its general altitude range is from about 300 to 900 m, but in some mountain ranges, such as the Sierra Nevada, Etna, and the Apennines, it is found at higher altitudes. It commonly occurs on acid soils, with a ground flora rich in acid-indicating species, but it can also grow on neutral or even basic soils, where it is often mixed with ash, lime, hornbeam, and beech.

Examples of chestnut woods in the southern Alps of northern Italy, have the dominant sweet chestnut, with the white oak, *Quercus pubescens*, and the silver birch, *Betula pendula*, in the tree layer. The shrub layer contains

Hazel	*Corylus avellana*
Sweet chestnut	*Castanea sativa*
Medlar	*Mespilus germanica*
Broom	*Cytisus scoparius*
Alder buckthorn	*Frangula alnus*.

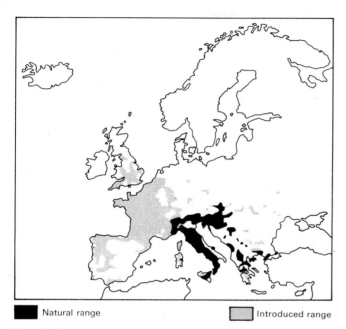

■ Natural range	▨ Introduced range

The field layer includes

Bilberry	*Vaccinium myrtillus*
	Luzula nivea
Fine-leaved sheep's-fescue	*Festuca tenuifolia*
Wavy hair-grass	*Deschampsia flexuosa*
Sweet vernal-grass	*Anthoxanthum odoratum*
	Agrostis capillaris
Purple moor-grass	*Molinia caerulea.*

Examples of sweet chestnut woods in Macedonia have a rich mixed tree layer, and well developed shrub and field layers.

Tree layer

Sweet chestnut	*Castanea sativa*
Walnut	*Juglans regia*
Hornbeam	*Carpinus betulus*
Oriental hornbeam	*Carpinus orientalis*
Beech	*Fagus sylvatica*
Sessile oak	*Quercus petraea*
Hop-hornbeam	*Ostrya carpinifolia*
Wild service-tree	*Sorbus torminalis*
Field maple	*Acer campestre*
Sycamore	*Acer pseudoplatanus*
Silver lime	*Tilia tomentosa*
Manna ash	*Fraxinus ornus*

Shrub layer

Hazel	*Corylus avellana*
Field rose	*Rosa arvensis*
Hawthorn	*Crataegus monogyna*
Cornelian cherry	*Cornus mas*

The field layer includes

Pink barren-strawberry	*Potentilla micrantha*
	Lathyrus venetus
Wood bedstraw	*Galium sylvaticum*
Wild basil	*Clinopodium vulgare*
Large skullcap	*Scutellaria columnae*
Wall lettuce	*Mycelis muralis*
Angular Solomon's-seal	*Polygonatum odoratum*
Southern wood-rush	*Luzula forsteri*
Wood melick	*Melica uniflora*
Slender false-brome	*Brachypodium sylvaticum*
Bracken	*Pteridium aquilinum*
Black spleenwort	*Asplenium adiantum-nigrum*
Brittle bladder-fern	*Cystopteris fragilis.*

Horse chestnut and walnut woods

Both these trees are native of the eastern Mediterranean region, and they form very local woods in restricted localities, often mixed with other trees (Fig. 33). However their natural range has been widely extended by cultivation and planting, so that they have become naturalized further west, and in Central Europe.

The horse chestnut, *Aesculus hippocastanum*, has two main areas of native distribution: in the mountains of Greece, Macedonia and Albania, and in Bulgaria in the northeastern part of the Balkan mountains (strictly part of the Central European region). The main woods of horse chestnut are found in damp mountain ravines and valleys, from about 350 to 1350 m in eastern Greece, southern Albania, and Macedonia. Horse chestnuts grow on stony well-drained soils, which can be either acidic or basic. An example in northern Greece has a mixed tree layer with

Horse chestnut	*Aesculus hippocastanum*
Walnut	*Juglans regia*
Alder	*Alnus glutinosa*
Hornbeam	*Carpinus betulus*
Beech	*Fagus sylvatica*
Oak species	*Quercus* sps.
Field maple	*Acer campestre*
Sycamore	*Acer pseudoplatanus*
Small-leaved lime	*Tilia cordata*
Silver lime	*Tilia tomentosa*
Ash	*Fraxinus excelsior*
Manna ash	*Fraxinus ornus.*

The shrub layer includes

Hazel	*Corylus avellana*
Hawthorn	*Crataegus monogyna*
Cornelian cherry	*Cornus mas*
Elder	*Sambucus nigra.*

Typical of the field layer are

Wood anemone	*Anemone nemorosa*
Herb robert	*Geranium robertianum*
Dog's mercury	*Mercurialis perennis*
Sanicle	*Sanicula europaea*

Ground elder	*Aegopodium podagraria*
Tuberous comfrey	*Symphytum tuberosum*
Jupiter's distaff	*Salvia glutinosa*
Common figwort	*Scrophularia nodosa*
Wall lettuce	*Mycelis muralis*
Large butcher's-broom	*Ruscus hypoglossum*
Angular Solomon's-seal	*Polygonatum odoratum*
Hard shield-fern	*Polystichum aculeatum.*

An example of a horse chestnut wood in Bulgaria has the aspen, *Populus tremula*, in the tree layer, as well as the field maple, sycamore, and hornbeam. The shrub layer has blackthorn, *Prunus spinosa*; dogwood, *Cornus sanguinea*; *Euonymus verrucosus* in addition to those species listed in the shrub layer of the previous community. The field layer includes

Wood anemone	*Anemone nemorosa*
Dog's mercury	*Mercurialis perennis*
Spotted deadnettle	*Lamium maculatum*
Danewort	*Sambucus ebulus*
Ramsons	*Allium ursinum.*

Fir woods of the Mediterranean region
(Plates 116, 117)

There are several species of fir which form forests in the Mediterranean montane region. The Greek fir, *Abies cephalonica*, takes the place of the silver fir in southern Greece. Included with this species is *Abies borisii-regis*, which in all probability is the hybrid between the Greek fir and the silver fir, *Abies alba*. This hybrid has a range through southern Albania, southern Bulgaria, and northern Greece. *Abies nebrodensis* is an endemic fir of Sicily, while the Spanish fir, *Abies pinsapo*, has a very limited range in southern Spain. The silver fir, *Abies alba*, has its main distribution in the southern part of Central Europe, but it also occurs in the Mediterranean–montane region in the northern Balkans, Italy, and southern France including the southeastern Pyrenees.

The Greek fir, *Abies cephalonica*, forms forests in southern Greece, particularly in the mountains of the Peloponnese, at altitudes of 800–1700 m, where montane humidity with cloud formation to some extent ameliorates the dry summers of this part of Greece. The soil is usually limestone.

The upper montane forests of Greek fir, where the trees are fully developed, have a very sparse undergrowth with only young trees of Greek fir below the dense canopy. Characteristic of the field layer are

	Cardamine graeca
	Corydalis solida
	Huetia cynapioides
Alpine squill	*Scilla bifolia.*

Examples of Greek fir forests at lower altitudes have in the tree layer the white oak, *Quercus pubescens*, and less commonly the prickly juniper, *Juniperus oxycedrus*, and in the shrub layer *Pyrus amygdaliformis* and Jerusalem sage, *Phlomis fruticosa*. In the field layer, in addition to those species listed in the previous example, occur the shining crane's-bill, *Geranium lucidum*; sowbread, *Cyclamen hederifolium*; and lesser calamint, *Calamintha nepeta*.

Abies borisii-regis grows further north, for example on Mount Olympus. It is often mixed with beech and Crimean pine, *Pinus nigra* ssp. *pallasiana*, and with shrub and field layer species of a more Central European character and distribution.

The Spanish fir, *Abies pinsapo*, is restricted to about four localities in the mountains around Ronda in southern Spain, at altitudes of 1000 to 1700 m. It is a dark-green tree with a wide

Fig. 33. Walnut trees *Juglans regia* in a rocky stream bed on Mount Olympus, N. Greece, with Black pine *Pinus nigra* above.

Plate 35. Mediterranean montane fir forests

1. Lesser calamint, *Calamintha nepeta*; **2**. *Huetia cynapioides*; **3**. *Phlomis purpurea*; **4**. *Cardamine graeca*; **5**. *Corydalis solida*; **6**. Spanish fir, *Abies pinsapo*; **7**. Grey-leaved cistus, *Cistus albidus*; **8**. Poplar-leaved cistus, *Cistus populifolius* (leaf); **9**. *Ulex parviflorus*; **10**. *Berberis hispanica*; **11**. Shining crane's-bill, *Geranium lucidum*; **12**. Sowbread, *Cyclamen hederifolium*; **13**. *Colchicum triphyllum*; **14**. Alpine squill, *Scilla bifolia*; **15**. Greek fir, *Abies cephalonica*.

121

pyramidal crown and thick needles; it grows to a height of about 20 m. The shrub layer, at lower altitudes, has

Holm oak	*Quercus ilex*
	Quercus fruticosa
	Ulex parviflorus
Hedgehog broom	*Erinacea anthyllis*
Mastic tree	*Pistacia lentiscus*
Grey-leaved cistus	*Cistus albidus*
Poplar-leaved cistus	*Cistus populifolius*
	Phillyrea latifolia
	Phlomis purpurea.

At higher altitudes, the shrub layer includes

	Berberis hispanica
	Astragalus granatensis
	Ononis reuteri
Spurge-laurel	*Daphne laureola*

with bulbous species such as

	Colchicum triphyllum
	Scilla sps.
	Hyacinthoides hispanica
	Narcissus pseudonarcissus
	ssp. *major.*

The Sicilian endemic, *Abies nebrodensis*, is in danger of extinction, though it is now being replanted. 150 years ago there were extensive forests of this fir in southern Sicily.

Cypress and juniper woods

These are very restricted in the Mediterranean region, though a number of species are commonly found as bushes in the different Mediterranean communities.

The funeral cypress, *Cupressus sempervirens* (Plate 118) and, in particular, the tall pyramidal form, forma *sempervirens*, is commonly cultivated in the Mediterranean coastal areas, and is distinctive and typical of this region. The funeral cypress is a native tree of Asia Minor, and further east; however it is probably also native in Crete and the montane region of mainland Greece, while it has in all probability been introduced by man into Albania and Yugoslavia. It very rarely forms natural woods, except in Crete where small woods occur mostly between 800 and 1500 m, usually with maquis undergrowth. Here cypress woods are best developed above about 1000 m, and they are often accompanied by the semi-evergreen maple, *Acer sempervirens*, which may become locally dominant. Associated shrubby species include

Kermes oak	*Quercus coccifera*
Hawthorn	*Crataegus monogyna* ssp.
	azarella
Greek spiny spurge	*Euphorbia acanthothamos*
	Rhamnus prunifolius
Jerusalem sage	*Phlomis fruticosa*
	Phlomis lanata
	Thymus capitatus
Three-lobed sage	*Salvia triloba.*

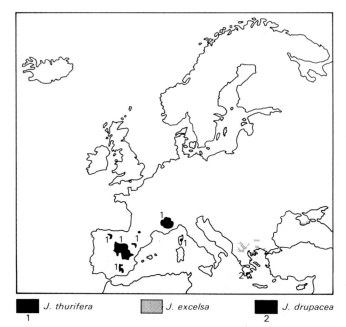

Map 38. The distribution of three Mediterranean junipers—Spanish juniper *Juniperus thurifers*, Syrian juniper *J. drupacea*, and Grecian juniper *J. excelsa*.

The following herbaceous species also occur

	Arabis verna
Shining crane's-bill	*Geranium lucidum*
	Asphodeline liburnica
Bulbous meadow-grass	*Poa bulbosa*
Cock's-foot	*Dactylis glomerata*
	Stipa bromoides.

There are three species of juniper which sometimes form woods and which grow to heights of 10 m, but they are very restricted in their distribution in Europe, though quite distinctive when encountered (Map 38). They are the Spanish juniper, *Juniperus thurifera* (Plate 119), which occurs in Spain and southern France, with outliers in Corsica; the Greek juniper, *Juniperus excelsa*, which is restricted in Europe to Macedonia, southern Bulgaria, Thrace, and some Greek islands (Thasos, Pyrgos, Euboea); and the Syrian juniper *Juniperus drupacea*, which just reaches the Peloponnese from Asia Minor, and forms very limited woods on Mount Parnon.

Another coniferous tree, the Barbary abor-vitae, *Tetraclinis articulata*, is a north African species which in Europe just reaches Malta and southeastern Spain in the dry coastal region of Cartagena. It forms scattered stands with a typical maquis shrub layer.

The Spanish juniper is found in the Iberian mountains of Spain, and in the southwestern Alps of France. The trees grow up to 10 m in height and they form dense dark woods without shrub or field layers, and with a thick layer of needles on the ground. However this closed community is very rare, and this juniper is likely to be encountered in a more open mixed community with holm oak, *Quercus ilex*, and Portugese oak,

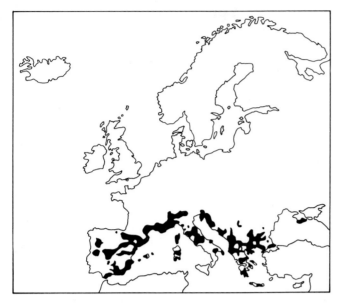

Map 39. Prickly juniper *Juniperus oxycedrus*.

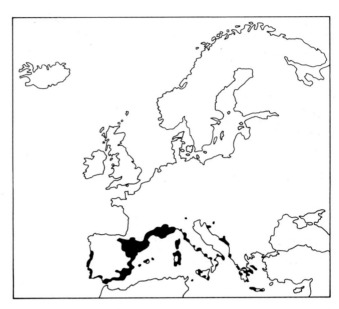

Map 40. Phoenician juniper *Juniperus phoenicea*.

Quercus faginea. In the shrub undergrowth, the prickly juniper, *Juniperus oxycedrus* (Map 39); Phoenician juniper, *Juniperus phoenicea* (Map 40), and savin, *Juniperus sabina*, occur together with the laurel-leaved cistus, *Cistus laurifolius.* The Spanish juniper may also form woods with the maritime pine, *Pinus pinaster;* the black pine, *Pinus nigra;* and the Scots pine, *Pinus sylvestris.*

The Greek juniper, *Juniperus excelsa*, is the tallest of the junipers, and may grow up to 20 m in height. It is a dark-green tree with a characteristic pyramidal crown. An example of a Greek juniper wood in Macedonia, has, in addition to the dominant juniper, the following in the lower tree and shrub layers

Prickly juniper	*Juniperus oxycedrus*
Joint-pine	*Ephedra fragilis*
	Ephedra major
Oriental hornbeam	*Carpinus orientalis*
White oak	*Quercus pubescens*
St. Lucie's cherry	*Prunus mahaleb*
	Prunus webbii
Christ's thorn	*Paliurus spina-christi*
Butcher's-broom	*Ruscus aculeatus.*

Sub-Mediterranean wet woods

These are largely dominated by the alder, and various oaks and willows. They are closely related to the wet woods of Central Europe. They occur along river valleys, or in low-lying waterlogged areas, scattered throughout the region.

An example of alder woods in southern France has *Alnus glutinosa*, with the narrow-leaved ash, *Fraxinus angustifolia*, and in addition in the tree layer

White willow	*Salix alba*
Crack willow	*Salix fragilis*
Purple willow	*Salix purpurea*
White poplar	*Populus alba*
Small-leaved elm	*Ulmus minor.*

The shrub layer consists of

	Salix elaeagnos
Hazel	*Corylus avellana*
Dewberry	*Rubus caesius*
Blackthorn	*Prunus spinosa*
Privet	*Ligustrum vulgare*
Dogwood	*Cornus sanguinea*
Elder	*Sambucus nigra.*

Field layer species include

Garlic mustard	*Alliaria petiolata*
Nettle	*Urtica dioica*
Enchanter's nightshade	*Circaea lutetiana*
Wild angelica	*Angelica sylvestris*
Hedge bindweed	*Calystegia sepium*
Hedge woundwort	*Stachys sylvatica*
Bittersweet	*Solanum dulcamara*
Common figwort	*Scrophularia nodosa*
Hedge bedstraw	*Galium mollugo*
Pendulous sedge	*Carex pendula*
False brome	*Brachypodium sylvaticum*

with the climbing hop, *Humulus lupulus.*

A swamp type of alder wood in Macedonia has the common alder, *Alnus glutinosa*, dominant. The shrub layer contains

123

Mediterranean vegetation

	Rosa corymbifera
Tartarian maple	*Acer tataricum*
Alder buckthorn	*Frangula alnus*
Privet	*Ligustrum vulgare*

with the climbers

Hop	*Humulus lupulus*
Grape vine	*Vitis vinifera*
Silk-vine	*Periploca graeca.*

The field layer includes

Purple loosestrife	*Lythrum salicaria*
Ivy	*Hedera helix*
Large bindweed	*Calystegia sylvatica*
Water germander	*Teucrium scordium*
Great horsetail	*Equisetum telmateia*
Royal fern	*Osmunda regalis*
Bracken	*Pteridium aquilinum*
Marsh fern	*Thelypteris palustris.*

Mixed oak woods are found in wet river valleys in a few places in the sub-Mediterranean region. For example in Istria, northern Yugoslavia, pedunculate oaks may be mixed with alder, hornbeam, white willow, small-leaved elm, and Caucasian ash, *Fraxinus angustifolia*.

Wet-valley willow woods are uncommon in the Mediterranean region. An example in France has the purple willow, *Salix purpurea*, mixed with white willow, *Salix alba*, with, in addition, the alder, black poplar, *Populus nigra*; white poplar, *Populus alba*; and small-leaved elm, *Ulmus minor*.

By the very nature of the Mediterranean and sub-Mediterranean climates, wet woodlands are restricted to river valleys which drain down into this region from Central Europe. Common woods are those of the white poplar, *Populus alba*, which form dense stands on deep alluvial soils, particularly in Eastern Europe. They tend to replace the willow–alder–ash woods of further north. These poplar woods also occur in northwestern Italy and southern France, especially along the Rhone valley, and along some of the Spanish valleys in the sub-Mediterranean zone.

White poplar woods in Languedoc, France contain in the tree layer

White willow	*Salix alba*
	Salix atrocinerea
Crack willow	*Salix fragilis*
Almond willow	*Salix triandra*
Black poplar	*Populus nigra*
Small-leaved elm	*Ulmus minor*
Box-elder	*Acer negundo*
Manna ash	*Fraxinus ornus.*

The shrub layer contains

Dewberry	*Rubus caesius*
Dog rose	*Rosa canina*
Hawthorn	*Crataegus monogyna*
Spindle	*Euonymus europaeus*
Elder	*Sambucus nigra*

and the Hop *Humulus lupulus*.

Typical field layer species include

Soapwort	*Saponaria officinalis*
Wood spurge	*Euphorbia amygdaloides*
Wild parsnip	*Pastinaca sativa*
Spreading hedge-parsley	*Torilis arvensis*
Hedge bedstraw	*Galium mollugo*
Hedge bindweed	*Calystegia sepium*
Hemp agrimony	*Eupatorium cannabinum*
Stinking iris	*Iris foetidissima*
Italian lords-and-ladies	*Arum italicum*
Pendulous sedge	*Carex pendula.*

Macedonian examples of a poplar wood have the grey poplar, *Populus canescens*, and the aspen, *Populus tremula*, mixed with the white poplar in the tree layer with, in addition, the white willow, *Salix alba*, and the small-leaved elm, *Ulmus minor*. The shrub layer includes

Traveller's-joy	*Clematis vitalba*
Dewberry	*Rubus caesius*
Dog rose	*Rosa canina*
Blackthorn	*Prunus spinosa*
Grape-vine	*Vitis vinifera*
	Asparagus acutifolius.

Another type of wet woodland is formed by the oriental plane, *Platanus orientalis* (Plate 121), which is a very distinctive tree. It forms small woods along the valleys of the Mediterranean region of the Balkan peninsula, on the gravelly river beds and flood plains. However most of them have been destroyed, and nowadays these woods are often much better developed higher up the river valleys in the sub-Mediterranean region. Good examples may be seen in the upper valleys of the Vardar and Struma, and near lake Dorjan, at 400–500 m, on the Greece–Yugoslavia border. Oriental planes grow quickly, and in about 80 years may reach a height of 40 m. They have a wide-spreading crown which casts considerable shade. In consequence of this they are often planted by springs, houses, and villages, thus casting a pleasant shade and providing meeting places for the inhabitants and performing a social role which is so distinctive in the villages of Greece, particularly in Macedonia. At low altitudes in dry river valleys, the oriental plane may form a bush community with such typical shrubs as *Tamarix tetrandra*; oleander, *Nerium oleander;* chaste-tree, *Vitex agnus-castus*; and *Rubus ulmifolius*. The field layer may include

	Alyssum murale
	Peltaria emarginata
	Euphorbia exigua
Field madder	*Sherardia arvensis*
Aromatic inula	*Dittrichia viscosa*
	Crepis neglecta
Dragon arum	*Dracunculus vulgaris.*

In river valleys which are infrequently flooded, for example in Albania, the oriental plane may be associated with the white willow, *Salix alba*; purple willow, *Salix purpurea*; and the

Rosemary Wise.

Plate 36. Mediterranean wet woodlands

1. Oriental plane, *Platanus orientalis*;   **2**. Aspen, *Populus tremula* (leaves) (inset×1);   **3**. White poplar, *Populus alba* (leaves);   **4**. Grey poplar, *Populus canescens* (leaves);   **5**. *Tamarix tetrandra*;   **6**. Small-leaved elm, *Ulmus minor* (leaves);   **7**. *Rubus ulmifolius*;   **8**. Soapwort, *Saponaria officinalis*;   **9**. Pendulous sedge, *Carex pendula*;   **10**. Oleander, *Nerium oleander*;   **11**. Tree medick, *Medicago arborea*;   **12**. Grape-vine, *Vitis vinifera*;   **13**. Chaste-tree, *Vitex agnus-castus*.

alder, *Alnus glutinosa*, in the tree layer. The shrub layer includes

	Rosa sempervirens
	Rubus ulmifolius
Judas tree	*Cercis siliquastrum*
Tree medick	*Medicago arborea*
Grape-vine	*Vitis vinifera*
Myrtle	*Myrtus communis*
Oleander	*Nerium oleander*
Chaste tree	*Vitex agnus-castus*.

The field layer includes some more southern species such as

Slender centaury	*Centaurium tenuiflorum*
Balm	*Melissa officinalis*
	Smilax aspera
Italian lords-and-ladies	*Arum italicum*
Hard Poa	*Desmazeria rigida*.

A type of oriental plane wood with the walnut, *Juglans regia*, is found in Macedonia with a rich tree layer containing

White willow	*Salix alba*
Black poplar	*Populus nigra*
Alder	*Alnus glutinosa*
Oriental hornbeam	*Carpinus orientalis*
Hornbeam	*Carpinus betulus*
Field maple	*Acer campestre*
Manna ash	*Fraxinus ornus*.

The shrub layer includes

Kermes oak	*Quercus coccifera*
Traveller's-joy	*Clematis vitalba*
	Rubus ulmifolius
Dog rose	*Rosa canina*
Hawthorn	*Crataegus monogyna*
Dogwood	*Cornus sanguinea*.

The field layer includes

	Rumex tuberosus
Celandine	*Ranunculus ficaria*
Soapwort	*Saponaria officinalis*
Wood avens	*Geum urbanum*
Cleavers	*Galium aparine*
Tuberous comfrey	*Symphytum tuberosum*
Wild basil	*Clinopodium vulgare*
Water mint	*Mentha aquatica*.

Mediterranean bush communities

The most distinctive and widespread semi-natural vegetation of the south are the bush and dwarf shrub communities which cover large areas of the coastal and hill regions in the Mediterranean and sub-Mediterranean zones. These are not the natural climax communities, but are the result of a number of factors which maintain them in this sub-climax state. The main factors are: the cutting of trees for fuel and charcoal; grazing by man's livestock (goats in particular); the clearance of land for cultivation—particularly for olive; and fires (Fig. 34). In consequence, the climax woods of evergreen and deciduous oaks, or coniferous woods, etc. are now very limited, and in their place bush and dwarf shrub communities prevail, depending on the degree of man's activities, and also the extent to which the top soil has been removed by erosion. The final state of depauperation of these natural woodland communities is a grass-like steppe, often with extensive areas of exposed rocky or stony ground. Here many interesting annuals, and bulbous and rhizomatous perennials can survive—flowering in early spring or late autumn, and dying down completely in the hot dry summer months.

The main types of bush communities are

Maquis: a dense shrub community of 1 m or more in height, consisting of evergreen sclerophyllous shrubs. In the sub-Mediterranean zone a largely deciduous shrub community

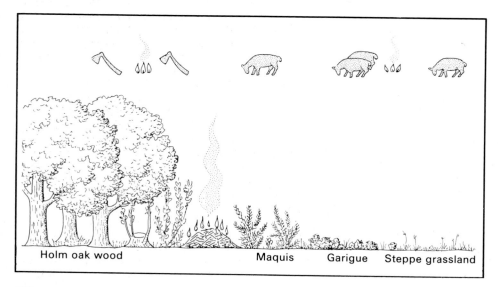

Fig. 34. Maquis. The original holm oak forest has been removed or converted into maquis, garigue, and dry grassland by felling, burning, and grazing.

Holm oak wood   Maquis   Garigue   Steppe grassland

takes its place—it is known as *šibljak* in the Balkan area, but this term will be extended to the western Mediterranean here. An intermediate type of shrub community, part evergreen and part deciduous, is often referred to as *pseudomaquis*.

Garigue: a general term for a dwarf shrub community usually less than 0.5 m high, which is rich in aromatic and mostly evergreen shrublets. This is termed *phrygana* in Greece, and there are many local names used to describe different variants of these dwarf shrub communities in the different countries bordering the Mediterranean. The local names may overlap in their meaning, or they may embrace more than one type of community; for example the widely used Spanish term *matorral* includes all evergreen bush communities, both maquis and garigue.

Maquis (Plate 122) (in our terminology) includes all types of evergreen bush communities growing to 1 m or more. It is particularly characteristic of the western Mediterranean region where the climate is comparatively moist; thus it is found largely on the western coasts of land masses, from sea-level up to about 600 m, or in local areas where the micro-climate is moist. It is less well developed on the eastern coasts of land masses where rain shadows prevail with the predominantly west-to-east circulating anticyclones. Further east, maquis is well developed along the Adriatic coast of Yugoslavia, and to a lesser extent in the Aegean region where it may be found at altitudes of 300 to 800 m, as for example in Euboea. In Crete, maquis is best developed up to 1100 m, where the cloud zone prevails. The activities of man, particularly charcoal-burning, firewood-cutting, and burning to give improved grazing, have very much reduced this natural maquis. Burnt maquis can however recover in as little as three years, but different species re-establish themselves at different rates. The strawberry tree, *Arbutus unedo*, is one of the quickest to recover; the tree heath, *Erica arborea* is one of the slowest. There are three main types of maquis, each characteristic of the different climatic zones in the Mediterranean. These zones are termed the olive–carob zone, the evergreen oak zone, and the sub-montane and montane zone.

The maquis of the olive–carob zone occurs in the driest climates. It is probable that a climax wood of mixed olive, *Olea europaea*, and carob, *Ceratonia siliqua*, prevailed in the past in this climatic zone up to altitudes of about 600 m, but the natural woods were virtually exterminated a long time ago by man, with the result that the soil has dried out and erosion has taken place, so that what soil still exists is very depauperate. However examples of these olive–carob woods still persist in, for example, the remoter parts of Sicily.

True maquis itself is very limited in this olive–carob zone, largely because of the long settlement by man, while the climate with its long periods of drought alternating with short heavy rainstorms causes maximum erosion. Much of the original maquis is now replaced by garigue, grass-steppe, or bare rock.

Fragments of maquis can still be found in Provence, in southern France. It has, in addition to bushes of the olive and carob,

Holm oak	*Quercus ilex*
Fragrant clematis	*Clematis flammula*
Spiny broom	*Calicotome spinosa*
Jupiter's beard	*Anthyllis barba-jovis*
Tree spurge	*Euphorbia dendroides*
	Cneorum tricoccon
Mediterranean buckthorn	*Rhamnus alaternus*
Myrtle	*Myrtus communis*
Wild madder	*Rubia peregrina*
	Lonicera implexa
	Asparagus acutifolius
	Smilex aspera

with the grass, *Brachypodium ramosum*. In Spain, a similar olive–carob maquis contains

	Osyris quadripartita
Mastic tree	*Pistacia lentiscus*
	Cneorum tricoccon
	Rhamnus lycioides
Myrtle	*Myrtus communis*
	Asparagus albus
Dwarf fan palm	*Chamaerops humilis.*

In Greece, in Euboea, the maquis of the olive–carob zone has in addition to the olive

Joint pine	*Ephedra fragilis*
White oak	*Quercus pubescens*
Golden alyssum	*Alyssum saxatile*
Judas tree	*Cercis siliquastrum*
	Pyrus amygdaliformis
Spiny broom	*Calicotome villosa*
Mastic tree	*Pistacia lentiscus*
Turpentine tree	*Pistacia terebinthus*
Mediterranean buckthorn	*Rhamnus alaternus*
	Phillyrea latifolia
Three-lobed sage	*Salvia triloba*
	Asparagus aphyllus
Butcher's-broom	*Ruscus aculeatus.*

The ground layer commonly has the Greek cyclamen, *Cyclamen graecum*, and the grasses, *Brachypodium ramosum* and tor-grass, *Brachypodium pinnatum*. This maquis forms a dense thick bush community, up to 3 m in height, with a poor field layer, but often it is heavily grazed and contains many phrygana and grass-steppe species.

Maquis of the evergreen oak zone (Plate 123). In the west of the Mediterranean this maquis is a degraded form of the holm oak woods which have already been described. This bush community, from which the holm oak trees have been removed, is very characteristic of southern France, the Italian coast, and Spanish Catalonia, and is composed of an assortment of species similar to those of the climax woods. These bush communities would in all probability with adequate protection redevelop into the climax woodland, though this is not certain.

Maquis of the eastern Mediterranean region (Fig. 35) are also in almost every case degraded forms either of evergreen oak woodlands, or olive–carob communities in the south. In Istria and along the Adriatic coast degraded oak woodlands are found, and an example on a promontory near Budva,

Plate 37. Mediterranean maquis—shrubs

1. *Calicotome spinosa*; **2**. Spiny broom, *Calicotome villosa*; **3**. Spanish broom, *Spartium junceum*; **4**. Myrtle, *Myrtus communis*; **5**. Jupiter's beard, *Anthyllis barba-jovis*; **6**. *Asparagus aphyllus*; **7**. *Asparagus acutifolius*; **8**. Carob, *Ceratonia siliqua*; **9**. Olive, *Olea europaea*; **10**. Wild madder, *Rubia peregrina*; **11**. Three-lobed sage, *Salvia triloba*; **12**. Bladder senna, *Colutea arborescens* (inset×1); **13**. Tree heath, *Erica arborea* (inset×6); **14**. *Erica manipuliflora* (inset×6); **15**. *Smilax aspera* (inset×1); **16**. *Phillyrea latifolia* (inset×1).

Yugoslavia, on well drained terra-rossa, is dominated by tall shrubs 1–2 m high which include

Prickly juniper	*Juniperus oxycedrus*
Mastic tree	*Pistacia lentiscus*
Strawberry tree	*Arbutus unedo*
Tree heath	*Erica arborea*
	Phillyrea latifolia.

Above these shrubs there are often widely spaced trees of holm oak, *Quercus ilex*. The field layer includes

Tunic flower	*Petrorhagia saxifraga*
	Teucrium polium
	Veronica austriaca

and the grass, *Brachypodium ramosum.*
Further south, the proportion of evergreen species increases, and bushes of kermes oak, *Quercus coccifera*, may become widespread. Other shrub species include

Laurel	*Laurus nobilis*
Spanish broom	*Spartium junceum*
Scorpion senna	*Coronilla emerus*
Christ's thorn	*Paliurus spina-christi*
	Frangula rupestris
Manna ash	*Fraxinus ornus.*

A type of maquis rich in kermes oak, *Quercus coccifera*, is developed predominantly in the Aegean region. It has a similar assortment of species to those listed in the olive–carob zone.

Maquis of the sub-montane and montane zone. This is usually found where the vegetation climax would otherwise be deciduous oak woods. However, this maquis still contains a number of evergreen species, due in part to their greater resistance to grazing. The most important and widespread evergreen species of this maquis (which is sometimes referred to as pseudomaquis) are the kermes oak, *Quercus coccifera*; box, *Buxus sempervirens*; and prickly juniper, *Juniperus oxycedrus*. Along the Adriatic coast, in southern Macedonia and Thrace, such sub-montane maquis contains the following shrubs

Prickly juniper	*Juniperus oxycedrus*
Oriental hornbeam	*Carpinus orientalis*
Turkey oak	*Quercus cerris*
Kermes oak	*Quercus coccifera*
Hungarian oak	*Quercus frainetto*
Hawthorn species	*Crataegus* sps.
St. Lucie's cherry	*Prunus mahaleb*
Scorpion senna	*Coronilla emerus*
Turpentine tree	*Pistacia terebinthus*
Christ's thorn	*Paliurus spina-christi*
Wall germander	*Teucrium chamaedrys*
Butcher's-broom	*Ruscus aculeatus*

and the sedge, *Carex hallerana*. In the Spanish montane maquis, additional species include

Spanish juniper	*Juniperus thurifera*
Portuguese oak	*Quercus faginea*
Snowy mespilus	*Amelanchier ovalis*
Portugal laurel	*Prunus lusitanica*
	Genista scorpius
Holly	*Ilex aquifolium*
	Acer granatense.

In the eastern Mediterranean the following additional shrubs may be found in the sub-montane and montane maquis

Macedonian oak	*Quercus trojana*
Cherry-laurel	*Prunus laurocerasus*

Fig. 35. Maquis with Kermes oak *Quercus coccifera*, Holm oak *Quercus ilex* and Wig tree *Cotinus coggygria* on a hillside in Greece.

Mediterranean vegetation

Spiny broom	*Calicotome villosa*
Bladder senna	*Colutea arborescens*
Holly	*Ilex aquifolium*
Wild jasmine	*Jasminum fruticans*
	Asparagus acutifolius.

Deciduous bush communities. These are termed *šibljak* in the Balkans, but comparable communities occur in central Spain, southern France, central Italy, Sardinia, and Sicily. They occur in sub-Mediterranean climates which have hot dry summers but cool continental winters, often with snow. In most cases šibljak is the sub-climax community which develops as a result of forest destruction. Along the northern coastal region of the Adriatic all stages of degradation can be seen, ranging from rare woodlands, to coppiced woods, true bush communities, and to grasslands, and these often occur as a mosaic. Alternatively, ascending from the coast to the hills inland, one passes through evergreen maquis to šibljak, with sometimes an intermediate middle zone dominated by kermes oak (pseudomaquis). An example of šibljak on the Croatian coast of Yugoslavia has a bushy vegetation up to about 3 m consisting of a rich mixture of shrubs, largely dominated by the oriental hornbeam, *Carpinus orientalis*; Montpellier maple, *Acer monspessulanum*; hop-hornbeam, *Ostrya carpinifolia*; and turpentine tree, *Pistacia terebinthus*. Other shrubs include

Turkey oak	*Quercus cerris*
White oak	*Quercus pubescens*
	Rubus ulmifolius
St. Lucie's cherry	*Prunus mahaleb*
Scorpion senna	*Coronilla emerus*
Wig tree	*Cotinus coggygria*
Field maple	*Acer campestre*

Rock buckthorn	*Rhamnus saxatilis*
Alder buckthorn	*Frangula alnus*
Christ's thorn	*Paliurus spina-christi*
Cornelian cherry	*Cornus mas*
Manna ash	*Fraxinus ornus*

and the climber traveller's joy, *Clematis vitalba*. The dwarf shrub and field layer includes

Hairy violet	*Viola hirta*
Ivy	*Hedera helix*
	Cnidium silaifolium
Betony	*Stachys officinalis*
Winter savory	*Satureja montana*
Wild basil	*Clinopodium vulgare*
	Asparagus acutifolius
Butcher's-broom	*Ruscus aculeatus*
The grass	*Sesleria autumnalis*
Tor-grass	*Brachypodium pinnatum.*

In Macedonia and Thrace, another type of šibljak has a rather similar bush composition with the oriental hornbeam, *Carpinus orientalis*; turpentine tree, *Pistacia terebinthus*; Judas tree, *Cercis siliquastrum*; and manna ash, *Fraxinus ornus*. Other shrubby species include

Prickly juniper	*Juniperus oxycedrus*
Turkey oak	*Quercus cerris*
White oak	*Quercus pubescens*
Hungarian oak	*Quercus frainetto*
Service tree	*Sorbus domestica*
Bladder senna	*Colutea arborescens*
Scorpion senna	*Coronilla emerus*
Box	*Buxus sempervirens*
Christ's thorn	*Paliurus spina-christi*

Fig. 36. Christ's thorn *Paliurus spina-christi* is common in coastal bush communities, sometimes with the Chaste tree *Vitex agnus-castus* (left) as here in Greece.

| Cornelian cherry | *Cornus mas* |
| Privet | *Ligustrum vulgare.* |

In the field layer, such distinctive species as the blue wood-anemone, *Anemone apennina*; Greek hellebore, *Helleborus cyclophyllus*; and *Silene viridiflora*, all of which flower in the spring, and the autumn-flowering sowbread, *Cyclamen hederifolium*, often occur.

Further south along the Adriatic coast, additional shrub species occur, such as the Dalmatian laburnum, *Petteria ramentacea*, and the Macedonian oak, *Quercus trojana*, while locally near Lake Skadar and the Gulf of Kotor, Yugoslavia, the pomegranate, *Punica granatum*, occurs in the šibljak. In the karst dolines a bush community occurs with more Central European species including the hornbeam, *Carpinus betulus*, together with the wood anemone, *Anemone nemorosa*; dewberry, *Rubus caesius*; sanicle, *Sanicula europaea*; and hart's-tongue fern, *Phyllitis scolopendrium*.

Mediterranean dwarf-shrub, or garigue

The term *garigue* is here used in its widest sense, to include all the different types of dwarf shrub communities found in the Mediterranean region; in the Balkans the equivalent term, again used in its widest sense, is *phrygana*. Local variants of dwarf-shrub communities are also individually named in different countries, particularly in Spain where each type of garigue is given a distinctive name.

Garigue is a more or less open shrub community, usually about 50 cm high, and rarely growing above 1 m. The dominant dwarf shrubs are often widely spaced, with a considerable amount of bare, stony ground between the clumps of shrubs, which, in the hot and very dry summers, gives the vegetation a very parched appearance. Many species are aromatic, and have small dry leathery leaves which often curl up in summer, thus reducing transpiration loss. Some shrublets are chemically unpalatable to grazing animals, others are very spiny and are thus partially protected (Fig. 37). Garigue is very widely distributed in the Mediterranean and sub-Mediterranean zone, particularly on very dry stony or rocky ground, where the maquis or woodlands have been destroyed, and where heavy grazing and burning is persistent, resulting in the erosion of soil to the bed-rock. However despite the difficult growing conditions, garigue is usually rich, and can be very rich, in species. These are largely annuals which flower during the spring or autumn rains, and die down completely in the summer. Others are geophytes, that is to say they are herbaceous species which have underground resting and storage organs, such as bulbs, tubers, or rhizomes, which lie protected in the soil during the unfavourable period, and yet as a result of the stored food, are able to grow very rapidly and flower as soon as the rains come early in spring, or in autumn. A number of plant families have species particularly well adapted to survive in these conditions; they include, in particular, the pea family, Leguminosae; spurge family, Euphorbiaceae; daphne family, Thymelaeaceae; mint family, Labiatae; daisy family, Compositae; lily family, Liliaceae, and orchid family, Orchidaceae.

There is great diversity in the different types of garigue, and they show many intermediate stages, which is understandable in view of the fact that garigues are largely dependent on grazing pressures, burning, soil erosion, etc. Regarding grazing, genera like spurge, *Euphorbia* and *Cistus* are chemically resistant to grazing, while many others, such as thistles, *Carlina*, and *Eryngium*, are very spiny and are largely avoided by grazing animals, but many more species are grazed and survive to a greater or lesser extent.

Twelve types of garigue have been selected and are briefly

Fig. 37. Grazing, particularly by goats as here in Greece, has a marked effect on vegetation maintaining dry grassland and phrygana (seen in the background).

Mediterranean vegetation

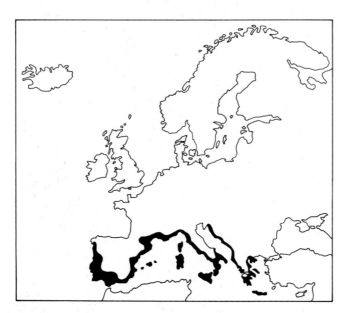

Map 41. Narrow-leaved cistus *Cistus monspeliensis*.

described, thus demonstrating the diversity of the Mediterranean dwarf-shrub communities. Characteristic of many of these types of garigue are such bulbous species as sea squill, *Urginea maritima*; *Ornithogalum collinum*; *Allium chamaemoly*, grape-hyacinths, *Muscari* species, as well as many tuberous and rhizomatous species such as anemone, cyclamen, asphodel, iris, arum, and orchids.

There are also many annual species characteristic of the garigue, which for short periods, mostly in early spring, give patches of colour to the otherwise rather monotonous terrain of parched shrublets and bare patches of stony ground. Typical annual species mostly belong to the following families: the pink family, Caryophyllaceae; buttercup family, Ranunculaceae; mustard family, Cruciferae; pea family, Leguminosae; grass family, Gramineae; etc.

Kermes oak garigue. This is the true garigue in the original sense (Spanish *goscojal*) and it is most frequently found on dry calcareous slopes in the western Mediterranean. It is also the degraded community of the Aleppo pine and holm oak woods in France, and of manna ash woods in the Balkans, and it is often maintained in this garigue state by grazing and/or by occasional burning by farmers.

Examples in southern France typically contain, among other species

Kermes oak	*Quercus coccifera*
	Dorycnium pentaphyllum
Large Mediterranean spurge	*Euphorbia characias*
	Euphorbia nicaeensis
Wall germander	*Teucrium chamaedrys*
	Aphyllanthes monspeliensis
	Asparagus acutifolius
	Brachypodium phoenicoides
	Brachypodium retusum.

In Spain, on the northern slopes of the Sierra Nevada, the kermes oak garigues contain the following species

Prickly juniper	*Juniperus oxycedrus*
	Berberis hispanica
	Genista cinerea
	Ulex parviflora
Mediterranean mezereon	*Daphne gnidium*
Rosemary	*Rosmarinus officinalis*.

Cistus garigue (Plate 124). This is widespread in the Mediterranean zone, replacing kermes oak as well as holm oak woods and white oak woods particularly after burning (Map 41). Cistus garigue in Languedoc, southern France contains such cistus species as

	Cistus crispus
Narrow-leaved cistus	*Cistus monspeliensis*
Sage-leaved cistus	*Cistus salvifolius*

and commonly in addition

Kermes oak	*Quercus coccifera*
	Calicotome spinosa
Bell heather	*Erica cinerea*
Heather	*Calluna vulgaris*
French lavender	*Lavandula stoechas*
	Bromus ramosus
	Brachypodium retusum
Silver hair-grass	*Aira caryophyllea*
Great quaking-grass	*Briza mazima*.

An example in Dalmatia, Yugoslavia has abundant *Cistus incanus* with

Prickly juniper	*Juniperus oxycedrus*
Spanish broom	*Spartium junceum*
Annual scorpion vetch	*Coronilla scorpioides*
	Erica manipuliflora
	Linaria microsepala
Sage	*Salvia officinalis*
	Brachypodium retusum

and some dwarfed maquis shrubs such as the kermes oak, mastic tree, *Pistacia lentiscus*, and *Phillyrea latifolia*.

Euphorbia garigue (Plate 127). Spurge species are very resistant to grazing, and they may come to dominate some garigues. The spiny spurge, *Euphorbia spinosa*, grows on dry stony soils, near the coast, from western Italy to northern Greece, while the much larger tree spurge, *Euphorbia dendroides*, which forms rounded bushes up to 2 m in height, dominates on rocky ground not far from the coast in the olive–carob zone. Examples in Italy of the tree spurge garigue contain

Holm oak	*Quercus ilex*
	Calicotome spinosa
Montpellier milk-vetch	*Astragalus monspessulanus*
Spanish broom	*Spartium junceum*
	Dorycnium hirsutum
	Dorycnium pentaphyllum
Mediterranean coriaria	*Coriaria myrtifolia*

Plate 38. Mediterranean garigue—shrubs and dwarf shrubs
1. Sage-leaved cistus, *Cistus salvifolius*; **2**. *Cistus crispus*; **3**. Narrow-leaved cistus, *Cistus monspeliensis*; **4**. *Erica multiflora*; **5**. Bell heather, *Erica cinerea*; **6**. Spiny spurge, *Euphorbia spinosa*; **7**. French lavender, *Lavandula stoechas*; **8**. Thyme *Thymus vulgaris*; **9**. Wall germander, *Teucrium chamaedrys*; **10**. *Genista sericea*; **11**. Dwarf fan palm, *Chamaerops humilis*; **12**. Thorny burnet, *Sarcopoterium spinosum*; **13**. *Thymus capitatus* (inset×5); **14**. Sage, *Salvia officinalis*.

Plate 39. Mediterranean low and dwarf shrubs

1. *Euphorbia acanthothamnos*;   **2**. Firethorn, *Pyracantha coccinea* (inset×3);   **3**. Mediterranean mezereon, *Daphne gnidium*;   **4**. Mediterranean coriaria, *Coriaria myrtifolia* (inset×2);   **5**. *Ephedra fragilis*;   **6**. *Cistus incanus*;   **7**. Tree spurge, *Euphorbia dendroides*;   **8**. *Hypericum empetrifolium*;   **9**. Felty germander, *Teucrium polium*;   **10**. Butcher's-broom, *Ruscus aculeatus* (inset×3);   **11**. *Petteria ramentacea*;   **12**. *Cneorum tricoccon* (inset×2);   **13**. Rosemary *Rosmarinus officinalis*;   **14**. *Staehelina uniflosculosa*;   **15**. Golden alyssum, *Alyssum saxatile* (inset×1).

Rosemary Wise

Plate 40. Mediterranean garigue and phrygana—dwarf shrubs and perennials

1. *Fumana laevipes*;   **2**. Mountain germander, *Teucrium montanum*;   **3**. *Fumana ericoides*;   **4**. *Adenocarpus complicatus*;   **5**. Shrubby gromwell, *Lithodora fruticosa*;   **6**. *Helichrysum stoechas*,   **7**. Montpellier milk-vetch, *Astragalus monspessulanus* (inset×1);   **8**. Aromatic inula, *Dittrichia viscosa*;   **9**. *Osyris alba* (inset×2);   **10**. *Dorycnium hirsutum*;   **11**. *Argyrolobium zanonii*;   **12**. *Prasium majus*;   **13**. Common rue, *Ruta graveolens*;   **14**. *Micromeria juliana* (inset×8);   **15**. *Lavandula latifolia*;   **16**. *Ballota acetabulosa*.

Mediterranean vegetation

Grey-leaved cistus	*Cistus albidus*
	Helianthemum sps.
Rosemary	*Rosmarinus officinalis*
	Helichrysum stoechas
	Hyoseris radiata.

In southern France, near Antibes, euphorbia garigue has both tree spurge and spiny spurge with

Fragrant clematis	*Clematis flammula*
Carob	*Ceratonia siliqua*
	Calicotome spinosa
Mastic tree	*Pistacia lentiscus*
Mediterranean buckthorn	*Rhamnus alaternus*
	Fumana laevipes
Olive	*Olea europaea*
	Smilax aspera.

In southern Macedonia, Greece, and the Greek Islands, the Greek spiny spurge, *Euphorbia acanthothamnos*, takes the place of the spiny spurge in the garigue.

Rosemary garigue. This occurs throughout the Mediterranean region on limestone, and is well developed in southern and eastern Spain, southern France, Italy, and islands off the Dalmatian coast. It replaces the Mediterranean coniferous and holm oak woods, though the former often re-invade such garigue. French examples in Bas-Languedoc have rosemary, *Rosmarinus officinalis*, and *Erica multiflora* commonly in the garigue with

Prickly juniper	*Juniperus oxycedrus*
Kermes oak	*Quercus coccifera*
	Genista scorpius
	Euphorbia nicaeensis
	Fumana ericoides
	Coris monspeliensis
Shrubby gromwell	*Lithodora fruticosa*
	Lavandula latifolia
	Staehelina dubia
	Aphyllanthes monspeliensis
	Brachypodium ramosum
Dwarf sedge	*Carex humilis.*

Examples of rosemary garigue in Dalmatia have rosemary associated with the heathers *Erica arborea*, *Erica multiflora*, and *Erica manipuliflora*, and the cistus species, *Cistus monspeliensis* and *Cistus salvifolius*, and other common shrubby species.

Tree heath garigue. This is known as *jaro* in Spain. It is predominantly a Western community, and occurs after burning of Atlantic heaths, or of maquis with tree heath. It is common in northern Spain and southwestern France, and extends eastwards along the northern Adriatic coast of Yugoslavia.

An example in the Cévennes, France at 250–450 m, has in addition to the tree heath, *Erica arborea*, the bell heather, *Erica cinerea*, and shrubs of

Holm oak	*Quercus ilex*
Broom	*Cytisus scoparius*
	Adenocarpus complicatus

Spotted rockrose	*Tuberaria guttata*
Heather	*Calluna vulgaris*
Wood sage	*Teucrium scorodonia*
Great quaking-grass	*Briza maxima*
Silver hair-grass	*Aira caryophyllea.*

On the Croatian coast, tree heath garigue is mixed with the sage-leaved cistus, *Cistus salvifolius*, with associated species such as

Prickly juniper	*Juniperus oxycedrus*
	Rosa sempervirens
	Dorycnium hirsutum
Spanish broom	*Spartium junceum*
Tor-grass	*Brachypodium pinnatum*
	Agrostis castellana
Glaucous sedge	*Carex flacca*
Bracken	*Pteridium aquilinum.*

Genista–Erica manipuliflora garigue. This is a montane garigue found in the coastal hills of Yugoslavia. It very often has the Dalmatian pine, *Pinus nigra* ssp. *dalmatica*, growing above it. Two species of genista, *Genista sericea* and the spiny *Genista sylvestris*, are characteristic, with the heather, *Erica manipuliflora*. Associated species include

Prickly juniper	*Juniperus oxycedrus*
Phoenician juniper	*Juniperus phoenicea*
	Argyrolobium zanonii
Spiny spurge	*Euphorbia spinosa*
	Fumana ericoides
	Asperula scutellaris
Mountain germander	*Teucrium montanum*
Common sage	*Salvia officinalis*
	Veronica species
	Brachypodium ramosum
	Chrysopogon gryllus
	Carex hallerana.

Thyme garigue. This is known as *tomillar* in Spain. The common thyme, *Thymus vulgaris*, is an important dominant of this type of garigue in the western Mediterranean, and it occurs as far east as western Italy. In Greece it is replaced by *Thymus capitatus* in this dwarf shrub community while, in central Spain, *Thymus zygis* is the main species. An example of thyme garique in Bordighera, Italy has in addition to common thyme

Prickly juniper	*Juniperus oxycedrus*
Mediterranean buckthorn	*Rhamnus alaternus*
Grey-leaved cistus	*Cistus albidus*
Rosemary	*Rosmarinus officinalis*
Aromatic inula	*Dittrichia viscosa*
Brown bee-orchid	*Ophrys fusca.*

Lavender garigue (Plate 125). The most important lavender species dominating considerable areas in the western Mediterranean region, and also in southern Greece, is the French lavender, *Lavandula stoechas*. It favours acid or neutral soils, and often associated with this lavender are

Plate 41. Mediterranean bulbous and rhizomatous perennials
1. Yellow bee orchid, *Ophrys lutea*; **2**. Mirror orchid, *Ophrys speculum*; **3**. Pink butterfly orchid, *Orchis papilionacea*; **4**. *Iris lutescens*; **5**. *Allium chamaemoly*; **6**. Friar's cowl, *Arisarum vulgare*; **7**. Sea squill, *Urginea maritima*; **8**. *Edraianthus tenuifolius*; **9**. *Ornithogalum collinum*; **10**. Mallow-leaved bindweed, *Convolvulus althaeoides*; **11**. *Coris monspeliensis*; **12**. *Serapias neglecta*; **13**. *Asphodelus ramosus*.

Mediterranean vegetation

Kermes oak	*Quercus coccifera*
Holm oak	*Quercus ilex*
Sage-leaved cistus	*Cistus salvifolius*
Narrow-leaved cistus	*Cistus monspeliensis*
Green heather	*Erica scoparia*
Heather	*Calluna vulgaris*
	Phillyrea angustifolia
Great quaking-grass	*Briza maxima*
	Brachypodium ramosum
	Agrostis castellana
	Carex oedipostyla
	Orchis morio ssp. *picta.*

Sage garigue. Several species of sage are common and distinctive in some garigues, particularly on the Balkan Adriatic coast on calcareous hills, eg. near Dubrovnik, and southwards to Greece. They also occur in the eastern Pyrenees and southern France, and are less common in Italy, Corsica, and Sardinia. The four most imortant species are common sage, *Salvia officinalis,* which occurs in northern Spain, southern France, southern Italy, and the Adriatic coast; *Salvia lavandulifolia* which is restricted largely to south–central and eastern Spain; silver sage, *Salvia argentea;* and the three-lobed sage, *Salvia triloba,* both of which have their main centres in the eastern Mediterranean.

An example of a sage garigue from the Dubrovnik region has common sage, *Salvia officinalis,* dominant with

Phoenician juniper	*Juniperus phoenicea*
	Osyris alba
	Alyssum montanum
	Genista sericea
	Dorycnium hirsutum
Common rue	*Ruta graveolens*
Mallow-leaved bindweed	*Convolvulus althaeoides*
Ground-pine	*Ajuga chamaepitys*
	Lavandula latifolia
Meadow clary	*Salvia pratensis*
	Micromeria juliana
	Edraianthus tenuifolius.

Eastern Thorny garigue (phrygana) (Plates 128–131). This is characteristic of the coastal and lower hills of the Balkan peninsula, particularly in Greece and Crete. It is a semi-natural dwarf shrub community which is maintained largely by burning, cutting, and very commonly by intensive grazing. It is a low scrub usually with a considerable amount of stony ground between the low cushion-like and mostly spiny shrubs; it rarely grows up to 1 m in height. Very characteristic of the Greek phrygana are the thorny burnet, *Sarcopoterium spinosum,* and the spiny dwarf shrub, *Anthyllis hermanniae.* In the Balearic Islands a comparable garigue has the endemic spiny astragalus, *Astragalus balearicus,* dominant, while in southern Italy, Sardinia, and Sicily, the spiny *Astragalus massiliensis* is common in the garigue. Southern Greek phryganas are characterized by the thorny burnet, *Sarcopoterium spinosum; Thymus capitatus;* and *Ballota acetabulosa.* Other typical species include

Spiny broom	*Calicotome villosa*
	Genista acanthoclada
	Anthyllis hermanniae
	Thapsia garganica
Felty germander	*Teucrium polium*
	Satureja thymbra
	Origanum onites
Aromatic inula	*Dittrichia viscosa*
Asphodel	*Ashpodelus aestivus*
Sea squill	*Urginea maritima*

and the grasses

Avena barbata
Piptantherum coerulescens
Stipa capensis
Hyparrhenia hirta.

In Euboea, a cistus-rich phrygana contains the sage-leaved cistus, *Cistus salvifolius,* and the pink-flowered cistus, *Cistus incanus* subsp. *creticus,* with *Erica manipuliflora,* and with other dwarf shrubs such as the thorny burnet, *Sarcopoterium spinosum; Anthyllis hermanniae;* felty germander, *Teucrium polium;* and *Hypericum empetrifolium.* Herbaceous species include

	Alyssum euboeum
	Stachys cretica
Greek cyclamen	*Cyclamen graecum*
	Piptantherum coerulescens
	Stipa bromoides.

Dwarf fan palm garigue (Plate 126). This is known as *palmito* in Spain and is dominated by the dwarf fan palm, *Chamaerops humilis.* It is predominantly a coastal community, spreading from the western Mediterranean to southwestern Italy, Sicily, and Sardinia. An example near Palermo, Sicily, where the palm grows to 2 m in height, is associated with such species as

	Anemone hortensis
Carob	*Ceratonia siliqua*
Bladder vetch	*Anthyllis tetraphylla*
Sowbread	*Cyclamen hederifolium*
Blue hound's-tongue	*Cynoglossum creticum*
Tree germander	*Teucrium fruticans*
	Prasium majus
	Verbascum sinuatum
	Orobanche lavandulacea
Hare's-foot plantain	*Plantago lagopus*
	Fedia cornucopiae
	Galactites tomentosa
White asphodel	*Asphodelus albus*
	Asphodelus ramosus
Sea squill	*Urginea maritima*
Friar's cowl	*Arisarum vulgare*
Hare's-tail	*Lagurus ovatus.*

Christ's thorn garigue. This garigue is largely dominated by the shrub, Christ's thorn, *Paliurus spina-christi,* which may grow 1–2 m high, with *Rhamnus intermedius* and the prickly juniper, *Juniperus oxycedrus*. It occurs on the coast of northern Yugoslavia and the Istrian peninsula. An example near Budva, Montenegro contains such additional shrubs as

Oriental hornbeam	*Carpinus orientalis*
Fragrant clematis	*Clematis flammula*
Blackthorn	*Prunus spinosa*
Spanish broom	*Spartium junceum*
Myrtle	*Myrtus communis*

while the dwarf shrubs and herbaceous species include

	Osyris alba
Tunic flower	*Petrorhagia saxifraga*
Large disk-medick	*Medicago orbicularis*
	Dorycnium hirsutum
	Linum nodiflorum
Upright yellow-flax	*Linum strictum*
Large-flowered orlaya	*Orlaya grandiflora*
	Putoria calabrica
Wall germander	*Teucrium chamaedrys*
Felty germander	*Teucrium polium*
Cut-leaved self-heal	*Prunella laciniata*
Spiny bear's breech	*Acanthus spinosus*
	Reichardia picroides
The grass	*Dichanthium ischaemum*.

Mediterranean and sub-Mediterranean grasslands

Dry grasslands and steppe-grasslands are widely distributed in the Mediterranean region. In the majority of cases they are semi-natural—that is to say that these communities are composed of native species—but they are maintained as grassland largely as the result of intense grazing by sheep and goats. In many cases they result from the destruction of woodlands, maquis, or garigue and consequently the land is often covered by a mosaic of scrub, bush, and grassland communities. However these steppe-grasslands can remain relatively stable for long periods under man's influence.

The feather-grasses, *Stipa* species, are characteristic of the driest communities where there is an annual rainfall of significantly less than 500 mm. Another grass of very dry, slightly salty soils, is the albardine, or esparto grass, *Lygeum spartum*, which forms steppe-grasslands in southeastern Spain, southwestern Italy, Sardinia, and Sicily.

Examples of Mediterranean steppe-grasslands from Spain in the coastal area have *Stipa tenacissima* dominant. It grows to 2 m in height and is harvested. Associated grasses include

Brachypodium retusum
Helictotrichon convolutum
Hyparrhenia hirta
Stipa capensis.

Typical herbaceous species found in the steppe-grasslands are *Thymus zygis*, *Evax pygmaea*, and *Asparagus stipularis*.

In Sicily and Sardinia, the feather-grasses, *Stipa capensis* and *Stipa offneri*, are important species. *Biscutella lyrata*, *Sedum caeruleum*, and *Anagallis monelli* are typically associated with them. Secondary steppe in southern Italy contains *Stipa*

capensis with the grass, *Lamarckia aurea*, and other species including

Ononis ornithopodioides
Linaria reflexa
Plantago afra
Fedia cornucopiae
Evax pygmaea.

Esparto-steppe is dominated by the albardine grass, *Lygeum spartum*. A Spanish example at Rio Dulce, northern Orihuela, contains

Joint-pine	*Ephedra fragilis*
Beet	*Beta vulgaris*
	Haloxylon articulatum
Shrubby orache	*Atriplex halimus*

which are mostly plants of somewhat saline soils, as well as

	Osyris alba
	Polygonum equisetiforme
	Anthyllis cytisoides
	Dorycnium pentaphyllum
Rush-like scorpion-vetch	*Coronilla juncea*
	Peganum harmala
Felty germander	*Teucrium polium*.

An example of an esparto-steppe in Sardinia has *Lygeum spartum* dominant with *Silene coeli-rosa*, *Thymelaea hirsuta*, *Thapsia garganica*, *Plantago bellardii*, *Ajuga iva*, and *Iris sisyrinchium*, among the non-graminaceous species.

Brachypodium retusum is an important Mediterranean grass and it is dominant over large areas in southern France, Spain, and Italy, and in parts of the eastern Mediterranean.

An example in southern France with *Brachypodium retusum* dominant has *Brachypodium distachyon* and bulbous meadow-grass, *Poa bulbosa*, with such species as

	Cerastium pumilum
	Sedum sediforme
	Trigonella gladiata
	Medicago minima
	Trifolium stellatum
	Euphorbia exigua
	Phlomis lychnitis
Thyme	*Thymus vulgaris*
	Valantia muralis
	Iris lutescens.

On calcareous soils, particularly on abandoned cultivated sites in southern France and Spain, *Brachypodium phoenicoides* is dominant, with such distinctive species as

Fennel	*Foeniculum vulgare*
	Echium pustulatum
	Salvia verbenaca
	Verbascum sinuatum
Mournful widow	*Scabiosa atropurpurea*
Globe-thistle	*Echinops ritro*
Salsify	*Tragopogon porrifolius*.

In the eastern Mediterranean, the dry grasslands are domi-

Plate 42. Mediterranean grasslands and steppe-grasslands
1. *Vulpia ciliata*; **2**. *Dichanthium ischaemum*; **3**. *Aira elegantissima*; **4**. *Chrysopogon gryllus*; **5**. *Koeleria glaucovirens*; **6**. *Stipa tenacissima*; **7**. *Brachypodium phoenicoides*; **8**. *Stipa offneri*; **9**. *Brachypodium retusum*; **10**. Great quaking-grass, *Briza maxima*; **11**. *Festuca rupicola*; **12**. *Hyparrhenia hirta*; **13**. Rough dog's-tail, *Cynosurus echinatus*; **14**. *Stipa capensis*; **15**. Albardine, *Lygeum spartum*.

nated largely by *Stipa capensis* and *Brachypodium retusum*, with other grasses including bulbous meadow-grass, *Poa bulbosa*; Bermuda-grass, *Cynodon dactylon*; great quaking-grass, *Briza maxima*; rough dog's-tail, *Cynosurus echinatus*; *Brachypodium distachyon*; timothy grass, *Phleum pratense*; *Hyparrhenia hirta*; and *Dichanthium ischaemum*. These grass-lands are rich in species and only the most distinctive are listed below

Proliferous pink	*Petrorhagia prolifera*
Spiny rest-harrow	*Ononis spinosa*
Cockscomb sainfoin	*Onobrychis caput-galli*
Field eryngo	*Eryngium campestre*
Mallow-leaved bindweed	*Convolvulus althaeoides*
	Myosotis incrassata
	Helichrysum italicum
Flat-topped carline-thistle	*Carlina corymbosa*
	Hedypnois cretica
	Romulea bulbocodium.

On dry sandy, more acid soils, which are uncommon in the eastern Mediterranean, the grasslands are dominated largely by *Vulpia ciliata* with *Aira elegantissima*, *Anthoxanthum ovatum*, and great quaking-grass, *Briza maxima*.

Sub-Mediterranean grasslands are quite extensive; they usually occur as a result of the clearance of deciduous woods, šibljak, and woody species. They are largely used for sheep grazing in both spring and autumn while, during the summer when the grasslands dry out, they are not grazed. There are a number of types of grassland each dominated by different grass species. In Provence in southern France, on rocky ground, at altitudes ranging from 700 to 1000 m, the blue moor-grass, *Sesleria caerulea*, may be dominant with

Wood pink	*Dianthus sylvestris*
	Iberis saxatilis
Burnet rose	*Rosa pimpinellifolia*
Hairy greenweed	*Genista pilosa*
	Helianthemum oelandicum
Common lavender	*Lavandula angustifolia*
	Plantago argentea.

In France above about 500 m, in the Cévennes and Corbières, *Koeleria vallesiana* becomes dominant, with sheep's-fescue, *Festuca ovina*, and with such herbaceous species as

	Sanguisorba minor
Common bird's-foot-trefoil	*Lotus corniculatus*
Kidney vetch	*Anthyllis vulneraria*
Horseshoe vetch	*Hippocrepis comosa*
Rock-rose species	*Helianthemum* sps.
Burnet-saxifrage	*Pimpinella saxifraga*
	Seseli montanum
Felty germander	*Teucrium polium*
Common globularia	*Globularia punctata*
Small scabious	*Scabiosa columbaria.*

A steppe-like grassland found locally from the western Alps to the Pyrenees, and well developed in the Causse region of France, at altitudes ranging from 500 to 1500 m, is dominated by the feather-grass, *Stipa pennata*. It contains the species

Grass-leaved buttercup	*Ranunculus gramineus*
Alpine cinquefoil	*Potentilla crantzii*
	Ononis striata
	Coronilla minima
	Euphorbia seguierana
	Helianthemum apenninum
Breckland thyme	*Thymus serpyllum*
	Inula montana
	Leontodon crispus.

In the eastern Mediterranean region the destruction of maquis and phrygana results in a steppe-grassland which is heavily grazed in spring and autumn. In the hills of Macedonia and Thrace, the grass, *Chrysopogon gryllus*, often dominates, and is mixed with other grasses, such as *Koeleria glaucovirens* and *Dichanthium ischaemum*. In lowland Macedonia *Festuca rupicola* is often dominant, forming a steppe-like grassland with

Poa bulbosa
Melica ciliata
Bromus cappadocicus
Koeleria glaucovirens
Phleum phleoides
Stipa lessingiana
Stipa pulcherrima
Chrysopogon gryllus
Dichanthium ischaemum.

The grasslands of the Adriatic coastal regions are well developed and are again maintained by regular grazing; other-wise woody species would quickly establish themselves. These grasslands are rich in attractive species which flower in the spring. *Festuca lemanii*, *Festuca valesiaca*, and *Koeleria splendens* are important species in lowland Dalmatia, with herbaceous species such as

Spanish catchfly	*Silene otites*
	Bupleurum baldense
Blue eryngo	*Eryngium amethystinum*
Felty germander	*Teucrium polium*
	Thymus longicaulis
	Plantago subulata
	Helichrysum italicum
	Carduus micropterus
	Centaurea spinosociliata
Flat-topped carline thistle	*Carlina corymbosa.*

A montane grassland of the coastal mountains of Croatia, has the upright brome, *Bromus erectus*; and *Koeleria splendens* as the important grasses, with the dwarf sedge, *Carex humilis*. Other species include

Pasque-flower	*Pulsatilla vulgaris*
Lesser meadow-rue	*Thalictrum minus*
	Genista sylvestris
	Asperula aristata
Mountain germander	*Teucrium montanum*

Mediterranean vegetation

Winter savory	*Satureja montana*
Spiked speedwell	*Veronica spicata*
Matted globularia	*Globularia cordifolia*
	Plantago argentea
	Jurinea mollis
	Centaurea rupestris
	Iris pallida.

Greek sub-alpine grasslands, which often replace Greek fir forests between about 1500 and 1700 m contain *Stipa* *pulcherrima* with *Melica ciliata* and *Festuca varia*. Characteristic additional species include

Cerastium candidissimum
Daphne oleoides
Morina persica
Pterocephalus perennis
Anthemis cretica.

11 Pannonic vegetation (Plates 43–47; Plates 132–138)

This is the western extension of the *steppe region* which spreads far into the Soviet Union to the east (Map 42). This western enclave is centred on the Hungarian Plain and takes its name from the Roman province of *Pannonia*. It stretches eastwards in southern Romania and northern Bulgaria as a narrow corridor of lowlands along either side of the Danube River to the steppes of the northwestern coast of the Black Sea.

It is distinguished by its low elevation with few hills, and its dry continental climate. The soils are mostly fertile and mineral-rich, and at the present day most of the area, something like 90 per cent, is under intensive cultivation, grazing, or re-afforestation. This region is perhaps the earliest settled area in the whole of Europe, beginning in the Neolithic and Bronze Ages, and consequently very little, if any, of the natural vegetation is present. In some parts of the Pannonic region, the natural climax communities were steppe-woodlands, dominated by various species of oak. Today only scattered remnants of the semi-natural oak woods are found, and many of these are degraded to bushlands. Characteristic of the region now are the steppe-grasslands known as *puszta* in Hungary. They are semi-natural and are maintained in this condition by grazing. Though they are composed of native species it is difficult to know the composition and structure of the original climax communities in these grassland areas.

Distinctive types of vegetation in this grass-steppe region are the salt-rich communities which possess many species of maritime and coastal origin, though they occur a long distance inland. As a result of excessive water evaporation in the dry climate, where evaporation exceeds precipitation, salts are brought up to the surface layers of the soil. Here the salts accumulate and consequently support saline-tolerant communities known as *salt-steppes*. Inland dune systems resulting from blown sand are also relatively frequent, and these support a distinctive dune vegetation. Around the margin of the Pannonic region, usually where there is higher ground, as for example in southwestern Hungary and northeastern Yugoslavia, more continental oak woods and beech woods may occur.

Wet woodlands, dominated largely by willows, poplars, and alder, which are subjected to annual flooding, often occur along the main river valleys, but these have been much reduced as a result of cultivation, drainage, and water-level regulation.

Steppe-woodlands

Fragments of woodlands, dominated largely by different species of oaks, may be found in the region but it is not known to what extent they are natural. In all probability most lowland sites on rich soil originally supported dry oak woods of one kind or another (Fig. 38). The most important oaks in the Pannonic region are the Turkey oak, *Quercus cerris;* Hungarian oak, *Quercus frainetto;* and the white oak, *Quercus pubescens*. Eastern species, such as *Quercus pedunculiflora*, form woods in Romania and Bulgaria, and other eastern species like *Quercus polycarpa* and *Quercus dalechampii* also occur in some woods. The more westerly European oaks such as the pedunculate oak, *Quercus robur*, and sessile oak, *Quercus petraea*, also form woods on moister sites in river valleys, and at higher altitudes. Other characteristic trees of these pannonic woodlands are the silver lime, *Tilia tomentosa;* hornbeam, *Carpinus betulus;* oriental hornbeam, *Carpinus orientalis;* and the tartarian maple, *Acer tataricum*, the latter also being important in the shrub layer.

In the more natural examples of the remaining steppe-woodlands, the tree layer is rather dense and casts much shade, so that the ground vegetation reaches its fullest development in the spring. However it is more usual to find woods where grazing or cutting has altered their structure. Typical woods now commonly consist of well separated groups of trees, interspersed with large clearings, in which the shrub layer is well developed and grows to a considerable height.

Oak woods dominated by the Turkey oak, *Quercus cerris*, are some of the most widespread in the Pannonic region. They often occur mixed with the white and sessile oaks. The following are also commonly found in the tree layer

Hornbeam	*Carpinus betulus*
Oriental hornbeam	*Carpinus orientalis*
Small-leaved elm	*Ulmus minor*
Wild pear	*Pyrus pyraster*
Snow pear	*Pyrus nivalis*
Crab apple	*Malus sylvestris*
	Malus dasyphylla
Wild service-tree	*Sorbus torminalis*
Field maple	*Acer campestre*
Tartarian maple	*Acer tataricum*
Small-leaved lime	*Tilia cordata*
Silver lime	*Tilia tomentosa*
Manna ash	*Fraxinus ornus.*

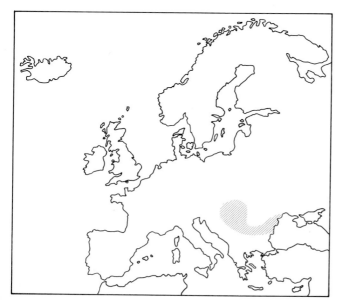

Map 42. Pannonic region.

Pannonic vegetation

Characteristic shrubs include

Juniper	*Juniperus communis*
Hazel	*Corylus avellana*
Ground cherry	*Prunus fruticosa*
Blackthorn	*Prunus spinosa*
Wig tree	*Cotinus coggygria*
Rough-stemmed spindle	*Euonymus verrucosus*
Bladder-nut	*Staphylea pinnata*
Cornelian cherry	*Cornus mas*
Wayfaring tree	*Viburnum lantana.*

The Hungarian oak, *Quercus frainetto*, is more widespread in the sub-Mediterranean region of the Balkan interior, but it does penetrate at fairly low altitudes into the Pannonic region in northern Bulgaria and northeastern Serbia. It is very often mixed with the Turkey oak and has similar tree and shrub layers. The field layer, from examples in Serbia, contains

	Helleborus odorus
Wild strawberry	*Fragaria vesca*
Wood avens	*Geum urbanum*
	Lathyrus pannonicus
Cypress spurge	*Euphorbia cyparissias*
Purple gromwell	*Buglossoides purpurocaerulea*
	Tanacetum corymbosum
Various-leaved fescue	*Festuca heterophylla*
Wood meadow-grass	*Poa nemoralis*
False brome	*Brachypodium sylvaticum.*

The white oak, *Quercus pubescens*, is important in the dry Pannonic oak woods. It occurs with the pedunculate oak in western Hungary and northern Yugoslavia, as well as on drier sites in the lower Danube plain, where it is mixed with oriental hornbeam, *Carpinus orientalis;* and manna ash, *Fraxinus ornus.*

Examples of white oak woods in the northern Dobrogea of Romania have, in addition to oriental hornbeam and manna ash, the sessile oak, *Quercus petraea; Quercus polycarpa;* and the service-tree, *Sorbus domestica.* The shrub layer includes

St. Lucie's cherry	*Prunus mahaleb*
Blackthorn	*Prunus spinosa*
Wild pear	*Pyrus pyraster*
Wig tree	*Cotinus coggygria*
Tartarian maple	*Acer tataricum*
Rough-stemmed spindle	*Euonymus verrucosus*
Cornelian cherry	*Cornus mas*
Privet	*Ligustrum vulgare.*

The field layer contains

Wild strawberry	*Fragaria vesca*
Bloody crane's-bill	*Geranium sanguineum*
Herbaceous periwinkle	*Vinca herbacea*
	Cynoglossum hungaricum
	Ajuga laxmannii
Wall germander	*Teucrium chamaedrys*
	Achillea coarctata

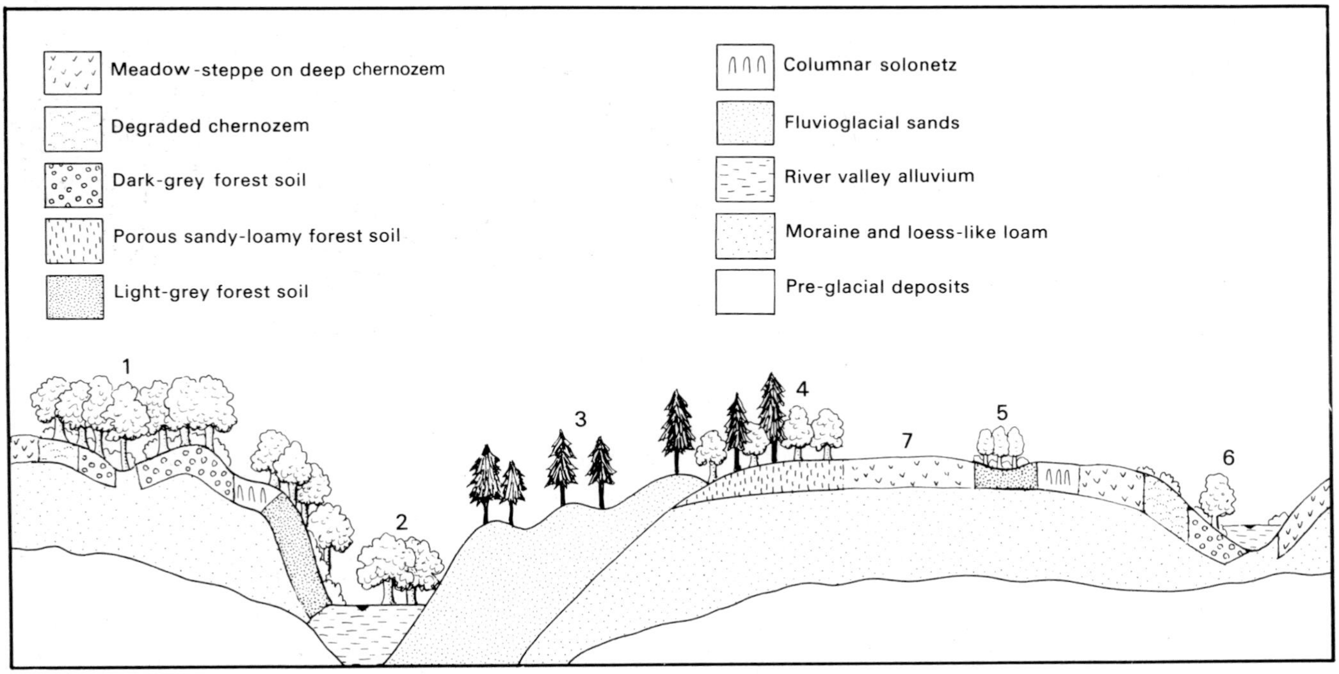

Fig. 38. Vegetation and soil in the steppe zone. This shows the relationship between vegetation, soil, and relief at the border between the steppe and forest regions. The black-earth soils (chernozem) would naturally support steppe communities. 1. Upland oak forest; 2. flood-plain forest (mainly oak); 3. pine forest on poor sands; 4. pine–oak forests; 5. aspen thickets in hollows; 6. ravine oak forest; steppe shrubs at upper margin; 7. arable land (formerly steppe). (Redrawn from Walter 1964.)

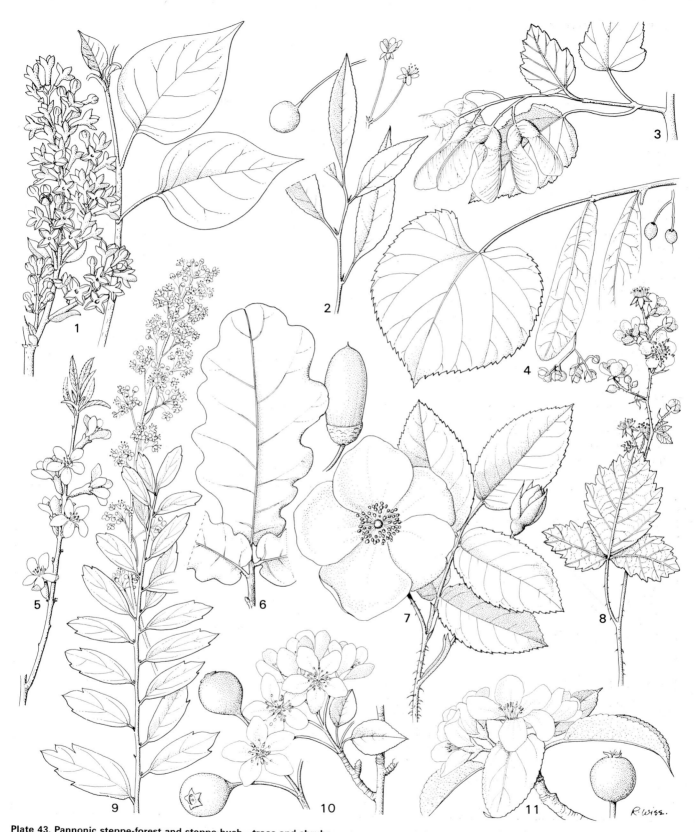

Plate 43. Pannonic steppe-forest and steppe-bush—trees and shrubs
1. Lilac, *Syringa vulgaris*; **2**. *Prunus fruticosa* (insets×2); **3**. Tartarian maple, *Acer tataricum*; **4**. Silver lime, *Tilia tomentosa*; **5**. *Prunus tenella*; **6**. *Quercus pedunculiflora*; **7**. *Rosa gallica*; **8**. *Rubus canescens*; **9**. *Spiraea media*; **10**. Wild pear, *Pyrus pyraster*; **11**. *Malus dasyphylla*.

Pannonic vegetation

	Tanacetum corymbosum
Wild Asparagus	*Asparagus officinalis*
	Asparagus verticillatus

with the grasses

	Festuca valesiaca
Narrow-leaved meadow-grass	*Poa angustifolia*
Cock's-foot	*Dactylis glomerata*
Hairy melick	*Melica ciliata*
Tor-grass	*Brachypodium pinnatum.*

The oak, *Quercus pedunculiflora*, is more closely associated with the Pannonic region, though it extends southwards into much of Bulgaria. An example of a wood in which this oak is co-dominant with the Hungarian oak, *Quercus frainetto*, contains in the tree layer

Turkey oak	*Quercus cerris*
White oak	*Quercus pubescens*
Small-leaved elm	*Ulmus minor*
Wild service-tree	*Sorbus torminalis*
Field maple	*Acer campestre*
Silver lime	*Tilia tomentosa*
Ash	*Fraxinus excelsior*

and in the shrub layer

Wild pear	*Pyrus pyraster*
Tartarian maple	*Acer tataricum*
Cornelian cherry	*Cornus mas*
Dogwood	*Cornus sanguinea*
Privet	*Ligustrum vulgare.*

The field layer includes

Rose campion	*Lychnis coronaria*
Dropwort	*Filipendula vulgaris*
Wood avens	*Geum urbanum*
Wild strawberry	*Fragaria vesca*
Purple gromwell	*Buglossoides purpurocaerulea*
	Tanacetum corymbosum

with the narrow-leaved meadow-grass, *Poa angustifolia;* wood meadow-grass, *Poa nemoralis;* cock's-foot, *Dactylis glomerata;* and others.

In the river valleys, the pedunculate oak, *Quercus robur*, is the most important oak, but it may also dominate in dry mixed oak woods. An example of a river-valley oak wood in Hungary has other oaks such as the white oak, Turkey oak, and sessile oak, with the small-leaved elm, *Ulmus minor.*

The shrub layer includes

Field rose	*Rosa arvensis*
Dog rose	*Rosa canina*
Provence rose	*Rosa gallica*
Wild pear	*Pyrus pyraster*
Blackthorn	*Prunus spinosa*
Field maple	*Acer campestre*
Tartarian maple	*Acer tataricum*
Spindle	*Euonymus europaeus*

Rough-stemmed spindle	*Euonymus verrucosus*
Buckthorn	*Rhamnus catharticus*
Dogwood	*Cornus sanguinea*
Privet	*Ligustrum vulgare*

and the field layer

Yellow adonis	*Adonis vernalis*
Celandine	*Ranunculus ficaria*
Wood avens	*Geum urbanum*
Burning bush	*Dictamnus albus*
Common vincetoxicum	*Vincetoxicum hirundinaria*
Purple gromwell	*Buglossoides purpurocaerulea*
Marjoram	*Origanum vulgare*
	Tanacetum corymbosum
Broad-leaved Solomon's-seal	*Polygonatum latifolium*
Variegated iris	*Iris variegata*
	Carex michelii

and the grasses *Festuca valesiaca; Poa angustifolia;* and cock's-foot, *Dactylis glomerata.*

The sessile oak sometimes dominates on the edge of the Pannonic region, in southwestern Hungary and in the foothills of the Transylvanian Alps in southern Romania.

Bush communities

These occur in the Pannonic region usually as a result of the destruction of the dry oak woodlands, or they may be a stage in the recolonization of abandoned pastures, ultimately to return to woodlands. Intense grazing is common in this region, and this may severely degrade the woodland to bush communities. The most typical bush community is that which is distinguished by the presence of the ground cherry, *Prunus fruticosa*. Other associated shrubby species include

	Spiraea media
	Rubus canescens
Provence rose	*Rosa gallica*
Wild pear	*Pyrus pyraster*
Hawthorn	*Crataegus monogyna*
Blackthorn	*Prunus spinosa*
Dwarf Russian almond	*Prunus tenella*
Tartarian maple	*Acer tataricum*
Buckthorn	*Rhamnus catharticus*
Cornelian cherry	*Cornus mas*
Privet	*Ligustrum vulgare.*

On steep rocky slopes, a bush community rather similar to the šibljak of Bulgaria occurs in southwestern Romania, western Bulgaria, and eastern Serbia. Characteristic shrubs are the oriental hornbeam, *Carpinus orientalis;* and the common lilac, *Syringa vulgaris*. Other shrubs include

White oak	*Quercus pubescens*
Wig tree	*Cotinus coggygria*
Rough-stemmed spindle	*Euonymus verrucosus*
Cornelian cherry	*Cornus mas*
Manna ash	*Fraxinus ornus.*

The field layer commonly includes

	Delphinium fissum
Orpine	*Sedum telephium*
Crown vetch	*Coronilla varia*
Hedge bedstraw	*Galium mollugo*
Wall germander	*Teucrium chamaedrys*
Small scabious	*Scabiosa columbaria*
	Campanula sibirica
Angular Solomon's-seal	*Polygonatum odoratum*

with the grasses hairy melick, *Melica ciliata;* tor-grass, *Brachypodium pinnatum;* and *Piptatherum virescens,* with the dwarf sedge, *Carex humilis,* amongst others. Where secondary scrub develops as the result of the destruction of woodland, the white oak, *Quercus pubescens,* is frequently the dominant bush with such plants as the burning bush, *Dictamnus albus;* bloody crane's-bill, *Geranium sanguineum; Artemisia pancicii; Scorzonera hispanica;* and variegated iris, *Iris variegata,* to list a few of the distinctive species.

Wet woodlands (Plate 138)

Wet woodlands were commonly developed along the margins of rivers on sites which were subjected to regular flooding. This flooding resulted in the annual addition of mineral salts to the soil from the flood water. However, today the regulation of the flow of most of the major rivers in the Pannonic region, the Danube and Tisza rivers in particular, has limited the extensive and regular floodings of the past and consequently reduced the extent of these wet woodlands considerably. There is also now a sharp margin between these wet woodlands and the adjoining agricultural land, where before there was a natural transition to dry Pannonic oak woods. Many of these wet woodlands have tall heavily crowned trees, and there is a rich ground vegetation including the common nettle, *Urtica dioica,* which perhaps here is in one of its few natural habitats in Europe. These woodlands also support such climbers as the wild grape-vine, *Vitis vinifera,* and the hop, *Humulus lupulus,* giving a dense almost tropical appearance to these woods. Many wet woods contain a mixture of dominant willow and poplar. The woods may be flooded for up to 90 days in the year, in which case willow is commonly dominant, while with less flooding poplar and willow may be mixed.

An example of the more flooded areas from Vojvodina, Yugoslavia, has a tree layer containing

White willow	*Salix alba*
Crack willow	*Salix fragilis*
Purple willow	*Salix purpurea*
Almond willow	*Salix triandra*
Osier	*Salix viminalis*
Black poplar	*Populus nigra*
Fluttering elm	*Ulmus laevis*
Narrow-leaved ash	*Fraxinus angustifolia.*

The shrub layer includes

| Dewberry | *Rubus caesius* |
| | *Crataegus nigra* |

| Alder buckthorn | *Frangula alnus* |
| Guelder-rose | *Viburnum opulus.* |

The more distinctive species of the field layer are

Nettle	*Urtica dioica*
Birthwort	*Aristolochia clematitis*
Bittersweet	*Solanum dulcamara*
Berry catchfly	*Cucubalus baccifer*
Creeping spearwort	*Ranunculus reptans*
Creeping yellow-cress	*Rorippa sylvestris*
Marsh spurge	*Euphorbia palustris*
Marsh woundwort	*Stachys palustris*
Creeping-jenny	*Lysimachia nummularia*
Common marsh-bedstraw	*Galium palustre*
Greater plantain	*Plantago major*
Skullcap	*Scutellaria galericulata*
Trifid bur-marigold	*Bidens tripartita*
Water-plantain	*Alisma plantago-aquatica*
Yellow iris	*Iris pseudacorus*

and the climbers, common grape-vine, *Vitis vinifera;* and hop, *Humulus lupulus.*

Other examples of wet woodlands in the Pannonic region contain the black poplar, *Populus nigra;* and white poplar, *Populus alba,* mixed with the white willow, *Salix alba;* and purple willow, *Salix purpurea.* The small-leaved elm, *Ulmus minor;* the fluttering elm, *Ulmus laevis;* the narrow-leaved ash, *Fraxinus angustifolia;* and the pedunculate oak, *Quercus robur,* are mixed in the tree layer.

Where there is less flooding, and where the water-table is lower, in localities usually situated further away from the rivers, wet oak–ash woods prevail. These woods are often heavily grazed, or the trees are felled, in which cases they are often replaced by willow–poplar woods.

Oak–ash woods in the Danube delta contain the following trees

White poplar	*Populus alba*
	Quercus pedunculiflora
Pedunculate oak	*Quercus robur*
Wild pear	*Pyrus pyraster*
Narrow-leaved ash	*Fraxinus angustifolia*

and have a shrub layer containing

Grey willow	*Salix cinerea*
Dewberry	*Rubus caesius*
Alder buckthorn	*Frangula alnus*
Dogwood	*Cornus sanguinea*
Guelder-rose	*Viburnum opulus.*

The field layer includes among others: marsh spurge, *Euphorbia palustris;* common comfrey, *Symphytum officinale;* marsh woundwort, *Stachys palustris;* and the climber, the common grape-vine, *Vitis vinifera.*

Steppe-grasslands (Plate 132)

Most of the dry grasslands which persist today in the Pannonic region are semi-natural and are maintained in this state by grazing. It is possible that some of the driest parts have remnants of the original steppe-grassland climax communities, where conditions are too dry for bush or woodlands to develop (Fig. 39). These dry grasslands are dominated by the feather-grasses, *Stipa* species, and occur typically on black-earth soils.

These semi-natural steppe-grasslands are quite widespread, and they probably have a greater variety of grasses and a larger number of distinctive flowering plants, than the original natural grasslands. Most grassland species flower in the spring, in regular monthly cycles, giving a distinctive appearance to the plains at different times of the year. For example in the Steppe Nature Reserves of Fîntînita–Murfatlar in Romania, in March the following will be flowering

Umbellate chickweed	*Holosteum umbellatum*
Common whitlowgrass	*Erophila verna*
	Viola suavis
Fingered speedwell	*Veronica triphyllos*
Meadow gagea	*Gagea pratensis*
	Gagea pusilla.

In April, flowering species will include

Yellow adonis	*Adonis vernalis*
	Adonis volgensis
Celandine	*Ranunculus ficaria*
Narrow-leaved peony	*Paeonia tenuifolia*
Annual androsace	*Androsace maxima*
Herbaceous periwinkle	*Vinca herbacea*
	Centaurea napulifera
	Taraxacum erythrospermum
	Colchicum triphyllum
	Ornithogalum refractum
	Hyacinthella leucophaea.

In early summer the maximum development of the steppe grassland occurs, and among others, the following will be in flower.

Minuartia viscosa
Ranunculus illyricus
Ranunculus oxyspermus

Erysimum diffusum
Astragalus glaucus
Euphorbia nicaeensis
Crepis sancta

and the distinctive grasses, *Stipa lessingiana* and *Festuca rupicola*. In summer the steppe becomes brown with the fruiting stems of feather-grases, umbellifers, composites, etc.; while by the autumn much of the above-ground vegetation dies right down.

The most natural steppe-grasslands are those dominated by the feather-grasses. They occur very scattered throughout the region and their composition depends largely on the grazing regimes.

Steppe communities from the Dobrogea, Romania, which are probably the most typical, have been rapidly destroyed in recent years, leaving only tiny remnants today. These remaining steppe-grasslands contain the following grasses

	Festuca pseudovina
	Festuca valesiaca
Bulbous meadow-grass	*Poa bulbosa*
	Bromus riparius
	Bromus squarrosus
Drooping brome	*Bromus tectorum*
	Agropyron cristatum
	Stipa capillata
	Stipa lessingiana
	Chrysopogon gryllus
	Dichanthium ischaemum.

The non-graminaceous species are numerous, particularly in protected reserve areas where there is little grazing. Conversely, overgrazing will alter the floral composition and destroy many species. Thus the problem of steppe conservation in Nature Reserves is an important one. The following are, among others, present in the protected steppes

Polycnemum arvense
Ceratocarpus arenarius
Herniaria incana
Minuartia viscosa

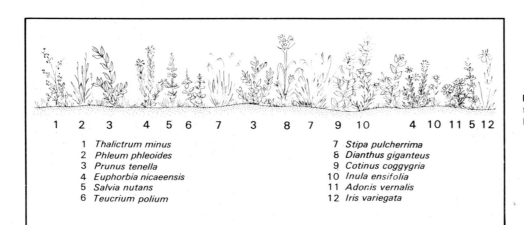

Fig. 39. Steppe grassland. Transect through a primary steppe community in Romania.

1 2 3 4 5 6 7 3 8 7 9 10 4 10 11 5 12

1 *Thalictrum minus*
2 *Phleum phleoides*
3 *Prunus tenella*
4 *Euphorbia nicaeensis*
5 *Salvia nutans*
6 *Teucrium polium*
7 *Stipa pulcherrima*
8 *Dianthus giganteus*
9 *Cotinus coggygria*
10 *Inula ensifolia*
11 *Adonis vernalis*
12 *Iris variegata*

Plate 44. Pannonic grasslands—herbaceous species

1. *Phlomis herba-venti* ssp. *pungens*; **2**. *Adonis volgensis*; **3**. *Androsace maxima*; **4**. *Chamaecytisus austriacus*; **5**. *Salvia nemorosa*; **6**. *Eryngium campestre*; **7**. *Gagea pusilla*; **8**. *Ranunculus illyricus*; **9**. *Ranunculus oxyspermus*; **10**. *Ornithogalum refractum*; **11**. *Vinca herbacea*; **12**. *Bunias orientalis*; **13**. *Astragalus glaucus*; **14**. *Astragalus exscapus*; **15**. *Verbascum phoeniceum*; **16**. *Dianthus pontederae*; **17**. *Paeonia tenuifolia*.

Plate 45. Pannonic grasslands—grasses

1. *Stipa lessingiana*;   **2**. Drooping brome, *Bromus tectorum*;   **3**. *Agropyron cristatum*;   **4**. Grey hair-grass, *Corynephorus canescens*;   **5**. *Stipa joannis*;   **6**. Soft-brome, *Bromus hordeaceus*;   **7**. *Bromus squarrosus*;   **8**. *Koeleria glauca*;   **9**. Bulbous meadow-grass, *Poa bulbosa*;   **10**. *Secale sylvestre*;   **11**. *Festuca valesiaca*;   **12**. *Festuca vaginata*;   **13**. *Agropyron cristatum* ssp. *pectinatum*;   **14**. Common bent, *Agrostis capillaris*;   **15**. *Festuca pseudovina*;   **16**. *Bromus riparius*.

Thyme-leaved sandwort	*Arenaria serpyllifolia*
Yellow adonis	*Adonis vernalis*
	Ceratocephalus sps.
	Alyssum desertorum
	Alyssum hirstum
Hoary cinquefoil	*Potentilla argentea*
Sainfoin milk-vetch	*Astragalus onobrychis*
Star-fruited fenugreek	*Trigonella monspeliaca*
Sickle medick	*Medicago sativa* ssp. *falcata*
Black medick	*Medicago lupulina*
Small medick	*Medicago minima*
	Medicago rigidula
	Euphorbia nicaeensis
	Euphorbia seguierana
Annual androsace	*Androsace maxima*
Felty germander	*Teucrium polium*
Branched horehound	*Marrubium peregrinum*
	Thymus pannonicus
Fingered speedwell	*Veronica triphyllos*
Micropus	*Bombycilaena erecta*
	Achillea setacea
	Artemisia austriaca
Annual cocklebur	*Xeranthemum annuum*
	Centaurea diffusa
Chondrilla	*Chondrilla juncea*
	Crepis tectorum.

Most steppe-grasslands however in the Pannonic region are secondary, and they largely replace the original woodlands. Grazing and fires are the two important factors which help now to maintain these secondary steppes. Though they are unstable, and will develop ultimately into different communities, they have been maintained in this state for long periods by traditional management. *Puszta* is the Hungarian name for these secondary steppes which are essentially a traditionally maintained artificial pasture in the lowlands of the Danube delta.

Secondary steppe occurring in moist localities, particularly in the areas of western Serbia and Bulgaria, is dominated by *Chrysopogon gryllus*, with other grasses such as the sweet vernal-grass, *Anthoxanthum odoratum*, and *Agrostis capillaris*. Other non-grass species include

	Dianthus pontederae
Dropwort	*Filipendula vulgaris*
Hop trefoil	*Trifolium campestre*
Mountain clover	*Trifolium montanum*
	Dorycnium pentaphyllum
Cut-leaved self-heal	*Prunella laciniata*
Ribwort plantain	*Plantago lanceolata*
	Valerianella dentata
Ox-eye daisy	*Leucanthemum vulgare*
Bug orchid	*Orchis coriophora.*

The fescue grass, *Festuca rupicola*, associated with the feather-grass, *Stipa joannis*, forms a community which is widespread in southern and eastern Romania, at lower altitudes, and it usually results from heavy grazing and subsequent erosion. Other grasses include *Agropyron cristatum* ssp *pectinatum*,

Stipa capillata, *Dichanthium ischaemum*, and bulbous meadow-grass, *Poa bulbosa*. Other species include

Dropwort	*Filipendula vulgaris*
Sainfoin milk-vetch	*Astragalus onobrychis*
Sickle medick	*Medicago falcata*
Sainfoin	*Onobrychis viciifolia*
Perennial flax	*Linum perenne*
Squinancywort	*Asperula cynanchica*
Lady's bedstraw	*Galium verum*
	Phlomis herba-venti
	ssp. *pungens*
	Salvia nemorosa
Ribwort plantain	*Plantago lanceolata.*

Wind-created, dry, sandy inland dunes occur in eastern Serbia and between the Danube and the Tisza rivers and also in the Danube delta. They are colonized by a quite distinctive vegetation, which may ultimately develop into lime–oak woods where there is sufficient ground-water. The pioneer communities of these sand dunes are dominated by annual species. This is followed by a brome-grassland dominated by the drooping brome, *Bromus tectorum*, with a rye-grass, *Secale sylvestre*, common, and typically with the bulbous meadow-grass, *Poa bulbosa*, and the soft-brome, *Bromus hordeaceus*. Other typical species are

	Kochia laniflora
Grey cinquefoil	*Potentilla cinerea*
Hare's-foot clover	*Trifolium arvense*
Field eryngo	*Eryngium campestre*
	Achillea ochroleuca
Round-headed club-rush	*Scirpus holoschoenus.*

Dunes which have become more stabilized, for example at Nagykörös, Hungary, have the fescue, *Festuca vaginata*, dominant, with *Koeleria glauca* abundant, and other grasses including

Bulbous meadow-grass	*Poa bulbosa*
	Festuca rupicola
Soft brome	*Bromus hordeaceus*
Drooping brome	*Bromus tectorum*
	Secale sylvestre
Grey hair-grass	*Corynephorus canescens*
	Stipa joannis

with such non-grass species as

Little mouse-ear	*Cerastium semidecandrum*
	Dianthus pontederae
	Dianthus serotinus
Cypress spurge	*Euphorbia cyparissias*
Field eryngo	*Eryngium campestre*
Lady's bedstraw	*Galium verum*
	Onosma arenaria
Perennial yellow wound-wort	*Stachys recta*
	Thymus glabrescens
	Achillea ochroleuca
Field southernwood	*Artemisia campestris.*

Plate 46. Pannonic inland salt communities
1. *Plantago tenuiflora*; **2**. Lesser sea-spurrey, *Spergularia marina*; **3**. Sea plantain, *Plantago maritima*; **4**. *Plantago schwarzenbergiana*; **5**. *Limonium gmelinii*; **6**. Glasswort, *Salicornia europaea*; **7**. *Atriplex tatarica*; **8**. Grass-leaved orache, *Atriplex littoralis*; **9**. *Salsola soda*; **10**. *Trigonella procumbens* (inset×6); **11**. Annual sea-blight, *Suaeda maritima* (inset×8); **12**. Wild chamomile, *Chamomilla recutita*; **13**. *Ranunculus pedatus*; **14**. *Ranunculus lateriflorus* (inset×2); **15**. *Aster sedifolius*; **16**. Oak-leaved goosefoot, *Chenopodium glaucum*.

Plate 47. Pannonic inland salt communities—grasses, sedges, rushes
1. Saltmarsh rush, *Juncus gerardii*; **2**. *Crypsis alopecuroides*; **3**. *Crypsis schoenoides*; **4**. Bermuda-grass, *Cynodon dactylon*; **5**. Creeping bent, *Agrostis stolonifera*; **6**. *Crypsis aculeata*; **7**. Divided sedge, *Carex divisa*; **8**. *Puccinellia festuciformis*; **9**. *Aeluropus littoralis*; **10**. Sea club-rush, *Scirpus maritimus*; **11**. *Beckmannia eruciformis*; **12**. *Koeleria macrantha*; **13**. *Pholiurus pannonicus*; **14**. Sea barley, *Hordeum marinum*; **15**. Reflexed saltmarsh-grass, *Puccinellia distans*; **16**. *Cyperus pannonicus*.

Pannonic vegetation

Rich meadows in the Pannonic region replace river-valley woodlands where moisture and fertility are maintained by regular flooding. These grasslands are restricted to the margins of the region and do not occur on the drier plains. They are similar to the rich meadow communities of Central and Western Europe. The dominant grasses in the wettest parts are tufted hair-grass, *Deschampsia cespitosa;* creeping bent, *Agrostis stolonifera;* and meadow foxtail, *Alopecurus pratensis,* while in the drier grasslands the dominants are smooth meadow-grass, *Poa pratensis;* cock's-foot, *Dactylis glomerata;* and false oat-grass, *Arrhenatherum elatius.*

Inland salt communities (Plates 133–137)

Salt-rich soils have increased considerably in the Pannonic region largely as a result of the regulation of the water levels, with the resulting lowering of the water-table. Where evaporation exceeds precipitation there is a natural accumulation of mineral salts in the surface soils. In the Pannonic region this salt accumulation has increased so that today in southeast Europe and Hungary, salt soils now cover something like one million hectares, and these are located particularly in the old flood plains of the major rivers. Salty soils are known as *szik* soils in Hungary, meaning soda, for the salts are largely carbonates and sulphates and not simply the chlorides of coastal soils. Thus in many cases the vegetation is different, with some quite distinct species, differing from those of the coast, yet other species are common to both salt-rich communities. Some characteristic plants of these continental salt communities are

	Ranunculus lateriflorus
	Ranculus pedatus
	Trigonella procumbens
	Limonium gmelinii
	Plantago cornuti
	Plantago schwarzenbergiana
	Plantago tenuiflora
	Aster sedifolius
Scented mayweed	*Matricaria recutita*
	Taraxacum bessarabicum

and the grasses and sedges

	Puccinellia distans
	Beckmannia eruciformis
	Pholiurus pannonicus
	Crypsis alopecuroides
	Crypsis schoenoides
	Cyperus pannonicus
Divided sedge	*Carex divisa.*

A small number of these species are exclusive to the Pannonic salt communities but, in addition, the following Mediterranean species, including succulent species, occur in the lowlands of the Danube

Sea purslane	*Halimione pedunculata*
	Halimione portulacoides
	Arthrocnemum glaucum
Water germander	*Teucrium scordium*

Tassel pondweed	*Ruppia spiralis*
Sea rush	*Juncus maritimus*
	Puccinellia festuciformis
Sea barley	*Hordeum marinum*
	Aeluropus littoralis
Long-bracted sedge	*Carex extensa.*

The most salt-rich sites (Plates 135–137) are usually dominated by relatively short-lived and frequently succulent species, and they have many similarities with the coastal communities. In Vojvodina, Yugoslavia the glasswort, *Salicornia europaea,* is abundant, with the lesser sea-spurrey, *Spergularia marina,* and the greater sea-spurrey, *Spergularia media,* and with the reflexed saltmarsh-grass, *Puccinellia distans;* and *Crypsis aculeata.* In Hungary where the soil is very rich in soda and may reach an alkalinity of between pH 8.5 and pH 11, the annual sea-blight, *Suaeda maritima;* with reflexed saltmarsh-grass, *Puccinellia distans;* and *Crypsis aculeata* may be abundant. Other species may include

Grass-leaved orache	*Atriplex littoralis*
	Atriplex tatarica
	Camphorosoma annua
Saltwort	*Salsola soda*
Sea aster	*Aster tripolium*
	Crypsis schoenoides
Sea club-rush	*Scirpus maritimus.*

Less salt-rich sites, particularly on damp sands, are dominated by grasses (Plates 133, 134) such as: *Puccinellia festuciformis;* creeping bent, *Agrostis stolonifera; Pholiurus pannonicus;* and *Crypsis schoenoides,* with *Festuca pseudovina* in less salt-rich localities. Other species include

Knotgrass	*Polygonum aviculare*
	Rumex stenophyllus
	Ranunculus lateriflorus
Celery-leaved buttercup	*Ranunculus sceleratus*
	Myosotis stricta
Sea plantain	*Plantago maritima*
Wild chamomile	*Chamomilla recutita*
Round-fruited rush	*Juncus compressus*
	Carex secalina.

On early flooded, slightly sandy soils, a community dominated by the 'eastern' galingale, *Cyperus pannonicus,* contains other salt-tolerant species including

Annual sea-blight	*Suaeda maritima*
Oak-leaved goosefoot	*Chenopodium glaucum*
Red goosefoot	*Chenopodium rubrum*
Orache species	*Atriplex* sp.
Trifid bur-marigold	*Bidens tripartita*
Reflexed saltmarsh-grass	*Puccinellia distans*
	Crypsis aculeata
	Crypsis schoenoides
Sea club-rush	*Scirpus maritimus.*

Some salty meadows, which usually develop under heavy grazing, are dominated by rushes such as the round-fruited rush, *Juncus compressus,* and the saltmarsh rush, *Juncus gerardii.* Commonly occurring with these rushes are the straw-berry clover, *Trifolium fragiferum;* common spike-rush, *Eleocharis palustris;* divided sedge, *Carex divisa;* and the creeping bent, *Agrostis stolonifera.*

Dry salt steppes are intermediate between dry grasslands and steppe. They are especially common in Hungary, but occur elsewhere in the Pannonic region. They are distin-guished by the abundance of the fescue, *Festuca pseudovina,* which is associated with other grasses such as

Perennial rye-grass	*Lolium perenne*
Soft-brome	*Bromus hordeaceus*
Common couch	*Agropyron repens*
Crested hair-grass	*Koeleria macrantha*
Bermuda-grass	*Cynodon dactylon*

Abundant species include

Grey cinquefoil	*Potentilla cinerea*
Star-fruited fenugreek	*Trigonella monspeliaca*
Sickle medick	*Medicago sativa* ssp. *falcata*
	Taraxacum serotinum
	Carex praecox
	Carex stenophylla.

Less common species include

Knotgrass	*Polygonum aviculare*
	Polycnemum arvense
Thyme-leaved sandwort	*Arenaria serpyllifolia*
	Cerastium dubium
Annual knawel	*Scleranthus annuus*
	Dianthus pontederae
Spiny restharrow	*Ononis spinosa*
Black medick	*Medicago lupulina*
Slender trefoil	*Trifolium micranthum*
Knotted clover	*Trifolium retusum*

Soft trefoil	*Trifolium striatum*
Crown vetch	*Coronilla varia*
Cypress spurge	*Euphorbia cyparissias*
Common stork's-bill	*Erodium cicutarium*
Wild thyme	*Thymus serpyllum*
Ribwort plantain	*Plantago lanceolata*
	Achillea collina
Chicory	*Cichorium intybus.*

Some notes on the vegetation of Neusiedlersee area (Seewinkel), Austria

This is an important region which contains many interesting inland salt plant communities. The reason for surface salt accumulation is the same as that mentioned already, but here much of the salt originates in local geological deposits rather than being transported to the area in rivers. As elsewhere, there is little sodium chloride, the important salts being sodium carbonate, bicarbonate, sulphate, as well as sulphates of magnesium and potassium. The surface is covered with pools of varying composition and water level—the smaller of these lakes dry out completely in summer.

Interesting species include *Lepidium cartilagineum* ssp. *crassifolium, Salicornia europaea, Suaeda maritima* ssp. *pannonica* (Plate 136), *Puccinellia festuciformis* ssp. *inter-media.* Where the soil is quite strongly salty but is not subjected to so much drying out, one finds the attractive *Aster tripolium* ssp. *pannonicus.* Less salty and rather moist soils support salt meadows with saltmarsh rush, *Juncus gerardii* and Sea Plantain, *Plantago maritima.*

Most of the damp salty sites used to grade into the formerly widespread *puszta* and, as in the puszta, the characteristic plant communities of these are now rare, preserved particu-larly in certain reserves, for example near to the villages of Illmitz and Apetlon.

12 Alpine plant communities (Plates 48–53; Plates 139–149)

These are considered azonal, for of the plant communities growing in the alpine regions, many are very similar throughout the different climatic zones of Europe. In general alpine plant communities may be defined as those occurring above the tree line. This naturally increases in altitude on progressing further south. For example in central Scandinavia the tree line is at about 1000 m, while in the Balkans it may be at nearly twice this altitude. In many cases man has lowered the natural tree line as a result of clearance, firing, and summer grazing by his flocks.

The main types of vegetation in the Alpine region are the low shrub communities usually found above the upper forests; the mountain grasslands and heaths; the distinctive communities of rocks, cliffs, screes, moraines, and exposed summits; and, locally, the wet flushes and streamsides and snow-melt areas.

The mountains of Europe, particularly in the south, act as refuges for many species of plants which would otherwise have been eliminated during the changes of climate and vegetational development that have taken place since glacial times. As a result of isolation and lack of interbreeding with related species and forms, they have slowly evolved their own distinctive characters. These endemic species which are found today only in a single area or mountain range, often occupy specific niches in the environment, such as mountain cliffs, where they are not in competition with the more vigorous community-forming species. Isolated mountains such as Mount Olympus in Greece have as many as 20 endemic species, which only occur on this mountain. The Sierra Nevada range in southwestern Spain has about 40 species, the exact number depending on different botanists' conception of 'species'.

Bush and tall-herb communities (Plate 140)

Above the tree line, and in mountain environments where trees are unable to grow and develop into woodlands, there are usually either bush communities or closely related tall-herb communities.

The most widespread of the bush communities are those of the dwarf mountain pine, *Pinus mugo* (Fig. 40), which is typically associated with the hairy alpenrose, *Rhododendron hirsutum;* and the spring heath, *Erica carnea.* These communities are found at altitudes ranging from about 1400 to 2000 m, and are characteristic of limestone soils; they have been described in the Central European coniferous section.

The green alder, *Alnus viridis* (Fig. 41), is another bush-forming species which usually replaces the pine on siliceous soils in the moister regions of the Alps, and on other main mountain ranges of southeastern Europe (Map 43). It occurs on steep, shady, north-facing slopes, from about 1200 to 2000 m, and is often associated with a rich assortment of tall herbs, including, for example, in the Alps (Fig. 42)

Monk's rhubarb	*Rumex alpinus*
Red campion	*Silene dioica*
Common monkshood	*Aconitum napellus*
Large white buttercup	*Ranunculus platanifolius*
Round-leaved saxifrage	*Saxifraga rotundifolia*
Lady's-mantle	*Alchemilla vulgaris*
Wood crane's-bill	*Geranium sylvaticum*
Yellow wood-violet	*Viola biflora*
Whorled-leaved willowherb	*Epilobium alpestre*
Hairy chervil	*Chaerophyllum hirsutum*
Masterwort	*Peucedanum ostruthium*
Hogweed	*Heracleum sphondylium*
Large-leaved sneezewort	*Achillea macrophylla*
Adenostyles	*Adenostyles alliariae*
Alpine sow-thistle	*Cicerbita alpina*
Great marsh-thistle	*Carduus personata*
White false-helleborine	*Veratrum album*
Alpine lady-fern	*Athyrium distentifolium.*

Further south, e.g. in Bosnia and Croatia, the tall-herb community contains in addition to the common adenostyles and alpine sow-thistle, many of the species listed from the Alps, with such additional species as

Wolfsbane	*Aconitum vulparia*
Scopolia	*Scopolia carniolica*
Alpine eryngo	*Eryngium alpinum*
Austrian leopard's-bane	*Doronicum austriacum.*

In Scotland, a related community, which does not, however, contain the green alder, occurs at altitudes ranging from 400 to 1000 m in the north and from 800 to 1200 m in the south. It is widespread but fragmentary, occurring on steep crags out of reach of grazing. The following are commonly present

Roseroot	*Rhodiola rosea*
Stone bramble	*Rubus saxatilis*
Water avens	*Geum rivale*
Wood crane's-bill	*Geranium sylvaticum*
Wild angelica	*Angelica sylvestris*

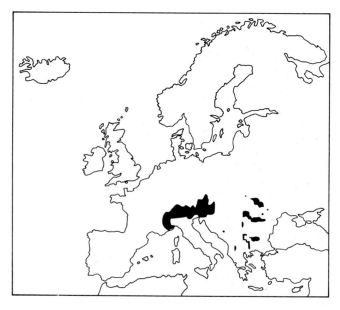

Map 43. Green alder *Alnus viridis.*

Fig. 40. Dwarf mountain pine *Pinus mugo* above the tree-line in the Tatra Mountains, Czechoslovakia.

Hogweed	*Heracleum sphondylium*
Bilberry	*Vaccinium myrtillus*
Alpine saw-wort	*Saussurea alpina*

and the grasses

Sheep's-fescue	*Festuca ovina*
Tufted hair-grass	*Deschampsia cespitosa*

Sweet vernal-grass	*Anthoxanthum odoratum*
Brown bent	*Agrostis canina.*

A number of shrubby willow species also form bushy thickets above the tree-line. In the Alps the widespread species are

Salix appendiculata
Salix breviserrata

Fig. 41. A north-facing slope in the Dolomites, N. Italy at about 2100 m. Green alder *Alnus viridis* bushes grow in the furrows and the Alpenrose *Rhododendron ferrugineum* on the dryer slopes. Spruce, larch, and arolla pine occur lower down.

157

Alpine plant communities

Blue-leaved willow	*Salix caesia*
	Salix glaucosericea
Large-stipuled willow	*Salix hastata*
Swiss willow	*Salix helvetica*
Tea-leaved willow	*Salix phylicifolia*
	Salix waldsteiniana

while in Scandinavia willows such as the following form low bushy thickets

Bluish willow	*Salix glauca*
Woolly willow	*Salix lanata*
Downy willow	*Salix lapponum.*

An example of an alpine willow bush community in southeastern Europe, in Croatia, contains *Salix appendiculata* and *Salix waldsteiniana,* and other shrubby and herbaceous species,

Dwarf juniper	*Juniperus communis* ssp. *nana*
Common columbine	*Aquilegia vulgaris*
Alpine rose	*Rosa pendulina*
Bilberry	*Vaccinium myrtillus*
Willow-leaved gentian	*Gentiana asclepiadea*
White false-helleborine	*Veratrum album*
	Calamagrostis arundinacea.

Alpine heaths (Plate 139)

Low-shrub communities, dominated by ericaceous species, occur widely in the Alpine regions of Europe, and in many ways they are associated with the heaths of lower altitudes described in the Atlantic, Central European, and Boreal regions.

In the Alps, at altitudes ranging from about 1500 to 2500 m, heaths are commonly dominated by the alpenrose, *Rhododendron ferrugineum*, or by the crowberry, *Empetrum nigrum* ssp. *hermaphroditum*, or less commonly by the trailing azalea, *Loiseleuria procumbens*.

In southeastern Europe, another member of the heath family, the spike-heath, *Bruckenthalia spiculifolia,* may become locally dominant in the mountainous regions.

On the higher mountains bordering the Mediterranean, quite distinctive hedge-hog heaths occur at high altitudes, dominated largely by spiny dwarf shrubs and unpalatable herbaceous perennials.

Rhododendron heaths occur on acid podsols, at altitudes between 1500 and 2300 m, in the Alps. They are dominated by the alpenrose, *Rhododendron ferrugineum* (Fig. 43), a small azalea-like shrub, usually less than 1 m high, and usually associated with

Dwarf juniper	*Juniperus communis* ssp. *nana*
Bilberry	*Vaccinium myrtillus*
Bog bilberry	*Vaccinium uliginosum*
Cowberry	*Vaccinium vitis-idaea*
Goldenrod	*Solidago virgaurea*
Alpine coltsfoot	*Homogyne alpina*
	Luzula sieberi
Wavy hair-grass	*Deschampsia flexuosa*
	Calamagrostis villosa

with the mosses, *Hylocomium splendens* and *Pleurozium schreberi*, and the lichens, *Cladonia arbuscula* and *Peltigera aphthosa*.

Crowberry heaths occur in the Alps and other ranges at

Fig. 42. Green alder *Alnus viridis*, with montane tall herb community with *Adenostyles* (right), Common monkshood *Aconitum napellus* (centre), Large white buttercup *Ranunculus platanifolius*, Wood ragwort *Senecio nemorensis*, Lady's-mantle *Alchemilla vulgaris*.

altitudes between about 1600 and 2400 m. Heaths at the higher altitudes in all probability represent the climatic climax vegetation, but crowberry-heaths often occur at lower altitudes largely as a result of forest clearance. The dominant shrublet is *Empetrum nigrum* ssp. *hermaphroditum* growing up to 50 cm in height. It is usually associated with similar species that are present in the rhododendron heath, but commonly with the addition of the alpine clubmoss, *Diphasium alpinum;* and the fir clubmoss, *Huperzia selago,* as well as the alpine bearberry, *Arctostaphylos alpinus.* Also present is the moss, *Rhytidiadelphus triquetrus,* and the lichens, *Cetraria islandica* and the reindeer moss, *Cladonia rangiferina.*

On high windswept localities, from altitudes of about 1750 to 2600 m, a *dwarf azalea–lichen heath* occurs in the Alps, containing (Fig. 44)

Trailing azalea	*Loiseleuria procumbens*
Bilberry	*Vaccinium myrtillus*
Bog bilberry	*Vaccinium uliginosum*
Cowberry	*Vaccinium vitis-idaea*

with the lichens

Cetraria cucullata
Cetraria islandica
Cetraria nivalis
Alectoria ochroleuca
Cladonia gracilis
Cladonia rangiferina
Cladonia arbuscula.

There are some examples of Alpine heath in which the dwarf juniper, *Juniperus communis* ssp. *nana,* is dominant, occurring in the higher regions of the Alps, and in the northern mountains of southwestern Europe and northwestern Scotland.

In southeastern Europe, alpine heaths are less widespread and more limited in extent. Rhododendron, dwarf juniper, and empetrum-heaths occur. However in the southern Carpathians, alpine heaths may be dominated by the spike-heath, *Bruckenthalia spiculifolia,* associated with

Dwarf juniper	*Juniperus communis* ssp. *nana*
Dyer's greenweed	*Genista tinctoria*
Bilberry	*Vaccinium myrtillus*
Bog bilberry	*Vaccinium uliginosum*
The grass	*Bellardiochloa violacea*
Wavy hair-grass	*Deschampsia flexuosa.*

Hedgehog heaths (Plates 142, 143)

The so-called hedgehog heaths, or cushion heaths, of the higher and much drier mountains of the Mediterranean region, have a very distinctive vegetation, which in some ways is more closely associated with similar communities in the mountains of north Africa. They contain a number of interesting endemic species which occur nowhere else in Europe. They include such low cushion-forming shrubs as *Acantholimon, Astragalus, Erinacea, Vella, Bupleurum, Ptilotrichum, Genista,* and *Echinospartum.* Many of these have protective spines on their stems or leaves, while other species, particularly some members of the daisy family, Compositae and the mint family, Labiatae, contain substances which make them unpalatable to grazing animals.

Spanish examples of such alpine hedgehog heaths (though they contain no ericaceous species and thus doubtfully qualify

Fig. 43. Alpine heath with Alpenrose *Rhododendron ferrugineum* and Bog bilberry *Vaccinium uliginosum* and Bilberry *Vaccinium myrtillus.* Dolomites, Italy.

Alpine plant communities

for the term heath) occur at altitudes ranging from about 1600 to 2000 m, on dry southern slopes in the Pyrenees, and in the Sierras of central and southern Spain. They are especially well developed on the Sierra Nevada where the following spiny shrublets are characteristic

	Berberis hispanica
	Ptilotrichum spinosum
	Vella spinosa
Hedgehog broom	*Erinacea anthyllis*
	Bupleurum spinosum.

In addition there are also species of *Astragalus, Genista,* and *Echinospartum,* some of which are spiny. In Italy in the central Apennines, at altitudes ranging from 1500 to 2000 m, the spiny shrublet *Anthyllis hermanniae* is dominant, while on Mount Etna in Sicily it occurs with tussock-forming *Astragalus granatensis* ssp. *siculus,* and the endemic *Viola aethnensis.* In the mountains of Corsica and Sardinia, and particularly in Crete, the low spreading shrub mountain cherry, *Prunus prostrata,* may be locally dominant.

Greek examples of hedgehog heaths for example on the mountains of Parnassus and Giona, at altitudes ranging from about 1700 to 2200 m, have the juniper, *Juniperus communis* ssp. *hemisphaerica;* the tussock-forming *Astragalus creticus* ssp. *rumelicus;* and the blue eryngo, *Eryngium amethystinum;* in addition they contain

	Galium thymifolium
	Marrubium velutinum
	Centaurea affinis
Alpine meadow-grass	*Poa alpina*
	Festuca varia
	Melica ciliata
	Bromus riparius
	Avenula compressa.

An example (c.f. Plate 141) of a similar community on Mount Olympus, Greece, has the abundant and very dense spiny cushion plant *Astragalus angustifolius,* and also commonly contains the box, *Buxus sempervirens; Marrubium velutinum;* and *Centaurea affinis.* Additional species present include

	Daphne oleoides
	Acinos alpinus
	Sideritis scardica
Wall germander	*Teucrium chamaedrys*
	Thymus sibthorpii

and the grasses

	Poa alpina
Alpine meadow-grass	*Festuca varia*
	Bromus riparius.

Snow-patch communities

Late-lying patches of snow, which are slow to melt in the alpine zone, have a considerable effect on the type of plant community. If the snow-free period is more than eight weeks long, dwarf willows may dominate but, if exposure is for shorter periods, mosses are commonly dominant. Snow not only stops growth, but it also protects the plant community from extremes of cold in winter and spring. Snow when it melts, also provides much water on the surface.

In the Alps, on acid soils, such snow-patch communities are commonly dominated by the dwarf willow, *Salix herbacea,* often with the two-flowered sandwort, *Arenaria biflora.* Also commonly present are

Sibbaldia	*Sibbaldia procumbens*
Dwarf snowbell	*Soldanella pusilla*
Alpine speedwell	*Veronica alpina*
Dwarf cudweed	*Omalotheca supina*
Alpine meadow-grass	*Poa alpina*

and the moss, *Polytrichum sexangulare.* Sometimes this snow-patch community is dominated by the wood-rush, *Luzula alpinopilosa.*

On calcareous soils where snow lies for 8 to 9 months, similar snow-patch communities commonly contain

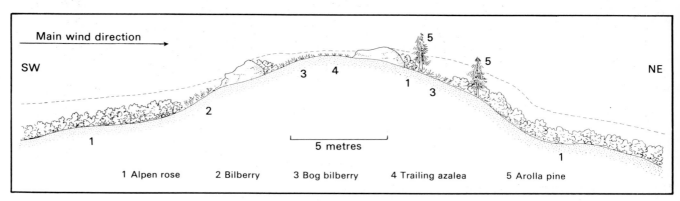

Main wind direction

SW

NE

5 metres

1 Alpen rose   2 Bilberry   3 Bog bilberry   4 Trailing azalea   5 Arolla pine

Fig. 44. Distribution of snow. Wind affects the distribution of snow, which in turn determines the distribution of plant communities. An example from the tree-line in the Alps (Oberengadin). 1. *Rhododendron ferrugineum;* 2. *Vaccinium myrtillus;* 3. Vaccinium uliginosum; 4. *Loiseluria procumbens;* 5. *Pinus cembra.* (Redrawn from Ellenberg 1982.)

Plate 48. Alpine bush, heath, and hedge-hog heath (s=southern species)
1. *Bupleurum spinosum* (s); **2**. *Daphne oleoides* (s); **3**. *Vella spinosa* (s); **4**. Hairy alpenrose, *Rhododendron hirsutum*; **5**. Spike-heath, *Bruckenthalia spiculifolia* (s); **6**. Mountain cherry, *Prunus prostrata* (s); **7**. Green alder, *Alnus viridis*; **8**. Hedgehog broom, *Erinacea anthyllis* (s); **9**. *Anthyllis hermanniae* (s); **10**. *Ptilotrichum spinosum* (s); **11**. *Eryngium amethystinum* (s); **12**. *Acantholimon androsaceum* (s); **13**. *Astragalus creticus* (s); **14**. *Astragalus angustifolius* (s); **15**. *Astragalus granatensis* (s).

Alpine plant communities

Alpine buttercup	*Ranunculus alpestris*
Blue arabis	*Arabis caerulea*
Scree saxifrage	*Saxifraga androsacea*
Alpine speedwell	*Veronica alpina*
	Omalotheca hoppeana
Alpine meadow-grass	*Poa alpina.*

Where the snow lies for shorter periods of up to 7–8 months in the Alps, these communities are dominated by the net-leaved willow, *Salix reticulata,* and the retuse-leaved willow, *Salix retusa.* Usually these willows are accompanied by

Alpine bistort	*Polygonum viviparum*
Alpine buttercup	*Ranunculus alpestris*
Mountain avens	*Dryas octopetala*
Alpine meadow-grass	*Poa alpina*
Black alpine-sedge	*Carex atrata.*

In the Boreal region of Scandinavia, the polar willow, *Salix polaris,* may be dominant, with the arctic mouse-ear, *Cerastium arcticum;* purple saxifrage, *Saxifraga oppositifolia;* and the mosses, *Distichium capillaceum* and *Pohlia commutata.*

Alpine grasslands (Plates 144 and 146)

Alpine grasslands occur throughout the higher mountains of Europe, either as dominant communities above the tree line, or as secondary communities as a result of de-afforestation, shrub clearance, and continued summer grazing.

There are two basic types of alpine grassland depending on the acidity or alkalinity of the bedrock or soil. Those developed on acid soils are often dominated by the mat-grass, *Nardus stricta,* or by sedges such as *Carex curvula.* On calcareous soils the fescues, *Festuca* sps. or different species of sedges are common.

Grasslands on the southern mountains of Europe again have different dominant grasses.

Acid alpine grasslands

Acid alpine grasslands in the mountains of both the north and south of Europe are dominated by the mat-grass, *Nardus stricta.* This is well developed under heavy grazing and in the wetter areas at altitudes ranging from about 1400 to 2400 m, in the Alps. Other important species include

Alpine avens	*Geum montanum*
Golden cinquefoil	*Potentilla aurea*
Tormentil	*Potentilla erecta*
Trumpet gentian	*Gentiana acaulis*
Bearded bellflower	*Campanula barbata*
Arnica	*Arnica montana*
Field wood-rush	*Luzula campestris*
Pale sedge	*Carex pallescens.*

High alpine grassland communities, at altitudes of about 1900–3000 m, on level or slightly sloping acid soils, are often dominated by the alpine sedge, *Carex curvula,* in the Alps. It is replaced in the Tatra mountains of Central Europe by communities of the three-leaved rush, *Juncus trifidus,* and the grass, *Oreochloa disticha,* whilst, in the Balkans, *Festuca paniculata* is important.

High alpine sedge communities dominated by *Carex curvula* (Plate 148), with *Oreochloa disticha,* include among other species

Alpine bistort	*Polygonum viviparum*
Least primrose	*Primula minima*
	Festuca halleri
Alpine meadow-grass	*Poa alpina*
	Agrostis rupestris.

Calcareous alpine grasslands

Calcareous alpine grasslands developed on cold, windy, rocky sites in the high Alps, at altitudes ranging from 2000 to 2800 m, are dominated by the grass, *Kobresia myosuroides.* This same grass also dominates extensive communities in the Arctic region, and thus in the Alps it may be considered as a relict of the once much more widespread tundra communities. Other species occurring in this high calcareous grassland include

Glacier pink	*Dianthus glacialis*
Snowy cinquefoil	*Potentilla nivea*
Unbranched lovage	*Ligusticum mutellinoides*
One-flowered fleabane	*Erigeron uniflorus*
Black alpine-sedge	*Carex atrata.*

On warmer south-facing calcareous slopes, at altitudes ranging from about 1600 to 2400 m, the moor-grass, *Sesleria albicans,* is dominant, with the following species common

Mountain avens	*Dryas octopetala*
Mountain milk-vetch	*Oxytropis jacquinii*
Edelweiss	*Leontopodium alpinum*
	Carex sempervirens.

Two sedges often form communities on calcareous soils. *Carex ferruginea* occurs on fresh soils at altitudes of about 1600–2000 m in the wetter parts of the Alps. With this sedge may occur

Alpine pasque-flower	*Pulsatilla alpina*
Leafy lousewort	*Pedicularis foliosa*
Alpine thistle	*Carduus defloratus*
Mountain hawk's-beard	*Crepis bocconi*
	Festuca violacea
	Phleum hirsutum.

Carex firma (Plate 149, Fig. 45) becomes important on steeper, stony sites, mostly at altitudes between 1700 and 2000 m. It forms hard evergreen cushions which can withstand wind, frost, and occasional drought. With this sedge a loose cushion community is commonly formed with

Alpine bistort	*Polygonum viviparum*
Moss campion	*Silene acaulis*
Blue saxifrage	*Saxifraga caesia*
Mountain avens	*Dryas octopetala*
Alpine rockrose	*Helianthum oelandicum* ssp. *alpestre*
Clusius's gentian	*Gentiana clusii*
False orchid	*Chamorchis alpina.*

In southern Europe, as for example in the mountains of

Plate 49. Alpine grasslands—grasses, sedges, rushes, etc.
1. *Festuca halleri*;   **2**. Mat-grass, *Nardus stricta*;   **3**. *Sesleria albicans*;   **4**. Three-leaved rush, *Juncus trifidus*;   **5**. *Kobresia myosuroides*;   **6**. *Alopecurus gerardii*;   **7**. *Carex curvula*;   **8**. *Phleum hirsutum*;   **9**. *Bellardiochloa violacea*;   **10**. *Oreochloa disticha*;   **11**. *Carex ferruginea*;   **12**. *Carex firma*;   **13**. Alpine meadow-grass, *Poa alpina*;   **14**. *Festuca violacea*;   **15**. *Festuca paniculata*.

Alpine plant communities

southern Greece, grasses such as *Bellardiochloa violacea* and *Alopecurus gerardii* become locally dominant, with

	Ranunculus demissus
	Trifolium parnassi
	Campanula radicosa
	Omalotheca hoppeana
Spring-sedge	*Carex caryophyllea.*

Alpine wet communities

These communities originate from two very different sources and both are local in the alpine region. There are mires, which develop over accumulating peat and which are either acid, neutral, or alkaline, depending on the substrata and on the water source. Secondly, there are the flush-communities which occur where there are springs or mountain streams, or where regular snow-melt water flows. These can also be acid, neutral, or alkaline.

Most *alpine mires* are developed in hollows, often in relation to the moraines of retreating glaciers, or in basins, where spring-water or snow-melt water can collect.

Raised bogs do occur, at altitudes ranging from about 800 to 1600 m, mostly below the treeline. They are probably relicts of bogs from the post-glacial period from about 2000–1000 BC, when they were actively developing and peat was still accumulating. Today the alpine raised-bogs are no longer active, in that under present conditions the growth period in most true alpine sites is not long enough to allow peat to build up.

Poor fens are frequent in the Alps and other mountain ranges. They cover small areas where the water-table is more or less at the surface and have a somewhat acid soil reaction; they are commonly dominated by sedges.

Examples of poor fens at altitudes of 1400 to 2400 m in the Alps commonly contain the common sedge, *Carex nigra,* with the common cottongrass, *Eriophorum angustifolium;* and the thread rush, *Juncus filiformis.* Additional species include

Marsh violet	*Viola palustris*
Deergrass	*Scirpus cespitosus*
White sedge	*Carex curta*
Star sedge	*Carex echinata*
Bog sedge	*Carex limosa.*

Another characteristic community developed over acid peat, where hollows are periodically flooded, or on peat surrounding small lakes or pools on siliceous rocks, is dominated by the cottongrass, *Eriophorum scheuchzeri.* Commonly associated with it is the thread rush, *Juncus filiformis;* the common sedge, *Carex nigra;* and the moss, *Depanocladus exannulatus.* Such mires occur above the tree-line at altitudes ranging from about 1700 to 2300 m in the Alps.

Calcareous fens are widespread and rich in species in the Alps and other mountains of Northern and Central Europe. Further south they become less common, as lower rainfall and higher temperatures do not allow the accumulation of fen peat over porous calcareous rocks. Examples of calcareous fens in the Alps, at altitudes ranging from about 1200 to 2300 m, commonly have the sedge, *Carex frigida,* abundant, with the common butterwort, *Pinguicula vulgaris,* and the tormentil, *Potentilla erecta.* Other species frequently present include

Yellow mountain saxifrage	*Saxifraga aizoides*
Grass-of-Parnassus	*Parnassia palustris*
Lady's-mantle	*Alchemilla vulgaris*
Alpine lovage	*Ligusticum mutellina*
Alpine snowbell	*Soldanella alpina*

Fig. 45. Alpine grassland with the sedge *Carex firma* forming cushions with Edelweiss *Leontopodium alpinum*. Julian Alps.

Plate 50. Alpine wet communities—mires and flushes (c=calcareous soil; a=acid soil)

1. Yellow mountain saxifrage, *Saxifraga aizoides* (c); **2**. Opposite-leaved golden saxifrage, *Chrysosplenium oppositifolium* (a); **3**. Starwort mouse-ear, *Cerastium cerastoides* (a); **4**. Bog stitchwort, *Stellaria alsine* (a) (inset×5); **5**. *Silene pusilla* (c) (inset×2); **6**. Tofield's asphodel, *Tofieldia calyculata* (c) (inset×5); **7**. Alpine snowbell, *Soldanella alpina* (c); **8**. Bladder gentian, *Gentiana utriculosa* (c); **9**. Blinks, *Montia fontana* (a); **10**. Lady's-mantle, *Alchemilla vulgaris* (c); **11**. Chickweed willowherb, *Epilobium alsinifolium* (a); **12**. Bird's-eye primrose, *Primula farinosa* (c); **13**. Marsh felwort, *Swertia perennis* (c); **14**. Soyer's rockcress, *Arabis soyeri* (c); **15**. False aster, *Aster bellidiastrum* (c).

Alpine plant communities

Alpine bartsia	*Bartsia alpina*
False aster	*Aster bellidiastrum*
Calycocorsus	*Calycocorsus stipitatus*
Tofield's asphodel	*Tofieldia calyculata*
Jointed rush	*Juncus articulatus*
Tufted hair-grass	*Deschampsia cespitosa*
Star sedge	*Carex echinata*
Large yellow-sedge	*Carex flava*
Lesser clubmoss	*Selaginella selaginoides.*

Flush communities (Plate 147) are developed where there is a continuous supply of ground water. They seldom freeze, although their temperature hardly ever rises above about 5°C, and they are often free from snow for most of the year. Characteristic species of such flushes are certain mosses, and such hygrophilous flowering plants as saxifrages, butterworts, and willowherbs, for example. There are two distinct communities—those developed in acid flushes, and those which are calcium-rich. Each community shows a remarkable similarity over a wide range of mountains, particularly from the Alps northwards.

In the Alps, at altitudes between about 1000 and 2500 m, and at lower altitudes further north, communities on *acid flushes* contain the following characteristic species

Starry saxifrage	*Saxifraga stellaris*
Nodding willowherb	*Epilobium nutans*

with the mosses, *Bryum schleicheri, Dicranella palustris,* and *Philonotis seriata.*

Also present are

Blinks	*Montia fontana*
Bog stitchwort	*Stellaria alsine*
Large bittercress	*Cardamine amara*
Opposite-leaved golden saxifrage	*Chrysosplenium oppositifolium*
Chickweed willowherb	*Epilobium alsinifolium*
Common sedge	*Carex nigra.*

In northern Wales, northern England, Scotland and Norway, such acid flush communities are common and the following mosses may be involved

Bryum pseudotriquetrum
Bryum weigelii (occasional in Scotland)
Dicranella palustris
Philonotis fontana
Pohlia albicans
Scapania undulata
Solenostoma cordifolium (liverwort).

Flowering plants characteristic of these northern mountain acid flushes include

Blinks	*Montia fontana*
Bog stitchwort	*Stellaria alsine*
Starwort mouse-ear	*Cerastium cerastioides*
Marsh marigold	*Caltha palustris*

Starry saxifrage	*Saxifraga stellaris*
Opposite-leaved golden saxifrage	*Chrysosplenium oppositifolium*
Chickweed willowherb	*Epilobium alsinifolium*
Alpine willowherb	*Epilobium anagallidifolium*
	Epilobium hornemannii
Tufted hair-grass	*Deschampsia cespitosa*
Creeping bent	*Agrostis stolonifera.*

Calcareous flushes have different communities of plants, and there are many similarities between such communities in the Alps, in Scotland, and Scandinavia. Water which drains from highly calcareous rocks usually with a pH of over 7, commonly is associated with the dominant moss, *Cratoneuron commutatum.* It is a distinctive species with its golden-green colour often grading into a rich orange-brown; it often forms chalky deposits, or tufa, within its cushions. Calcareous flush communities in the Alps occur between 750 and 2500 m, and are widespread particularly at lower levels; in upland Britain they occur at altitudes ranging from 300 to 900 m. The yellow mountain saxifrage, *Saxifraga aizoides,* and the chickweed willowherb, *Epilobium alsinifolium,* are common to both these mountainous regions. In the Alps *Silene pusilla* and Soyer's rockcress, *Arabis soyeri,* may also occur. In Scotland the cuckooflower, *Cardamine pratensis,* and the red fescue, *Festuca rubra,* are commonly present.

Alpine communities on rocks, cliffs, and screes
(Plate 149)

These natural rock gardens have a very distinctive and surprisingly stable composition, despite the instability of the terrain. The conditions under which the plants grow are often extreme, with heavy exposure to frost and wind and very variable moisture conditions. Plant growth is often restricted to fissures and ledges in cliffs, or to the more consolidated patches of the smaller-stoned screes, and in sandy moraines. Because of their exposed positions high in the mountains, snow may not lie for long periods, but elsewhere, in sheltered screes and moraines, late snow patches may lie for many months. In consequence of such extremes of growing conditions, many species are dwarf cushion-forming perennials, with wintering buds at or below the surface. Many others are herbaceous perennials, which die right down in the cold weather. Mosses and lichens are also characteristic, and it is interesting to record that in the colonization of these rocks and screes, it is the flowering plants which colonize such environments first, followed by the mosses and lichens. This is in direct contrast to the succession first of lichens, then bryophytes, and then higher plants, which colonize newly exposed rocks and sands in most lowland regions. The vegetation of the alpine rocks and screes probably represents the climax vegetation, or they are at least communities of long duration; the extreme climate is the primary factor preventing further development into more complex communities.

Rock and cliff communities growing on siliceous rocks are more frequent in Southern Europe. In the montane zone the following plants are characteristic

Orpine	*Sedum telephium*
Black spleenwort	*Asplenium adiantum-nigrum*
Forked spleenwort	*Asplenium septentrionale*
Maidenhair spleenwort	*Asplenium trichomanes*
Oblong woodsia	*Woodsia ilvensis.*

In the high Alpine zone, for example in the central Alps, the following are typical

	Minuartia cherlerioides
White musky-saxifrage	*Saxifraga exarata*
	Androsace vandellii
King of the Alps	*Eritrichium nanum*
Dwarf rampion	*Phyteuma humile*
Yellow genipi	*Artemisia umbelliformis*
	Hieracium intybaceum.

In southeastern Europe, in Bulgaria and Yugoslavia, *Silene lerchenfeldiana* occurs in similar communities.

Plant communities on calcareous rocks are more frequent and richer in species. Examples from the Alps, on steep rocks at altitudes ranging from about 1800 to 3000 m, commonly contain

	Draba tomentosa
Kernera	*Kernera saxatilis*
Auricula	*Primula auricula*
Swiss rock-jasmine	*Androsace helvetica*
	Festuca alpina.

In southeastern Europe these calcareous rock communities are often rich in endemics—that is species which have evolved in isolation, or are relict species from the last ice ages, which are now only found in restricted localities.

On Mount Olympus, Greece (Plate 145) there are something like 12 endemic species found only on the mountain at the alpine level above 2400 m, and perhaps 70 species which also only occur on other mountains in Greece or southeastern Europe. Rock-crevice plants found in the summit area of Mount Olympus include such attractive species as *Viola delphinantha*, *Omphalodes luciliae*, and *Campanula oreadum*, as well as several saxifrages including *Saxifraga spruneri*, *S. scardica*, *S. sempervivum*, *S. grisebachii*, and *S. moschata*. Other distinctive rock-crevice species on Olympus are

Arenaria cretica
Arabis bryoides
Aubrieta gracilis
Draba athoa
Aethionema saxatile
Potentilla deorum
Edraianthus graminifolius
Festuca olympica.

On Mount Vermion in northern Greece the following are present on cliffs and rocks

Vernal sandwort	*Minuartia verna*
	Saxifraga grisebachii
	Ramonda nathaliae
	Campanula formanekiana
	Jurinea consanguinea

and the widely distributed ferns

Wall rue	*Asplenium ruta-muraria*
Maidenhair spleenwort	*Asplenium trichomanes*
Rusty-back fern	*Ceterach officinarum.*

The Sierra Nevada in southern Spain, which is the highest mountain range in the southwest of Europe, has about 40 alpine species which are endemic to Spain, most of them occurring on the exposed screes round the summits. The limited limestone outcrop on the Sierra Nevada has the endemics

Helianthemum pannosum
Scabiosa pulsatilloides
Santolina elegans.

Mountain screes and gravel slopes, which are siliceous and result from the disintegration of rocks from glacier moraines or from alpine river deposits, have a distinctive flora. These are often pioneered by the mountain sorrel, *Oxyria digyna*. Commonly associated with it are

Glacier crowfoot	*Ranunculus glacialis*
Mignonette-leaved bittercress	*Cardamine resedifolia*
Alpine rockcress	*Arabis alpina*
Creeping avens	*Geum reptans.*

In the Alps, at higher alpine levels up to 3400 m, the alpine rock-jasmine, *Androsace alpina*, is characteristic. It occurs with

Mossy cyphel	*Minuartia sedoides*
Moss campion	*Silene acaulis*
Glacier crowfoot	*Ranunculus glacialis*
Mignonette-leaved bittercress	*Cardamine resedifolia*
Mossy saxifrage	*Saxifraga bryoides*
Purple saxifrage	*Saxifraga oppositifolia*
Bavarian gentian	*Gentiana bavarica*
Alpine moon-daisy	*Leucanthemopsis alpina*
	Luzula spicata
	Poa laxa
	Oreochloa disticha.

In the Sierra Nevada, above 2500 m, on the dark slaty screes about the summit, the following alpine species grow

Arenaria tetraquetra
Dianthus subacaulis ssp. *brachyanthus*
Ranunculus demissus
Ptilotrichum purpureum
Biscutella glacialis
Lotus glareosus
Anthyllis vulneraria ssp. *atlantis*
Linaria aeruginea
Linaria glacialis
Sideritis glacialis
Plantago nivalis

Rosemary Wise

Plate 51. Alpine rocks, screes in southeast Europe
1. *Draba athoa*;   **2**. *Arenaria cretica*;   **3**. Burnt candytuft, *Aethionema saxatile*;   **4**. *Viola delphinantha*;   **5**. *Jurinea consanguinea*;   **6**. *Omphalodes luciliae*;   **7**. *Minuartia verna*;   **8**. *Drypis spinosa*;   **9**. *Saxifraga sempervivum*;   **10**. *Saxifraga grisebachii*;   **11**. *Campanula formanekiana*;   **12**. *Achillea atrata*;   **13**. *Potentilla deorum*;   **14**. *Arabis bryoides (inset×2)*.

Plate 52. Alpine rocks, screes, cliffs—mostly acidic

1. Creeping avens, *Geum reptans*;   **2**. Oblong woodsia, *Woodsia ilvensis*;   **3**. Mignonette-leaved bittercress, *Cardamine resedifolia*;   **4**. Alpine moon-daisy, *Leucanthemopsis alpina*;   **5**. Dwarf rampion, *Phyteuma humile*;   **6**. Bavarian gentian, *Gentiana bavarica*;   **7**. *Androsace vandellii*;   **8**. *Artemisia umbelliformis*;   **9**. King-of-the-Alps, *Eritrichium nanum*;   **10**. Pyramidal saxifrage, *Saxifraga cotyledon*;   **11**. Mossy saxifrage, *Saxifraga bryoides*;   **12**. White mossy saxifrage, *Saxifraga exarata*;   **13**. Parsley fern, *Cryptogramma crispa*;   **14**. Mountain sorrel, *Oxyria digyna* (inset×3);   **15**. Orpine, *Sedum telephium*;   **16**. *Silene lerchenfeldiana*.

Rosemary Wise.

Plate 53. Alpine rocks, screes in southwest Europe
1. *Sideritis glacialis*;   **2**. *Lotus glareosus*;   **3**. *Dianthus subacaulis*;   **4**. *Senecio boissieri*;   **5**. *Scabiosa pulsatilloides*;   **6**. *Plantago nivalis*;   **7**. *Ptilotrichum purpureum*;   **8**. *Linaria aeruginea*;   **9**. *Arenaria tetraquetra*;   **10**. *Biscutella glacialis*;   **11**. *Ranunculus demissus*;   **12**. *Erigeron frigidus*;   **13**. *Santolina elegans*.

Erigeron frigidus
Senecio boissieri.

Of the above species, nine are endemic to Spain and three restricted to the Sierra Nevada.

On gravels, usually associated with alpine streams, pioneer communities of the following species often occur in the Alps

	Epilobium dodonaei
Alpine willowherb	*Epilobium fleischeri*
French figwort	*Scrophularia canina*
	Tolpis staticifolia
	Chondrilla chondrilloides
	Calamagrostis
	pseudophragmites.

On calcareous screes in the Alps and other ranges, a widely distributed community, occurring up to about 2700 m, is dominated by the French sorrel, *Rumex scutatus*, and the following usually occur with it

Bladder campion	*Silene vulgaris*
White stonecrop	*Sedum album*
Herb Robert	*Geranium robertianum*
Large pink hemp-nettle	*Galeopsis ladanum.*

On damp calcareous screes at about the treeline the alpine butterbur, *Petasites paradoxus*, is conspicuous, particularly where there is snow-melt water. With it commonly occur

Bladder campion	*Silene vulgaris*

Fairy thimble	*Campanula cochleariifolia*
	Adenostyles alpina
Tolpis	*Tolpis staticifolia.*

On limestone or dolomite scree in the calcareous Alps between about 1800 and 2600 m, a community with round-leaved pennycress, *Thlaspi rotundifolium* (Fig. 46) makes an attractive sight a few weeks after the snow has melted. Other species include *Saxifraga aphylla; Papaver sendtneri;* creeping sandwort, *Moehringia ciliata;* and alpine toadflax, *Linaria alpina.*

On the calcareous screes of the mountain ranges of southeastern Europe, the following are characteristic, for example in the northern Pindus ranges of Greece

	Silene caesia
	Silene multicaulis
Bladder campion	*Silene vulgaris*
	Drypis spinosa
	Ranunculus brevifolius
	Euphorbia deflexa
	Euphorbia herniariifolia
	Huetia cynapioides
	Sclerochorton junceum
	Asperula purpurea
	Nepeta sibthorpii
Harebell	*Campanula rotundifolia*
	Sesleria nitida.

Fig. 46. Alpine scree at about 2500 m with Round-leaved Pennycress *Thlaspi rotundifolium*. Dolomites, Italy.

Alpine plant communities

Further south in Greece, in the mountains of Taygetos and Killini, *Minuartia juniperina* is characteristic of the scree community, with *Galium incanum;* aethionema, *Aethionema saxatile; Drypis spinosa;* and *Euphorbia herniariifolia.* Other characteristic species of the high screes of the mountains of the Greek peninsula, at altitudes above 2300 m, include such attractive plants as

Minuartia stellata
Cerastium candidissimum

Corydalis bulbosa ssp. *blanda*
Erysimum pusillum
Thlaspi graecum
Astragalus angustifolius
Astragalus hellenicus
Daphne oleoides
Acantholimon androsaceum
Rindera graeca
Veronica thessalica
Veronica thymifolia.

13 Fresh-water wetland communities   (Plates 54–57; Plates 150–159)

Like the alpine and coastal vegetation, the fresh-water wetland communities show a similar composition and structure throughout Europe, and they are therefore described in this azonal section.

The term wetland is not precise. Water levels can change at different seasons, but in general those communities growing in soils which are more or less permanently flooded are described here. In contrast, mires, which are described in their different phytogeographical regions, may be flooded only for relatively short periods, but usually occur above the water level most of the year.

The main types of wetlands that will be described are rivers, both fast and slow; still waters such as lakes, ponds, and ditches; swamps which are permanently flooded or saturated with water; and marshes which may be partially flooded for long periods.

Rivers

These have a continuous flow of water which consequently has a profound effect on the type of vegetation that is developed, not only in the water, but on the surface, and on plants with aerial parts growing above the surface. Rivers vary in the rate and strength of the current—from rushing mountain torrents, with water speeds of up to 4 feet per second or more, to sluggish lowland rivers with a flow of perhaps less than 5 inches per second. Not only the speed of the current, but the source of the water, whether acid, neutral, or alkaline, and the rocks or substrata over which the rivers flow, have a very marked impact on the vegetation that develops. Plant communities growing in fast or slow currents, or growing in acid or alkaline waters are composed of quite different species. Other important physical variables are the width and depth of the channels, the amount of light reaching and penetrating the water, and the amount of dissolved oxygen in the water.

The submerged and floating vegetation of relatively fast eutrophic rivers—that is to say rivers with abundant nutrients—may contain the following species (Fig. 47)

River water-crowfoot	*Ranunculus fluitans*
Intermediate water-starwort	*Callitriche hamulata*
Lodden pondweed	*Potamogeton nodosus*
Fennel pondweed	*Potamogeton pectinatus*
Opposite-leaved pondweed	*Groenlandia densa.*

Fringe species, growing on the margins of these fast mineral-rich rivers or in shallow runs, commonly include some or all of the following species

	Nasturtium microphyllum
Water-cress	*Nasturtium officinale*
Lesser water-parsnip	*Berula erecta*
Fool's water-cress	*Apium nodiflorum*
Water forget-me-not	*Myosotis scorpioides*
Water mint	*Mentha aquatica*
Monkeyflower	*Mimulus guttatus*
Blue water-speedwell	*Veronica anagallis-aquatica*
Brooklime	*Veronica beccabunga*
Pink water-speedwell	*Veronica catenata*
Arrowhead	*Sagittaria sagittifolia*

Unbranched bur-reed	*Sparganium emersum*
Branched bur-reed	*Sparganium erectum.*

Verges of fast rivers which are subjected to regular flooding, as for example in the Alps and foothills, where river-gravels and sands accumulate, are colonized by a shrub vegetation composed usually of willow and less commonly of the German tamarisk, *Myricaria germanica* (Fig. 48). These colonizing willows include the violet willow, *Salix daphnoides;* hoary willow, *Salix elaeagnos;* and the purple willow, *Salix purpurea* (Fig. 49). Associated herbaceous species include *Tolpis staticifolia,* and such grasses as the creeping bent, *Agrostis stolonifera;* and *Calamagrostis pseudophragmites.* In fast-flowing oligotrophic rivers and streams which are deficient in minerals, arising from or passing over resistant rocks or acid sandstones, the following largely submerged species are commonly present

Common water-crowfoot	*Ranunculus aquatilis*
Pond water-crowfoot	*Ranunculus peltatus*
Alternate water-milfoil	*Myriophyllum alterniflorum*
Intermediate water-starwort	*Callitriche hamulata.*

On the margins of these rivers, or in shallow waters, the jointed rush, *Juncus articulatus;* soft rush, *Juncus effusus;* hemlock

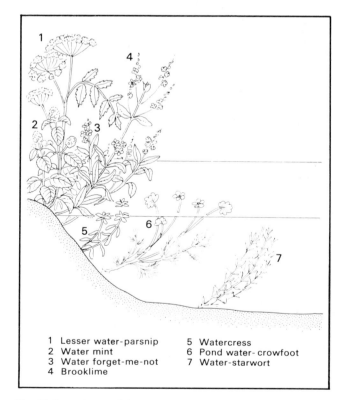

1	Lesser water-parsnip	5	Watercress
2	Water mint	6	Pond water-crowfoot
3	Water forget-me-not	7	Water-starwort
4	Brooklime		

Fig. 47. Some characteristic species. 1. *Berula erecta*; 2. *Mentha aquatica*; 3. *Mysotis scorpioides*; 4. *Veronica beccabunga*; 5. *Rorripa nasturtium-aquaticum*; 6. *Ranunculus peltatus*; 7. *Callitriche* sp. (Redrawn from Haslam 1978.)

Plate 54. Wetlands—rivers
1. Water-cress, *Nasturtium officinale*;   **2**. Amphibious bistort, *Polygonum amphibium*;   **3**. Brooklime, *Veronica beccabunga*;   **4**. Blue water-speedwell, *Veronica anagallis-aquatica* (leaf);   **5**. Arrowhead, *Sagittaria sagittifolia*;   **6**. Flowering rush, *Butomus umbellatus*;   **7**. Perfoliate pondweed, *Potamogeton perfoliatus*;   **8**. Broad-leaved pondweed, *Potamogeton natans*;   **9**. Narrow-leaved water-plantain, *Alisma lanceolatum*;   **10**. Branched bur-reed, *Sparganium erectum*;   **11**. Spiked water-milfoil, *Myriophyllum spicatum*;   **12**. Great yellow-cress, *Rorippa amphibia*;   **13**. Fool's water-cress, *Apium nodiflorum*;   **14**. Water-plantain, *Alisma plantago-aquatica* (leaf);   **15**. River water-crowfoot, *Ranunculus fluitans*;   **16**. Pond water-crowfoot, *Ranunculus peltatus*;   **17**. Fine-leaved water-dropwort, *Oenanthe aquatica*;   **18**. Water mint, *Mentha aquatica*.

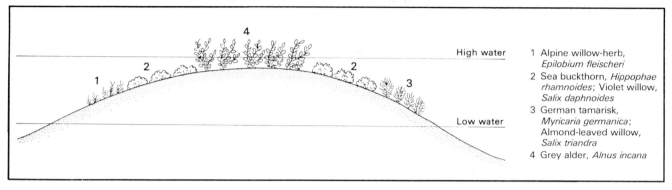

High water

1 Alpine willow-herb,
  Epilobium fleischeri
2 Sea buckthorn, *Hippophae*
  rhamnoides; Violet willow,
  Salix daphnoides
3 German tamarisk,
  Myricaria germanica;
  Almond-leaved willow,
  Salix triandra

Low water

4 Grey alder, *Alnus incana*

Fig. 48. Interesting 'de-Alpine' species become established on unstable gravel. (Redrawn from Wilmanns 1973.)

water-dropwort, *Oenanthe crocata;* branched bur-reed, *Sparganium erectum;* and others are commonly present.

In slow-running rivers and streams, different plant communities occur. Some of the most distinctive and dominant species include the reed canary-grass, *Phalaris arundinacea;* plicate sweet-grass, *Glyceria plicata;* and the reed sweet-grass, *Glyceria maxima.* In the slowest rivers and dykes, the common reed, *Phragmites australis,* and the bulrush, *Typha latifolia,* or less commonly the lesser bulrush, *Typha angustifolia,* are present (Fig. 50). Submerged or floating species include

Fennel pondweed	*Potamogeton pectinatus*
Yellow water-lily	*Nuphar lutea*
Unbranched bur-reed	*Sparganium emersum.*

Other emergent species like the arrowhead, *Sagittaria sagittifolia;* common club-rush, *Scirpus lacustris;* and water-plantain,

Alisma plantago-aquatica, may have narrower submerged or floating leaves and broader emergent leaves.

Other emergent or verge species bordering slow rivers commonly include

Amphibious bistort	*Polygonum amphibium*
Creeping buttercup	*Ranunculus repens*
Creeping yellow-cress	*Rorippa sylvestris*
Lesser water-parsnip	*Berula erecta*
Water mint	*Mentha aquatica*
Brooklime	*Veronica beccabunga*
Flowering-rush	*Butomus umbellatus*
Yellow iris	*Iris pseudacorus*
Swamp meadow-grass	*Poa palustris*
Whorl-grass	*Catabrosa aquatica*
Creeping bent	*Agrostis stolonifera*
Branched bur-reed	*Sparganium erectum.*

Fig. 49. River gravels colonized by willows on the margin of the Isar river, Bavaria, W. Germany.

Fresh-water wetland plant communites

In southern Europe in the Mediterranean region, slow rivers, streams, dykes, etc. may have a different assortment of species, though many of the aquatic species are very widespread. One example is the great reed, *Arundo donax*, which often occurs mixed with the common reed, *Phragmites australis*. It colonizes shallow slow-moving waters and marshes in the southern parts of the Mediterranean region. Although it is not a native species, it has been widely planted and is now commonly naturalized. Commonly associated with these reeds are

Purple loosestrife	*Lythum salicaria*
Water germander	*Teucrium scordium* ssp. *scordioides*
Water mint	*Mentha aquatica*
Horse mint	*Mentha longifolia*
Blue water-speedwell	*Veronica anagallis-aquatica*
Narrow-leaved water-plantain	*Alisma lanceolatum*
Water-plantain	*Alisma plantago-aquatica*
Flowering rush	*Butomus umbellatus*
Whorl-grass	*Catabrosa aquatica*
Great fen-sedge	*Cladium mariscus*
Branched bur-reed	*Sparganium erectum*
	Typha domingensis
Slender club-rush	*Scirpus cernuus*

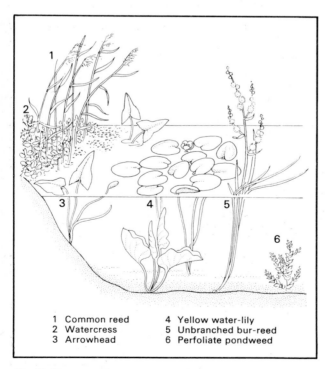

1 Common reed   4 Yellow water-lily
2 Watercress   5 Unbranched bur-reed
3 Arrowhead   6 Perfoliate pondweed

Fig. 50. Slow water. Some species typical of slow eutrophic silt streams. 1. *Phragmites australis*; 2. *Nasturtium officinale*; 3. *Sagittaria sagittifolia*; 4. *Nuphar lutea*; 5. *Sparganium emersum*; 6. *Potamogeton perfoliatus*. (Redrawn from Haslam 1978.)

Round-headed club-rush	*Scirpus holoschoenus*
Common spike-rush	*Eleocharis palustris*.

The following are commonly found submerged in the water

Rigid hornwort	*Ceratophyllum demersum*
Common water-crowfoot	*Ranunculus aquatilis*
Spiked water-milfoil	*Myriophyllum spicatum*
Water-starwort sps.	*Callitriche* sps.
Broad-leaved pondweed	*Potamogeton natans*
Loddon pondweed	*Potamogeton nodusus*
Fennel pondweed	*Potamogeton pectinatus*
Bog pondweed	*Potamogeton polygonifolius*
Lesser pondweed	*Potamogeton pusillus*
Horned pondweed	*Zannichellia palustris*.

Surface-floating plants are the common duckweed, *Lemna minor*, and fat duckweed, *Lemna gibba*. Less common in the south are the galingale species which are dominant in slow rivers and ditches in the lowlands. Such species as *Cyperus flavescens; Cyperus michelianus;* the brown galingale, *Cyperus fuscus;* and the common galingale, *Cyperus longus*, are found in southeastern Yugoslavia with such assorted species as

Jersey cudweed	*Gnaphalium luteo-album*
	Crypsis alopecuroides
Cockspur	*Echinochloa crus-galli*
	Fimbristylis bisumbellata.

Still waters   Plates 150–159)

The vegetation of still waters—of lakes, ponds, pools, ditches, and dykes—show many similarities of composition throughout Europe and though there are some different species which occur only in the north or south of Europe, many more are widespread. Again the composition of the communities is determined by the mineral content of the waters, the acidity and alkalinity, the oxygen content, and the amount of light penetrating the waters. In addition, the extent and the depth of water is important. Different species are adapted to different conditions (Fig. 51): some grow completely submerged; others are rooted in the soil and often have submerged as well as broad floating leaves; others again float freely on the surface; while many grow in shallower waters with emergent leaves and flowering shoots. These latter show all the stages of transition towards swamp and marsh vegetation.

Only selected communities of open waters are listed below. Free-floating surface communities are largely composed of duckweed species; the following are widespread in still, sheltered waters of ponds and ditches

Fat duckweed	*Lemna gibba*
Common duckweed	*Lemna minor*
Ivy-leaved duckweed	*Lemna trisulca*
Greater duckweed	*Spirodela polyrhiza*.

Less common is the smallest of all flowering plants, the rootless duckweed, *Wolffia arrhiza*. In Central and Southern Europe the floating ferns, *Salvinia natans, Azolla filiculoides*, and *Azolla caroliniana* are found—the latter two naturalized from America.

In deep eutrophic water in depths of up to 7 m the water is largely dominated by the pondweeds (Plate 152), such as the shining pondweed, *Potamogeton lucens*, which may become dominant in unpolluted waters. Other submerged species commonly include

Rigid hornwort	*Ceratophyllum demersum*
Spiked water-milfoil	*Myriophyllum spicatum*
Canadian waterweed	*Elodea canadensis*
Broad-leaved pondweed	*Potamogeton natans*
Perfoliate pondweed	*Potamogeton perfoliatus*

and the stonewort, *Chara fragilis.*

In northern and alpine regions, in deep clear and cool waters, on sand or mud, the following pondweeds may be abundant

Red Pondweed	*Potamogeton alpinus*
Slender-leaved pondweed	*Potamogeton filiformis*
Long-stalked pondweed	*Potamogeton praelongus*

with the stonewort, *Chara hispida.*

The common reed, *Phragmites australis,* is also common in waters up to 50 cm in depth. The water violet, *Hottonia palustris,* may dominate in some still or gently flowing eutrophic waters. This latter species is commonly associated with water-starwort, *Callitriche palustris;* common duckweed, *Lemna minor;* and the water-plantain, *Alisma plantago-aquatica.*

Mesotrophic waters, for example in Central Europe, have the frogbit, *Hydrocharis morsus-ranae,* and the water-soldier, *Stratiotes aloides,* abundant, commonly in association with

Yellow water-lily	*Nuphar lutea*
Canadian waterweed	*Elodea canadensis*
Broad-leaved pondweed	*Potamogeton natans*
Common duckweed	*Lemna minor*
Water horsetail	*Equisetum fluviatile.*

In pools, in the Danube delta, the following species are present

Abundant species

Water chesnut	*Trapa natans*

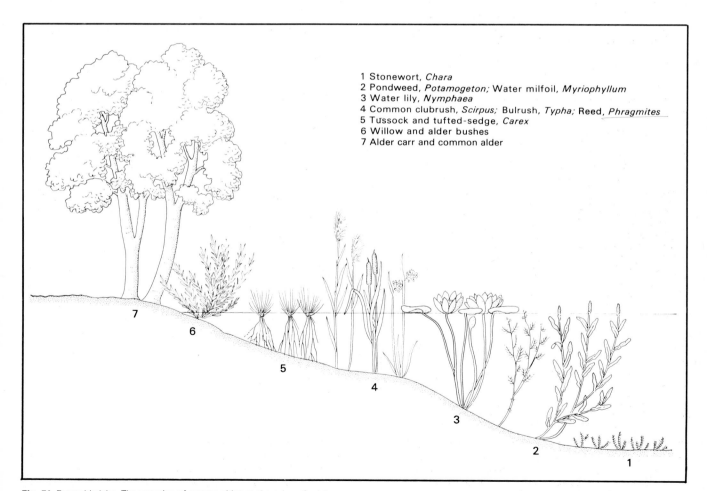

1 Stonewort, *Chara*
2 Pondweed, *Potamogeton;* Water milfoil, *Myriophyllum*
3 Water lily, *Nymphaea*
4 Common clubrush, *Scirpus;* Bulrush, *Typha;* Reed, *Phragmites*
5 Tussock and tufted-sedge, *Carex*
6 Willow and alder bushes
7 Alder carr and common alder

Fig. 51. Eutrophic lake. The zonation of communities at the edge of a lake.

Plate 55. Wetlands—Lakes, pools, and ditches

1. Canadian waterweed, *Elodea canadensis*;   **2**. *Salvinia natans* (×3);   **3**. Common water-starwort, *Callitriche palustris*;   **4**. Shining pondweed, *Potamogeton lucens*;   **5**. Slender-leaved pondweed, *Potamogeton filiformis*;   **6**. Water chestnut, *Trapa natans*;   **7**. Water-violet, *Hottonia palustris*;   **8**. Ivy-leaved duckweed, *Lemna trisulca*;   **9**. Greater duckweed, *Spirodela polyrhiza*;   **10**. Shoreweed, *Littorella uniflora*;   **11**. Quillwort, *Isoetes lacustris*;   **12**. Water fern, *Azolla filiculoides*;   **13**. Common water-crowfoot, *Ranunculus aquatilis*;   **14**. Water lobelia, *Lobelia dortmanna*;   **15**. Rigid hornwort, *Ceratophyllum demersum*;   **16**. Fringed water-lily, *Nymphoides peltata*;   **17**. Water-soldier, *Stratoites aloides*.

Frogbit	*Hydrocharis morsus-ranae*
Water-soldier	*Stratiotes aloides*
	Salvinia natans

Less common species

White water-lily	*Nymphaea alba*
Rigid hornwort	*Ceratophyllum demersum*
Whorled water-milfoil	*Myriophyllum verticillatum*
Cowbane	*Cicuta virosa*
The bladderwort	*Utricularia neglecta*
Arrowhead	*Sagittaria sagittifolia*
Shining pondweed	*Potamogeton lucens*
Ivy-leaved duckweed	*Lemna trisulca.*

A local insectivorous submerged plant, *Aldrovanda vesiculosa*, related to the sundews, may also be found in the above community.

Still-water communities dominated by the water-lilies are very distinctive and widespread in Europe (Plates, 150, 153). The commonest type is dominated by the white water-lily, *Nymphaea alba*, often with the yellow water-lily, *Nuphar lutea*, and least water-lily, *Nuphar pumila*, usually in sheltered eutrophic waters, with the following characteristic species

Spiked water-milfoil	*Myriophyllum spicatum*
Whorled water-milfoil	*Myriophyllum verticillatum*
Amphibious bistort	*Polygonum amphibium*
Broad-leaved pondweed	*Potamogeton natans.*

A less common community of ditches and dykes (Plate 151) is dominated by the fringed water-lily, *Nymphoides peltata;* the following often occur with it

Amphibious bistort	*Polygonum amphibium*
Yellow water-lily	*Nuphar lutea*
Rigid hornwort	*Ceratophyllum demersum*
Shining pondweed	*Potamogeton lucens*
Common duckweed	*Lemna minor.*

The water chestnut, *Trapa natans* (Plate 159) is another floating species which may become locally dominant in the shallow waters of lake verges and ditches, mostly in south-eastern Europe.

Aquatic communities dominated by the carnivorous bladderworts are local and limited. They occur in shallow fen pools, and in transitional mires. The widespread species in Northern and Central Europe are the intermediate bladderwort, *Utricularia intermedia;* lesser bladderwort, *Utricularia minor;* and the greater bladderwort, *Utricularia vulgaris;* while *Utricularia australis* is commoner in southern Europe.

Aquatic communities characteristic mainly of northwestern Europe are distinguished by the presence of two distinctive northern species, the water lobelia, *Lobelia dortmanna* (Plate 154) and the shoreweed, *Littorella uniflora.* Both grow in shallow water in oligotrophic lakes or ponds with a sandy substratum, from just above the water surface to about 4 m below. Commonly associated with these species are

Creeping spearwort	*Ranunculus reptans*
Awlwort	*Subularia aquatica*
Bulbous rush	*Juncus bulbosus*

Common spike-rush	*Eleocharis palustris*
Quillwort	*Isoetes lacustris.*

Swamp or marsh communities (Plates 151 and 155–157)

These are usually developed along the margins of open water, bordering rivers, lakes, or pools. They are closely related to mires, and they show all stages of transition from one to the other. Swamps also often occur as a transitional stage in the development of mires; as soil accumulates and the water-table falls below the surface of the soil, the vegetation of mires becomes dominant. Swamp vegetation exists on permanently or seasonally submerged soils, and it is typically composed of herbaceous plants with emergent leaves and inflorescences, and with perennating buds lying in the mud below the water surface for much of the year. In the summer the water-table may vary in depth from as much as 1 m or more above the soil surface, to a few centimetres below.

One of the most characteristic and widespread plants of periodically flooded soils, particularly in the western part of Europe, is the toad rush, *Juncus bufonius.* With it occur many mosses and liverworts which reach their fullest development in late summer and autumn. Examples in southeastern England have the following species associated with them

Amphibious bistort	*Polygonum amphibium*
Grass poly	*Lythrum hyssopifolia*
Scentless mayweed	*Matricaria perforata*

with the following mosses and liverworts

Bryum klinggraefii
Dicranella varia
Phascum cuspidatum
Pottia starkeana
Riccia cavernosa
Riccia glauca.

German examples of the toad rush, *Juncus bufonius*, community may be associated with the following

Water-pepper	*Polygonum hydropiper*
Procumbent pearlwort	*Sagina procumbens*
Chaffweed	*Anagallis minima*
Greater plantain	*Plantago major*
Marsh cudweed	*Filaginella uliginosa*

and the mosses, *Anthoceros punctatus* and *Anthoceros laevis.* In Vojvodina, Yugoslavia, similar toad rush communities include

Pale persicaria	*Polygonum lapathifolium*
	Potentilla supina
Grass poly	*Lythrum hyssopifolia*
Water purslane	*Lythrum portula*
	Lythrum tribracteatum
Greater plantain	*Plantago major*
Marsh cudweed	*Filaginella uliginosa*
Jersey cudweed	*Gnaphalium luteo-album*
Trifid bur-marigold	*Bidens tripartita*
Cockspur	*Echinochloa crus-galli*
Brown galingale	*Cyperus fuscus.*

Plate 56. Wetlands—swamps and marshes

1. Trifid bur-marigold, *Bidens tripartita*; **2**. *Potentilla supina*; **3**. Chaffweed, *Angallis minima* (inset×5); **4**. Ivy-leaved crowfoot, *Ranunculus hederaceus*; **5**. Nodding bur-marigold, *Bidens cernua*; **6**. Greater plantain, *Plantago major*; **7**. Grass-poly, *Lythrum hyssopifolia* (inset×5); **8**. Marsh cudweed, *Filaginella uliginosa*; **9**. Creeping yellow-cress, *Rorippa sylvestris*; **10**. *Riccia glauca* (×1½); **11**. Greater spearwort, *Ranunculus lingua*; **12**. Water-pepper, *Polygonum hydropiper* (inset×2); **13**. *Galium elongatum*; **14**. Creeping-jenny, *Lysimachia nummularia*; **15**. Creeping buttercup, *Ranunculus repens* (inset×2).

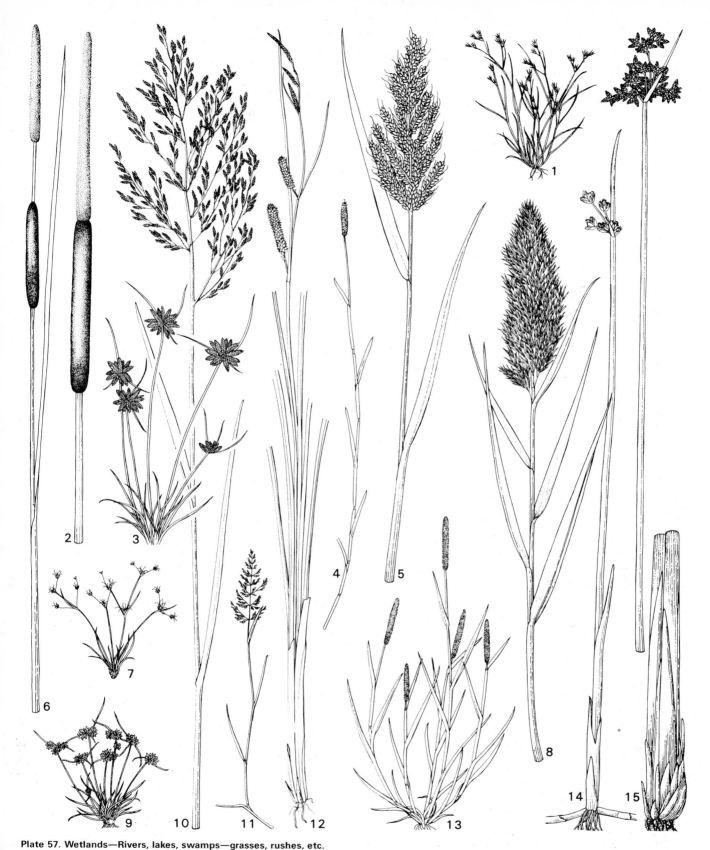

Plate 57. Wetlands—Rivers, lakes, swamps—grasses, rushes, etc.
1. Toad rush, *Juncus bufonius*; **2**. Bulrush, *Typha latifolia*; **3**. *Cyperus flavescens*; **4**. Marsh foxtail, *Alopecurus geniculatus*; **5**. Cockspur, *Echinochloa crus-galli*; **6**. Lesser bulrush, *Typha angustifolia*; **7**. Bulbous rush, *Juncus bulbosus*; **8**. Common reed, *Phragmites australis*; **9**. Brown galingale, *Cyperus fuscus*; **10**. Reed sweet-grass, *Glyceria maxima*; **11**. Whorl-grass, *Catabrosa aquatica*; **12**. Bottle sedge, *Carex rostrata*; **13**. Orange foxtail, *Alopecurus aequalis*; **14**. Triangular club-rush, *Scirpus triqueter*; **15**. Common club-rush, *Scirpus lacustris*.

Fresh-water wetland plant communites

In nitrogen-rich mud, on river banks and around cattle-drinking areas, other herbaceous plants such as the following are commonly present

Water-pepper	*Polygonum hydropiper*
Small water-pepper	*Polygonum minus*
Tasteless water-pepper	*Polygonum mite*
Great yellow-cress	*Rorippa amphibia*
Nodding bur-marigold	*Bidens cernua*
	Bidens connata
	Bidens radiata
Trifid bur-marigold	*Bidens tripartita*
Whorl-grass	*Catabrosa aquatica*
Orange foxtail	*Alopecurus aequalis.*

A common grass dominant of wet ground in the flood meadows of older rivers is the marsh foxtail, *Alopecurus geniculatus.* With it are often associated

Curled dock	*Rumex crispus*
Creeping buttercup	*Ranunculus repens*
Creeping yellow-cress	*Rorippa sylvestris*
Silverweed	*Potentilla anserina*
Creeping-jenny	*Lysimachia nummularia.*

Swamp communities (Plate 155) dominated by larger emergent species are widespread in Europe; owing to their large size many are very distinctive and easily recognized. However, many are also common to mires and there is no clear cut distinction between swamp and mire communities. The commonest dominants include the common reed, *Phragmites australis;* the reed canary-grass, *Phalaris arundinacea;* the common club-rush, *Scirpus lacustris;* and less commonly the bulrushes, *Typha latifolia* and *Typha angustifolia.* A number of sedge species may also dominate, such as the bottle sedge, *Carex rostrata;* the tufted sedge, *Carex elata;* and others; also the reed sweet-grass, *Glyceria maxima.* A typical community (Plate 156) with abundant *Scirpus* and *Phragmites* has the following species

Greater spearwort	*Ranunculus lingua*
Common reed	*Phragmites australis*
Branched bur-reed	*Sparganium erectum*
Lesser bulrush	*Typha angustifolia*
Common bulrush	*Typha latifolia*
Common club-rush	*Scirpus lacustris*
Water horsetail	*Equisetum fluviatile.*

In the Danube delta region, the common reed, *Phragmites australis,* is the dominant species, and it can form floating rafts of vegetation (*plaur*) which may drift under the influence of the wind and currents. These rafts may be as much as 2 m below the surface of the water, while the accumulated soil may rise up to 4 cm above the water level.

Other plants associated with these rafts are

Eared willow	*Salix aurita*
Cowbane	*Cicuta virosa*
Hemp agrimony	*Eupatorium cannabinum*
Greater tussock-sedge	*Carex paniculata*
Cyprus sedge	*Carex pseudocyperus*
Greater pond-sedge	*Carex riparia*
Marsh fern	*Thelypteris palustris.*

Another widespread emergent rush growing in reed-swamps and shallow lakes in the common club-rush, *Scirpus lacustris,* which is often associated with the bulbous rush, *Juncus bulbosus,* and the least bur-reed, *Sparganium minimum,* in the north; in southeastern Europe, in the Danube delta for example, the common club-rush occurs with

Sea rush	*Juncus maritimus*
Common reed	*Phragmites australis*
Branched bur-reed	*Sparganium erectum*
Bulrush	*Typha latifolia*
Sea club-rush	*Scirpus maritimus*
Triangular club-rush	*Scirpus triqueter.*

Sedge communities are commonly found in still mesotrophic waters, that is waters with a moderate mineral content, on mud or sand, in depths of up to 35 cm of water, to about 20 cm above water-level. In the northern and central montane regions of Europe, the bottle sedge, *Carex rostrata,* is common or dominant, associated with such species as

White water-lily	*Nymphaea alba*
Marsh cinquefoil	*Potentilla palustris*
Bogbean	*Menyanthes trifoliata*
Broad-leaved pondweed	*Potamogeton natans*
Soft rush	*Juncus effusus*
Water horsetail	*Equisetum fluviatile.*

The tufted-sedge, *Cares elata,* is another species that covers large areas of the lower river valleys in southern and southeastern Europe. An example from Yugoslavia has the following plants associated with this sedge

Purple loosestrife	*Lythrum salicaria*
Marsh pennywort	*Hydrocotyle vulgaris*
Yellow loosestrife	*Lysimachia vulgaris*
Water mint	*Mentha aquatica*
	Galium elongatum
Water-plantain	*Alisma plantago-aquatica*
Yellow iris	*Iris pseudacorus*
Common reed	*Phragmites australis*
Lesser bulrush	*Typha angustifolia*
Common spike-rush	*Eleocharis palustris*
Galingale	*Cyperus longus.*

14 Coastal plant communities (Plates 58–60; Plates 160–170)

The vegetation of the coasts of Europe is very distinct. It is composed largely of species which can tolerate salinity and which are widespread on the coasts, but which otherwise are not found inland unless there is an accumulation of salt. Certain genera have become adapted to live in such conditions of high salt content and some of them, like the saltworts, sea-purslane, and sea sandworts, have succulent fleshy leaves.

The main types of coastal communities are those found on mudflats and in saltmarshes; those in brackish estuaries and brackish pools and ponds; communities of dunes and shingle shores; communities of coastal rocks and cliffs. Within each of these above environments there are many similarities over the whole of Europe, both in species and structure of the community; for example, the saltmarshes of the northern Atlantic coasts may have a considerable number of species in common with salt marshes of the Mediterranean region, and the shingle beaches of the Atlantic coasts may share some species with shingle beaches of the Black Sea, though naturally the presence of northern or southern species may give a somewhat different aspect to these communities. Not only does the environment help to maintain these similarities, but maritime transport of seeds, fruits, and vegetative shoots, by tides and currents, brings about a wider dispersal of these coastal plants, as also does the movement of the coastal populations of sea birds.

Mudflats and saltmarshes

The most distinctive feature of communities growing on mudflats and saltmarshes is the zonation of the vegetation. Thus the species which are continuously submerged, or partially submerged, or occasionally submerged, differ as one passes inland. Relatively few species can survive these extreme conditions of salinity and exposure, but a few species have become adapted to them, and can flourish in large numbers over wide areas of the European coasts.

A characteristic zone structure is shown in the diagram (Fig. 52). Very few flowering plants can survive under continuous submersion by sea water. In Europe, the plants which do are eelgrasses, *Zostera marina* and *Zostera noltii*, which are widespread in the North Sea and the Baltic, occurring where there is a muddy substrate which may be exposed for several hours during the lowest tides. In the Mediterranean however they grow in water up to 1–10 m deep, on mud which is rarely exposed above water. In these communities only marine algae are associated with the eelgrasses. In the Mediterranean another unusual and distinctive flowering plant is found commonly growing on sand, often in deeper water than the former species. It is *Posidonia oceanica*, which produces fibres from the bases of old leaves, which during unsettled weather are often washed ashore in large numbers as compact balls on some of the sandy beaches of the Mediterranean. It also occurs less commonly along the Atlantic coast of southwestern Europe.

The first true land plants which are found on coastal mudflats and saltmarshes are the glassworts, *Salicornia* species (Plate 162), which dominate wide expanses of exposed mud to the exclusion of other flowering plants. They are low fleshy plants, often growing in close proximity, and thus tending to trap mud and silt round their stems. In consequence they very slowly bring about the raising of the level of the soil, and ultimately make way for the establishment of the next community—that dominated by the saltmarsh-grasses, *Puccinellia* species.

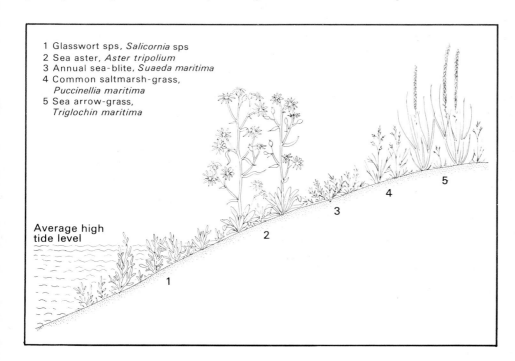

1 Glasswort sps, *Salicornia* sps
2 Sea aster, *Aster tripolium*
3 Annual sea-blite, *Suaeda maritima*
4 Common saltmarsh-grass, *Puccinellia maritima*
5 Sea arrow-grass, *Triglochin maritima*

Average high tide level

Fig. 52. Saltmarsh. Typical zonation of saltmarsh plants in relation to average high tide level. (Redrawn from Ellenberg 1982.)

Coastal plant communities

There are several species of glassworts, but they are all very similar and difficult to identify, even for the expert. They are annual plants, and some of them turn pink in the autumn giving the mudflats a distinctive colour.

Closely related dwarf shrubby species are the perennial glassworts, *Arthrocnemum perenne* (Plate 161), *A. fruticosum*, and *A. glaucum*, which are characteristic of the coastal mudflats of Southern and Western Europe. On the coasts of northwestern Europe another plant, Townsend's cord-grass, *Spartina x townsendii* (Plate 160) has, since about 1927, become a well established and very active colonizer of partially exposed tidal mudflats and, being a robust loosely clumped species, it also accumulates sediment and builds up the soil level. Once the plants are established they can form a dense sward in about 25 years. They can tolerate high salinity and can spread both by seed and vegetatively from fragmented shoots, but the accumulation of their own detritus is often the limiting factor in the further growth of the community. This cord-grass arose spontaneously, as a hybrid between a North American and a European species. It has been widely planted in many parts of the world to encourage stabilization and binding of coastal mudflats.

The next zoned community above these two partially submerged communities are the grasslands dominated by the common saltmarsh-grass, *Puccinellia maritima*. It covers a wide range of saltmarsh habitats, is tolerant of periodic water-logging and high salinity and, like the cord-grass, produces a high rate of silt accumulation, of up to 5 cm per year in some cases. It largely spreads vegetatively, and in the pioneer stages may have creeping stolons of up to 50 cm long. These grasslands commonly occur in a zone from about 15 cm below to 25 cm above the average high water level. They are often intensively grazed, and consequently form a tight mat-like turf, little more than 1 cm high. Typical species in these grasslands are

Sea-purslane	*Halimione portulacoides*
Annual sea-blite	*Suaeda maritima*
Greater sea-spurry	*Spergularia media*
English scurvygrass	*Cochlearia anglica*
Sea-milkwort	*Glaux maritima*
Common sea-lavender	*Limonium vulgare*
Sea plantain	*Plantago maritima*
Sea aster	*Aster tripolium*
Sea wormwood	*Artemisia maritima*
Sea arrowgrass	*Triglochin maritima.*

Often the red fescue, *Festuca rubra*, occurs with the common saltmarsh-grass at the upper, less flooded levels. Above these grasslands may be a zone of rushes, or rushes may develop at the same level as the grasslands. The saltmarsh rush, *Juncus gerardii*, is commonly dominant. It usually occurs with

Sea-milkwort	*Glaux maritima*
Thrift	*Armeria maritima*
Common sea-lavender	*Limonium vulgare*
Sea plantain	*Plantago maritima*
Sea arrow-grass	*Triglochin maritima*

Red fescue	*Festuca rubra*
Creeping bent	*Agrostis stolonifera.*

This community usually occurs from about 10 to 30 cm above mean high-water level. On light sandy soils, in areas covered by seawater only during storms, the sea wormwood, *Artemisia maritima*, may dominate.

The sea rush, *Juncus maritimus*, is another dominating and characteristic species of this upper zone, occurring further inland from the sea at about 20–40 cm above the mean high-water mark. Species similar to those in the saltmarsh rush community occur in this sea rush community, including the saltmarsh rush, *Juncus gerardii*; in addition, the spear-leaved orache, *Atriplex hastata*, and the wild celery, *Apium graveolens* are often present.

In brackish waters near the sea, in pools and ditches, other species are dominant such as the beaked tasselweed, *Ruppia maritima*; the spiral tasselweed, *Ruppia cirrhosa*; the horned pondweed, *Zannichellia palustris*; and the holly-leaved naiad, *Najas marina*, which is less common. All these species are widespread in Europe and occur in brackish waters throughout. Other flowering plants commonly associated with them are the brackish water-crowfoot, *Ranunculus baudotii*; blunt-fruited water-starwort, *Callitriche obtusangula*; and the fennel pondweed, *Potamogeton pectinatus*.

In estuaries, where there is an increase of fresh water inland, the creeping bent, *Agrostis stolonifera*, becomes increasingly dominant in grazed communities. In the absence of grazing, or cutting, the common reed, *Phragmites australis*, commonly becomes dominant. Neither of these species can tolerate a salinity of much more than 1 per cent.

Sand dune and shingle communities

These are both very distinctive communities. Like other coastal communities throughout Europe, they have a characteristic assemblage of species, many of which are restricted to the different zones of vegetation. In general the environment is difficult; there is the instability of the substrates, lack of mineral nutrients in most sand and gravels, and usually lack of soil moisture owing to excessive drainage, as well as exposure to wave-action and winds. Consequently, the initial establishment of plant communities is often very difficult, or impossible. However the accumulation of tidal litter in strandlines at high water at the top of the foreshore, makes plant establishment possible, both on sand and shingle.

Sand dune communities

On sand, the plant communities which develop on the fore-shore initiate the development of sand dunes inland (Fig. 53). Thus a series of zones of vegetation occurs inland from the foreshore. These communities are commonly termed: the foreshore communities; the primary dunes or fore-dunes; the white dunes; the grey dunes; and furthest inland and the oldest dune community, the brown dunes. The foreshore or strand communities develop above the high-water level, usually where the waves and currents have deposited some detritus. This detritus traps variable quantities of viable seeds, and

Rosemary Wise.

Plate 58. Coastal mudflats, saltmarshes

1. Greater sea-spurrey, *Spergularia media*;   **2**. Sea aster, *Aster tripolium*;   **3**. *Arthrocnemum fruticosum*;   **4**. Holly-leaved naiad, *Najas marina*;   **5**. Common sea-lavender, *Limonium vulgare* (inset×3);   **6**. Beaked tasselweed, *Ruppia maritima*;   **7**. Horned pondweed, *Zannichellia palustris* (inset×5);   **8**. Sea arrowgrass, *Triglochin maritima*;   **9**. English scurvy-grass, *Cochlearia anglica*;   **10**. *Posidonia oceanica*;   **11**. Eelgrass, *Zostera marina*;   **12**. Townsend's cord-grass, *Spartina x townsendii*;   **13**. Sea wormwood, *Artemisia maritima*;   **14**. Sea-purslane, *Halimione portulacoides* (inset×4);   **15**. Sea-milkwort, *Glaux maritima*;   **16**. Common saltmarsh-grass, *Puccinellia maritima*.

Coastal plant communities

supplies some nutrients to the sand, as well as increasing its ability to retain moisture.

Such foreshore communities are open and short-lived; on the northern Atlantic coasts, the following species are characteristic

Knotgrass	*Polygonum aviculare*
Ray's knotgrass	*Polygonum oxyspermum* ssp. *raii*
Sea beet	*Beta vulgaris* ssp. *maritima*
Babington's orache	*Atriplex glabriuscula*
Spear-leaved orache	*Atriplex hastata*
Grass-leaved orache	*Atriplex littoralis*
Common orache	*Atriplex patula*
Prickly saltwort	*Salsola kali*
Scurvygrass sps.	*Cochlearia sps.*
Sea rocket	*Cakile maritima*
Sea-kale	*Crambe maritima*
Sea radish	*Raphanus maritimus*
Oysterplant	*Mertensia maritima.*

In the eastern Mediterranean, in Greece for example, widespread foreshore species are

Prickly saltwort	*Salsola kali*
Sea rocket	*Cakile maritima*
Purple spurge	*Euphorbia peplis*
Spiny cocklebur	*Xanthium spinosum*
Cocklebur	*Xanthium strumarium*
Sand couch	*Elymus farctus*
Marram	*Ammophila arenaria*
Bermuda-grass	*Cynodon dactylon*
	Cyperus rotundus.

Primary dunes or fore-dunes develop as the result of the colonization and establishment of two distinctive and specialized grasses, which by their form of growth are able to flourish in conditions of high salinity, and where wind-blown sand from the foreshore accumulates inland. These grasses are the sand couch, *Elymus farctus* (commonly referred to as *Agropyron junceum* or *Agropyron junciforme*), and the lyme-grass, *Leymus arenarius* (commonly known as *Elymus arenarius*). Both grasses are perennial, and are able to produce long lateral and vertical shoots which can grow through the accumulating sand grains to a depth of up to 30 cm each year. The aerial parts of these grasses hold up the dry sand-grains as they are blown inland from the foreshore and bring about the accumulation of sand to heights of up to 5–7 m in a few years. However their growth is limited, and as these grasses cannot grow through the sand indefinitely the fore-dunes remain consequently relatively low and the dry sand is blown further inland. Species commonly present on these primary dunes in the northern Atlantic region typically include

Knotgrass	*Polygonum aviculare*
Curled dock	*Rumex crispus*
Spear-leaved orache	*Atriplex hastata*
Prickly saltwort	*Salsola kali*
Sea sandwort	*Honkenya peploides*
Sea campion	*Silene vulgaris* ssp. *maritima*
Sea rocket	*Cakile maritima*
Sea mayweed	*Matricaria maritima.*

Comparable communities on the much less widespread primary dunes of the Mediterranean, are commonly dominated by the sand couch, *Elymus farctus* (Plate 163), with the following species

| Prickly saltwort | *Salsola kali* |
| Sea medick | *Medicago marina* |

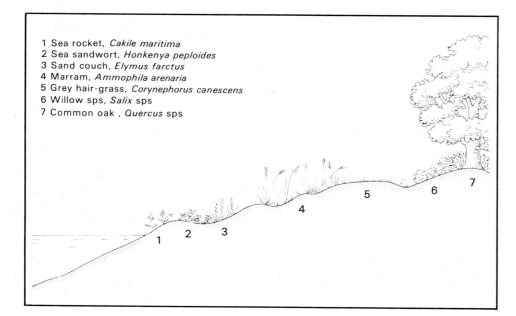

1 Sea rocket, *Cakile maritima*
2 Sea sandwort, *Honkenya peploides*
3 Sand couch, *Elymus farctus*
4 Marram, *Ammophila arenaria*
5 Grey hair-grass, *Corynephorus canescens*
6 Willow sps, *Salix* sps
7 Common oak , *Quercus* sps

Fig. 53. Sand dunes. Typical zonation of sand dune plants. (Redrawn from Wilmanns 1973.)

R. Wise.

Plate 59. Coastal sand dunes, shingle

1. Cocklebur, *Xanthium strumarium*;   **2**. Spiny cocklebur, *Xanthium spinosum*;   **3**. Sea pea *Lathyrus japonicus*;   **4**. Sea sandwort, *Honkenya peploides* (inset×3);   **5**. Sea mayweed, *Matricaria maritima* ssp. *maritima*;   **6**. Sea-holly, *Eryngium maritimum*;   **7**. Sea medick, *Medicago marina*;   **8**. Sea campion, *Silene vulgaris* ssp. *maritima*;   **9**. Sea daffodil, *Pancratium maritimum*;   **10**. Yellow horned-poppy, *Glaucium flavum*;   **11**. Sea bindweed, *Calystegia soldanella*;   **12**. Sea-buckthorn, *Hippophae rhamnoides*;   **13**. Sea spurge, *Euphorbia paralias* (inset×3);   **14**. Purple spurge, *Euphorbia peplis* (inset×3);   **15**. Prickly saltwort, *Salsola kali* (inset×5);   **16**. Sea rocket, *Cakile maritima*;   **17**. Curled dock, *Rumex crispus* (inset×4).

Coastal plant communities

Sea spurge	*Euphorbia paralias*
Purple spurge	*Euphorbia peplis*
Sea-holly	*Eryngium maritimum*
Woolly chamomile	*Anthemis tomentosa*
The grass	*Sporobolus pungens.*

White dunes are the secondary ranks of higher dunes which occur inland from the primary dunes. They are, like the primary dunes, open and still very mobile, so that blow-outs and crests develop as the mobile sand is carried further inland by the inshore winds. The main colonizer and stabilizing plant of the white dunes is the marram grass, *Ammophila arenaria* (Plate 164), which has the almost unique ability of growing vertically and horizontally through many metres of sand, provided the accumulation is not too rapid. This community remains open, that is to say that separate tufts of marram grass occur with areas of loose dry sand between. These sandy areas may be subsequently colonized by other species, including some non-maritime plants. Characteristic of marram grass dunes are stretches of bare sand on the slopes and in the hollows alternating with tufts of marram on the crests. The first species to commence the consolidation of the sandy areas is often the red fescue, *Festuca rubra,* and in relatively sheltered places the sand sedge, *Carex arenaria.* The following plants commonly grow on these white dunes, usually amongst the clumps of marram grass

Sea mouse-ear	*Cerastium diffusum*
Sea pea	*Lathyrus japonicus*
Small-flowered evening primrose	*Oenothera parviflora*
Sea-holly	*Eryngium maritimum*
Sea bindweed	*Calystegia soldanella*
Lyme-grass	*Leymus arenarius*
Grey hair-grass	*Corynephorus canescens.*

In addition non-maritime species such as the following are often present

Hedge bedstraw	*Galium mollugo*
Common ragwort	*Senecio jacobaea*
Perennial sow-thistle	*Sonchus arvensis*
Creeping thistle	*Cirsium arvense*
Spear thistle	*Cirsium vulgare.*

Grey dunes (Plates 165 and 166) are often described as fixed dunes, in contrast to the previous mobile dunes. The sandy dune surfaces gradually become stabilized, at first by the flowering plant cover, and then they are further consolidated by mosses and lichens which form a closed turf. This turf may take 10 to 20 years to develop. For example on the west coasts of England, the most important pioneer moss is *Tortula ruraliformis.* This forms rich moss-carpets over the sheltered sandy areas, while on the east coast, between the clumps of marram grass, the following species develop moss-carpets

Brachythecium albicans
Bryum sps.
Ceratodon purpureus
Hypnum cupressiforme.

These moss-carpets are gradually colonized by lichens which slowly take over and become abundant, thus ultimately turning the dunes grey in appearance, and so giving them their name. *Cladonia furcata* becomes abundant, with *Cladonia foliacea* and *Cetraria aculeata* which are common; these are later followed by *Cladonia arbuscula.* Large circular growths of *Parmelia physodes* and *Parmelia saxatilis* are also conspicuous.

Many species of flowering plants grow on these grey dunes; only a few of the more widespread and attractive species are listed in the examples given below. Grey dunes in eastern Friesland, Germany, typically contain the following species

Creeping willow	*Salix repens*
Silverweed	*Potentilla anserina*
Common bird's-foot-trefoil	*Lotus corniculatus*
Grass-of-Parnassus	*Parnassia palustris*
Fairy flax	*Linum catharticum*
Seaside centaury	*Centaurium littorale*
Autumn gentian	*Gentianella amarella*
Dune gentian	*Gentianella uliginosa*
Red bartsia	*Odontites verna*
Autumn hawkbit	*Leontodon autumnalis*
	Juncus anceps
Saltmarsh rush	*Juncus gerardii*
Creeping bent	*Agrostis stolonifera*
Glaucous sedge	*Carex flacca*
Marsh helleborine	*Epipactis palustris.*

The most characteristic species of the grey dunes in the eastern Mediterranean in the Peloponnese, Greece include

Silene nicaeensis
Euphorbia terracina
Hedypnois cretica

and grasses such as *Bromus rigidus;* hare's-tail, *Lagurus ovatus;* and *Vulpia fasciculata.* Another example, further north on the north Aegean coast typically contains

Joint-pine	*Ephedra distachya*
Striated catchfly	*Silene conica*
Forked catchfly	*Silene dichotoma*
Sea medick	*Medicago marina*
Common fumana	*Fumana procumbens*
Felty germander	*Teucrium polium*
	Verbascum pinnatifidum
	Jasione heldreichii
	Centaurea cuneifolia

with the grasses, *Corynephorus divaricatus* and the sand cat's-tail, *Phleum arenarium.*

Brown dunes are the oldest in the natural succession, and they usually occur furthest inland. Here acid peat accumulates near the surface to a limited extent—hence the colour of the dunes. They are commonly colonized by ericaceous shrublets, such as heather, *Calluna vulgaris;* and crowberry, *Empetrum nigrum;* they are associated with other heathland species and are commonly called dune heaths (Plate 167). In northwestern Germany, Holland, Denmark, etc. these brown dunes support a community dominated by such shrubs as the sea-buckthorn,

Plate 60. Coastal grasses, sedges, rushes, etc.
1. *Cyperus rotundus*; **2**. Sand couch, *Elymus farctus*; **3**. *Bromus rigidus*; **4**. Sand cat's-tail, *Phleum arenarium*; **5**. Round-headed club-rush, *Scirpus holoschoenus*; **6**. *Vulpia fasciculata*; **7**. *Sporobolus pungens*; **8**. *Corynephorus divaricatus*; **9**. Marram, *Ammophila arenaria*; **10**. Lyme-grass, *Leymus arenarius*; **11**. Sand sedge, *Carex arenaria*; **12**. *Juncus anceps*; **13**. Hare's-tail, *Lagurus ovatus*; **14**. Sea rush, *Juncus maritimus*.

Coastal plant communities

Hippophae rhamnoides, and the creeping willow, *Salix repens.* Commonly present with these shrub communities are

Common toadflax	*Linaria vulgaris*
Hedge bedstraw	*Galium mollugo*
Red fescue	*Festuca rubra*
Lyme-grass	*Leymus arenarius*
Marram	*Ammophila arenaria.*

Other shrubs may also occur less commonly on these brown dunes, such as

Juniper	*Juniperus communis*
Blackberry sps.	*Rubus* sps.
Rose sps.	*Rosa* sps.
Hawthorn	*Crataegus monogyna*
Gorse	*Ulex europaeus.*

In the Mediterranean region the prickly juniper, *Juniperus oxycedrus* ssp. *macrocarpa,* often occurs on old dunes. Eventually these dune shrub communities may be colonized by oaks, to form acid oak woods, or further east in Poland by the Scots pine, *Pinus sylvestris.* In the southwestern Atlantic region, the maritime pine, *Pinus pinaster,* colonizes mature dunes and forms forests. It is also widely planted elsewhere to consolidate sand dunes and poor acid soils.

Dune-slacks are the damp hollows left between the dune ridges, and here the ground water may reach the surface, forming permanent or semi-permanent shallow pools which are colonized by pondweeds, *Potamogeton* species, and the horned pondweed, *Zannichellia palustris,* with mosses commonly of *Hypnum* species. Round the margins of these pools usually occur typical species of inland swamps such as the marsh pennywort, *Hydrocotyle vulgaris;* yellow iris, *Iris pseudacorus;* reed sweet-grass, *Glyceria maxima;* reed canary-grass, *Phalaris arundinacea;* soft rush, *Juncus effusus;* and many other species. Elsewhere, where the water table is below the surface in the dune-slacks, dune marshes develop with a typical assortment of inland marsh species. A few maritime or sub-maritime species, such as the sea-milkwort, *Glaux maritima;* sea rush, *Juncus maritimus;* brookweed, *Samolus valerandi;* and the round-headed club-rush, *Scirpus holoschoenus,* may also be present.

Dune meadows may develop on old dune ridges, particularly on sandy calcareous soils which are formed from shell deposits. These develop a rich grassland vegetation which is usually heavily grazed. In western and northern Scotland these pastures are known as *machair,* and the important grasses include the smooth meadow-grass, *Poa pratensis,* and the red fescue, *Festuca rubra.* In the machair a large number of attractive flowering species occur, including such orchids as

Dark-red helleborine	*Epipactis atrorubens*
Common twayblade	*Listera ovata*
Fragrant orchid	*Gymnadenia conopsea*
Frog orchid	*Coeloglossum viride*
Common spotted-orchid	*Dactylorhiza fuchsii*

Northern marsh-orchid	*Dactylorhiza majalis* ssp. *purpurella*
Early-purple orchid	*Orchis mascula.*

Shingle shorelines (Plate 168) are developed where strong onshore currents cause the accumulation of rounded eroded pebbles along the coast. Only above highwater where the shingle has become stabilized are plant communities able to develop. Shingle communities are open and there are no really dominant species by comparison with sand dunes, but a few species are quite characteristic of shingle beaches, notably shrubby sea-blight, *Suaeda vera;* sea sandwort, *Honkenya peploides;* sea campion, *Silene vulgaris* ssp. *maritima;* yellow horned-poppy, *Glaucium flavum;* curled dock, *Rumex crispus;* sea pea, *Lathyrus japonicus;* and bittersweet, *Solanum dulcamara.*

Stony beaches in Greece, for example, have some typical widely distributed coastal species, such as the sea rocket, *Cakile maritima;* prickly saltwort, *Salsola kali;* rock samphire, *Crithmum maritimum;* sea-holly, *Eryngium maritimum;* and the common sea-lavender, *Limonium vulgare.* However more southerly species also occur such as

	Camphorosma monspeliaca
Stalked orache	*Halimione pedunculata*
Saltwort	*Salsola soda*
Shrubby glasswort	*Arthrocnemum fruticosum*
	Arthrocnemum glaucum
Three-horned stock	*Matthiola tricuspidata*
Small-headed blue-eryngo	*Eryngium creticum*
	Limonium cancellatum
Golden samphire	*Inula crithmoides*
	Artemisia caerulescens.

Coastal cliff (Plate 169) communities are locally well developed along the exposed coasts of the Atlantic. Communities of flowering plants occur above the highest tides, but are subjected to sea-spray. Consequently many species are halophytic plants, that is to say salt-tolerant, and are the same as are found in salt marshes and elsewhere on the coasts.

Characteristic species of many western Atlantic cliffs (Plate 170) are

Sea beet	*Beta vulgaris*
Sea campion	*Silene vulgaris* ssp. *maritima*
Hoary stock	*Matthiola incana*
Common scurvy-grass	*Cochlearia officinalis*
Wild cabbage	*Brassica oleracea*
English stonecrop	*Sedum anglicum*
Tree mallow	*Lavatera arborea*
Rock samphire	*Crithmum maritimum*
Fennel	*Foeniculum vulgare*
Thrift	*Armeria maritima*
Common sea-lavender	*Limonium vulgare*
Buck's-horn plantain	*Plantago coronopus*
Sea plantain	*Plantago maritima*

Sea mayweed	*Matricaria maritima* ssp. *maritima*
Crested hair-grass	*Koeleria macrantha.*

Comparable Mediterranean spray-exposed cliffs have, among other species

Biting stonecrop	*Sedum acre*
Rock samphire	*Crithmum maritimum*
	Limonium cancellatum
	Plantago subulata
	Reichardia picroides.

PART III

The National Parks and Nature Reserves of Europe

Europe

This map and the maps overleaf show the distribution of National Parks and Nature Reserves in Europe. The key is arranged alphabetically by country beginning on page 197.

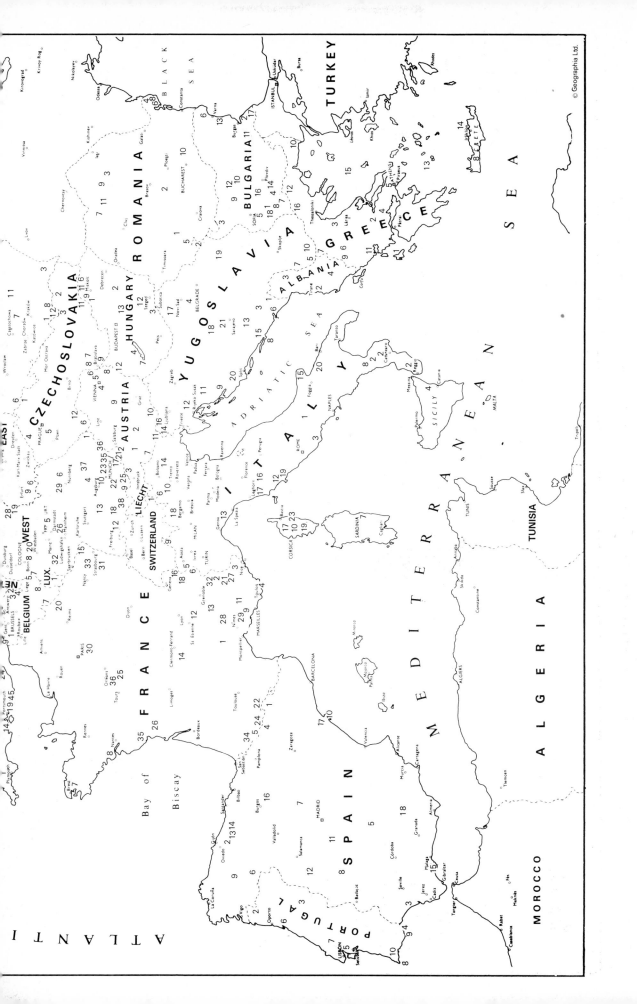

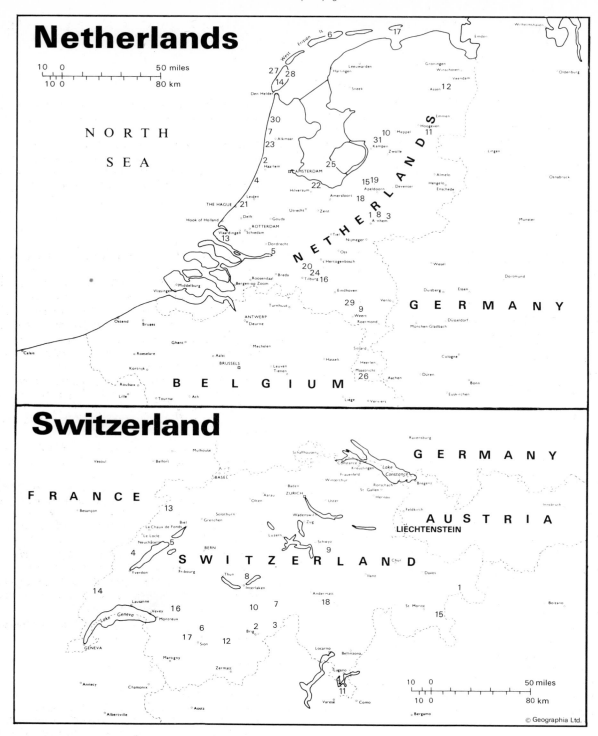

Netherlands

Switzerland

A selection of the most important National Parks and Nature Reserves and other reserves listed for each country (excluding the USSR), with an indication of the most distinctive plant communities present. In general only the larger reserves over 5 sq km have been included, but some exceptions have been made where the plant communities are of particular interest.

The locations are given on the map on pages 194–6.

ALBANIA

National Parks

1. Dajti. Area, 30 km²; altitude, 400–1611 m. Mountain ridge with beech and Balkan pine woods; also maquis with holm oaks.
2. Divjaka. Area, 10 km²; altitude, to 10 m. Dunes on Adriatic shore, with Aleppo and stone pine woods.
3. Lura. Area, 30 km²; altitude, 1000–2246 m. Two mountain ranges with beech and Bosnian, Macedonian, Scots, and black pines with firs. Alpine meadows and many mountain lakes.
4. Tomori. Area, 30 km²; altitude, 800–2400 m. Mountains with beech and Bosnian pine, and alpine meadows.

AUSTRIA

National Parks

1. Hohe Tauern (proposed). Area 2500 km²; altitude, to 3797 m. This includes the 3797 m high Grossglockner and the Krimmel waterfalls. Alpine and sub-alpine vegetation.
2. Niedere Tauern (proposed). Area 750 km²; altitude to 2863 m. Mountains of crystalline rocks and acid soils, with many lakes and waterfalls. Pastures and peat-bogs, and stands of Arolla pine up to an altitude of 2000 m.

Nature Reserves, Nature Protection Areas

3. Karwendel. Area, 720 km²; altitude, 1100–2756 m. Largest Nature Protection area in Austria, with beautiful scenery. Calcareous alps, with alpine pastures, lakes, and streams. Ancient sycamore woodland.
4. Lainzer Tiergarten. Area, 25 km²; altitude, 250–514 m. Nature Protection Area with Central European forest, meadows, and streams.
5. Lobau Nature Reserve. Area, 20 km². On Danube: river-valley woods; reed-swamps, grasslands, and ponds.
6. Marchauen. Area, 11 km². Nature Protection Area with Central European river-valley woods, alder carr, meadows, bogs, ponds.
7. Mussen. Area, 3.8 km². Nature protection Area in the Dolomites with alpine pastures with a rich and interesting flora; also quartz and schist areas with lime-free flora.
8. Neusiedlersee and Seewinkel Nature Reserve and Landscape Protection Area. Area, 500 km²; altitude, 180 m. One of the finest wetlands in Europe. Inland salt communities, lakes, reed beds, steppe-like grasslands or 'puszta'.
9. Rothwald Nature Reserve. Area, 6 km²; altitude, 1000–1500 m. One of the finest and largest virgin forests in Central Europe, with mixed forests of beech, fir, and spruce.
10. Trögener Klamm. Area, 2.5 km²; altitude, c.700 m. Nature Protection Area with a limestone gorge with alpine and sub-alpine plants, mixed with karst woodland species. Relict woods of coniferous and deciduous trees.

BELGIUM

Nature Reserves

1. De Blankaart. Area, 4 km²; altitude, 5 m. Large lake with reed swamps and alder carr.
2. De Mechelse Heide. Area, 3.9 km². Heathland, marsh, and relicts of oak–birch woodlands on sandy hills.
3. De Teut. Area, 2.1 km². Acid soils and peat; heathlands with planted conifers; boggy and dry areas; ponds; copses.
4. Genk. Area, 2.2 km²; altitude, 50 m. Thirty-five small artificial lakes with some interesting water plants. Also heaths, sand dunes, bogs.
5. Hautes Fagnes. Area, 40 km²; altitude, 500–700 m. Situated in the Hautes Fagnes–Eifel Natural Park of 500 km². Semi-natural woods of oak, beech, and birch, also planted spruce. Plateau of grassy heaths and acid bogs in high rainfall area.
6. Kalmthoutse Heide. Area, 8.5 km²; altitude, 20–25 m. Heathlands, sphagnum bogs, marshes, sand dunes, small lakes.
7. Lesse et Lomme. Area, 9.7 km²; altitude, 155–314 m. Dry calcareous grasslands with juniper and cornelian cherry; also rock-faces and caves. Interesting flora of over 600 species. Also oak and beech woods and cultivated land. Most northerly site of white oak.
8. Plateau des Tailles. Area, 2.0 km². Raised bogs, marshes, wet and dry heaths, surrounded by planted spruce.
9. Westhoek. Area, 5.4 km²; altitude, 0–24 m. Dunes and dune-slacks with interesting flora; shore vegetation. Fringing woodlands of oak and alder.
10. Zwin. Area, 1.25 km²; altitude, 0–5 m. Saltmarsh and mud-flat vegetation.

BULGARIA

National and People's Parks

1. Pirin. Area, 265 km²; altitude, 1100–2914 m. High mountain area with 60 peaks over 2600 m with many rivers, lakes, gorges, rocks, alpine grasslands, and snow-beds. Coniferous forests of Macedonian and Bosnian pines and, at lower altitudes, woods with beech, fir, and spruce; also stands of Grecian juniper. It includes three Nature Reserves.
2. Ropotamo. Area, 8.5 km²; altitude, to 30 m. Sand dunes with saline vegetation at the mouth of Ropotamo River as it enters the Black Sea, with marshes and vast meadows with interesting flora. Also old littoral forest of ash and elm with distinctive woody climbing species. It includes four reserves.

3. Roussenskilom. Area, 22 km². Part of the Danubian plain, with a canyon-like valley, rocks, caves, and karst scenery. Deciduous woods of oriental hornbeam; white, Turkey, Hungarian, and sessile oaks; lime and lilac, with much Christ's thorn bush.

4. Steneto National Park and Biosphere Reserve. Area, 39 km²; altitude, 800–1523 m. Ancient coniferous and deciduous woods, with nearly virgin woods of beech and fir; also with hornbeam and hop-hornbeam, spruce, sycamore, elm, and ash. Karst landscape with steep valleys and waterfalls, with caves, precipices, gorges, and rocks. Many shrub species including cherry laurel.

5. Vitocha. Area, 121 km²; altitude, 700–2290 m. Mountainous region with coniferous and deciduous woods of spruce, Scots pine, beech, birch, hornbeam and maple. Mountain pastures, screes, and streams, with a rich spring flora. Over 2700 species of plants recorded. It includes the Bistrishko Branishte Biosphere Reserve, area, 97 km².

6. Zlatni Pyassatsi. Area, 8.3 km²; altitude, 20–277 m. Black Sea climatic region, with forests of Hungarian, Turkey, white, sessile, and pedunculate oaks; hornbeams; silver lime; field maple; smooth-leaved elm; common and manna ash; rowan; white poplar; wild cherry; with a rich shrub layer with many woody climbers. Coastline with sandy beaches.

Nature Reserves and Biosphere Reserves

7. Alibotouch Biosphere Reserve. Area, 5.7 km²; altitude, 1140–2212 m. Karst area with Balkan pine forest, and mixed forests of beech and black pine; also mountain pastures. Over 1400 plant species, including 20 Bulgarian endemics recorded in the reserve.

8. Baevi Doupki Biosphere Reserve. Area, 8.5 km²; altitude 1300–2100 m. Mountainous region with limestone ravines, with varied coniferous woods of Macedonian and Bosnian pines and other pine species, and spruce, showing distinctive altitude zonation. Alpine calcareous meadows rich in species, rocks, and snow-beds.

9. Boatin Nature and Biosphere Reserve. Area, 12 km²; altitude 900–1500 m. Rugged hills with valleys and rivers, with old beech forests covering 90 per cent of the area, with spruce woods higher up.

10. Djendema Biosphere Reserve. Area, 18 km²; altitude 1500–2000 m. Mountainous region with deep gorges with beech and Greek fir woods; also meadows and rocks, with over 100 waterfalls and spectacular landscapes.

11. Dolna Topchia. Area, 5 km²; altitude, c.120 m. River valley, with undisturbed wet woodlands with distinctive woody climbers.

12. Doupkata Nature and Biosphere Reserve. Area, 12 km²; altitude, 1100–1645 m. Narrow gorge of the Devinska river in the Rhodope mountains, with woods of Scots pine, spruce, fir, beech, aspen, and *Quercus dalechampii*; partly influenced by Mediterranean climate.

13. Kamtchia Biosphere Reserve. Area, 5 km²; altitude, 0–4 m. Ancient Black Sea coastal flood-plains with deciduous woods of narrow-leaved ash, *Quercus pedunculiflora*, and elm, and with over 40 species of shrubs including distinctive woody climbers. Lakes and wetland vegetation. The reserve has a rich flora.

14. Koupena Biosphere Reserve. Area, 10 km²; altitude 650–1410 m. Typical Rhodope forest with woods largely of beech and fir, and with 10 other tree species.

15. Mantaritza Biosphere Reserve. Area, 6 km²; altitude, 1400–1900 m. Old forests of spruce, and also mixed beech and fir woods.

16. Maritchini Ezera (Maritza lakes) Nature and Biosphere Reserve. Area, 15 km²; altitude, 1900–2925 m. Mountain lakes in Rila mountains, with old forests of fir, Macedonian pine, and dwarf mountain pine. High alpine meadows and rocks, with several Bulgarian endemic species.

17. Ouzounbodjak (Lopouchuna) Nature and Biosphere Reserve. Area, 24 km²; altitude, 150–200 m. Hilly region of the Strandja mountains, with ancient forests of oriental beech, *Quercus hartwissiana*, *Quercus polycarpa* and Turkey oak, with shrubs of rhododendron, medlar, holly, *Daphne ponticum*, and *Vaccinium arctostaphylos*.

18. Parangalitza Biosphere Reserve. Area, 15 km²; altitude, 1400–2434 m. Primary mixed forests in Rila mountains largely of spruce and Scots pine; also with fir, beech, aspen, and Macedonian pine. Also alpine and sub-alpine grasslands rich in species.

19. Tsaritchina Biosphere Reserve. Area, 11.4 km²; altitude, 1200–1950 m. Forests of beech, fir, Macedonian pine, and spruce.

CZECHOSLOVAKIA

National Parks

1. Krkonose. Area, 380 km²; altitude, to 1603 m. Part of the mountains of Bohemia with extensive forests, alpine grasslands, peat-bogs, and mires; some endemic species.

2. Pieniny. Area, 21 km²; altitude, to 982 m. Limestone region with typical flora and vegetation. Adjoins the Pieniny National Park of Poland.

3. High Tatra. Area, 500 km²; altitude, to 2663 m. Crystalline and calcareous rocks with characteristic alpine landscape with lakes, moraines, and glaciers. Forests of beech, birch, spruce, larch, and Arolla and dwarf mountain pines.

Nature Reserves

4. Adršpašsko Teplické Skály. Area, 18 km². Cretaceous sandstone with canyons, with characteristic vegetation.

5. Karlštejn and Koda. Area, 20 km². Limestone karst terrain with thermophilous steppe vegetation and steppe-woodland flora.

6. Maly a Velky Tisy. Area, 6 km². Two lakes with marshes.

7. Súlovské Skály. Area, 6 km². Calcareous vegetation.

8. Šúr. Area, 5 km². Relict virgin alder wood in low-lying basin.

9. Zlatna na Ostrove. Area, 90 km². Cultivated steppes.

Biosphere Reserves

10. Kivoklatsko. Area, 628 km²; altitude, 221–611 m. Highlands with deep valleys, gorges, and hollows, with natural, semi-natural, and planted deciduous forests, with juniper, yew, silver fir, wild service tree, common whitebeam, cornelian cherry, and *Cotoneaster integerrimus,* etc.

11. Slovak Karst. Area, 361 km²; altitude, 200–924 m. Karstic country with plateaux, deep gorges, also areas of slates and sandstones. Interesting and rich flora (the richest in Central Europe) with over 900 species of vascular plants. Oak woods, marshes, meadows, rock-steppe, cliffs, chasms, and caves.

12. Trebon Basin. Area, 700 km²; altitude 400–500 m. Flat terrain with sandy, clay, and peat soils, half forest-covered, and also including arable land and meadows. Semi-natural and planted woods of Scots pine, silver fir, and spruce. Also river flood-plains with pedunculate oak, limes, elms, and maples, and fen woodland with alder. Peatlands with mountain pine and shrubs. Wetlands, wet meadows, and dry grasslands.

DENMARK

Nature Parks

1. Farum. Area, 50 km². Scattered lakes with marshes, swamps, dense thickets, and meadows.

2. Gudenåens og Skjern Åens Kilder. Area, 6.5 km². Lakes and swamps, springs, meadows, heaths, woodlands, and oak scrub.

3. Hindsholm. Altitude, *c.*36 m. Shrubberies of hazel, hawthorn, and blackthorn, and deciduous woodland on the small Romsø island. In the north is the Fyns Hoved Nature Reserve.

4. Høje Møn. Area, 21 km². A series of parallel chalk cliffs, with sink-holes with fens on isle of Møn.

5. Mols Bjerge. Area, 27.5 km². Sandy heathlands and grasslands, with about 600 species of flowering plants. Rocky hills with deciduous woods; some wet grasslands.

6. Nordbornholm. Area, 25 km². Northern part of the island of Bornholm. Bedrock of gneiss and granite, with rift valleys. Woods, heathlands, and grassy areas; also rock-crevice plants.

7. Rands Fjord. Area, 11 km². At the present day a fresh-water lake with marshes, reedbeds, and meadows.

8. Rømø. Area, 121 km². Offshore island with vast sand dunes and tidal flats, also saltmarshes, reed beds, and coastal meadows. In the south of Rømø is the important Vadehavet Bird Sanctuary.

9. Tystrup–Bavelse Søerne. Area, 37.5 km². Glaciated valley with two lakes, hilly moraines, forests, fens, springs, and surrounding cultivation.

10. Ulvshale og Nyord. Area, 11 km². Flint beaches with beach vegetation; also oak heathlands, meadows, and shallow ponds with reeds.

Nature Reserves

11. Anholt. Area, 8 km². Island vegetation.

12. Draved Skov. Area, 2.3 km². A probably ancient mixed wood of small-leaved lime with ash, oak, elm, and aspen.

13. Esrum sø og Gribskov. Area, 73 km². Large lake rich in aquatic life. Hilly country in west with small lakes and fens, and a large forest of beech and conifers.

14. Flyndersø and Stubbergardsø. Area, 167 km². Heathlands and oak woods.

15. Hansted. Area, 38 km²; altitude, 50 m. Large area of dunes, heathlands, and lakes rich in diatoms.

16. Harrild Hede. Area, 12 km². Heathlands.

17. Hjerl Hede. Area, 9 km². Heathlands.

18. Nekselø. Area, 2.2 km². The remains of a glacial moraine with heathlands, meadows, and small ponds, and with a little deciduous woodland.

19. Nordsamsø og Stavnsfjord. Area, 17 km². Islands with sandy and gravel moraines with areas of grassland with distinctive flora, also cultivated areas, and conifer plantations.

20. Råbjerg. Area, 17 km². Dunes and heathlands.

21. Randbøl Hede. Area, 7.4 km². Sand dunes and heathlands.

22. Selsø–Lindholm–Bognaes. Area, 20 km². Heathlands and meadows.

23. Skagen. Area, 43 km². Dune vegetation, heathlands; northern point of Jutland.

24. Skallingen. Area, 23 km². Sandy beaches and dune vegetation, with marshes and meadows inland.

FINLAND

National Parks

1. Lemmenjoki. Area, 2800 km²; altitude, 123–601 m. Broad plains with mountains, open fell country, valleys with lakes, gorges and canyons, streams and waterfalls. Pine and birch forests; mountain heaths with aapa mires. Northern limit of spruce and pine.

2. Liesjärvi. Area, 6.3 km²; altitude, 108–120 m. Uplands with extensive pine and spruce forests; mires and lake-shore plants.

3. Linnansaari. Area, 38.2 km²; altitude, 76–100 m. Large lake with a group of islands with coves, cliffs, ravines, and glens, with pine forests and deciduous woods of birch, alder, and lime.

4. Oulanka. Area, 202 km²; altitude, 145–380 m. Deep valleys with rock walls and steep banks; rivers with gravel banks, rapids, and waterfalls. Pine heaths, peat-bogs, grasslands with interesting arctic flora.

5. Pallas–Ounastunturi. Area, 500 km²; altitude, 272–807 m. Large areas of mountain plateaux with lakes, peat-bogs, tundra fell-field vegetation. Canyons with streams. Extensive spruce and pine woods, with birch woods above, and with arctic–alpine vegetation on the plateaux.

6. Petkeljärvi. Area, 6.3 km²; altitude, 140–150 m. Eskers and lakes with pine forests, pine-heaths, bogs, and marshy meadows.

7. Pyhähäkki. Area, 13.5 km^2; altitude, 159–193 m. The largest natural forests of spruce and pine. Also mires and small ponds; open peatland areas.

8. Pyhätunturi. Area, 41 km^2; altitude, 200–540 m. Five mountain peaks with poor fell-field vegetation. Below, pine and spruce forests and extensive aapa mires.

9. Rokua. Area, 4.2 km^2; altitude, 130–197 m. Stabilized old coastal dunes and flat sandy heathlands with tarns and mires. Pine woods with distinctive lichens.

Nature Reserves

10. Häädetkeidas. Area, 5.5 km^2; altitude, 149–155 m. Raised bogs.

11. Kevo. Area, 703 km^2; altitude, 74–551 m. Scrub birch woods; some far north pine woods.

12. Malla. Area, 30 km^2; altitude, 463–927 m. Arctic–alpine fell vegetation, with a rich flora on local calcareous outcrops.

13. Maltio. Area, 147 km^2; altitude, 240–400 m. Wooded hills, fells, and peat-bogs.

14. Paljakka. Area, 26.6 km^2; altitude, 300–384 m. Rich spruce forest; mires.

15. Pisavaara. Area, 50 km^2; altitude, 135–220 m. Probably ancient pine and spruce forests; also extensive peat bogs. A rich flora, particularly along streams.

16. Raunkaus. Area, 61 km^2; altitude, 99–160 m. Aapa mires.

17. Salamanpera. Area, 12 km^2; altitude, 180–195 m. Pine woods; mires.

18. Sompio. Area, 181 km^2; altitude, 245–544 m. Pine and spruce forest; aapa mires.

19. Ulvinsalo. Area, 25 km^2; altitude, 200–250 m. Pine and spruce forest; mires.

20. Veskijärvi. Area, 8 km^2; altitude, 53–61 m. Raised bogs.

FRANCE

National Parks

1. Cévennes. Area, 840 km^2; altitude, to 1700 m. Limestone plateau with deep gorges, with rich and interesting flora, including over 40 species of orchids. Limestone grasslands, forests of oak, sweet chestnut, and beech, and planted conifers. Also granite and schist peaks with distinctive and different flora.

2. Ecrins. Area, 1080 km^2; altitude to 4102 m. Largest French National Park, with superb alpine scenery, with four mountains over 3900 m, and lakes, gorges, cirques, and glaciers. Larch forests; a rich flora. Six Nature Reserves are included in the Park.

3. Mercantour. Area, 700 km^2; altitude, 1500–3061 m. Alpine country of crystalline and sedimentary rocks which is influenced somewhat by Mediterranean climate at lower levels, with a rich flora. Larch, spruce, and pine woods; alpine meadows; lakes.

4. Port-Cros. Area, 16 km^2; altitude, 0–200 m. Mediterranean island with original dense Aleppo pine and holm oak woods and maquis; also extensive marine vegetation.

5. Pyrenées. Area, 457 km^2; altitude to 3298 m. With the Spanish National Park of Ordesa it covers a large mountain area, rich in alpine species, and it includes the Nature Reserves of Néouville, Mont Valier, and Carlitte. There are cliffs, rocky outcrops, waterfalls, cirques, lakes, with beech and mountain pine woods.

6. Vanoise. Area, 528 km^2; altitude, 1250–3852 m. Very fine scenery with 107 mountain summits over 3000 m, with valleys and lakes, of both calcareous and metamorphic rocks. It borders the Gran Paradiso National Park of Italy. Coniferous forests of larch, fir, spruce, and Scots, mountain, and Arolla pines. Alpine meadows on calcareous soils are rich in species.

Regional Parks

7. Armorique. Area, 650 km^2. Largest protected area in Brittany, with offshore islands. Woodlands, heaths, bogs, and small hedged fields.

8. Brière. Area, 400 km^2. One of the largest areas of marshes and lagoons in France, with heaths and boggy meadows.

9. Camargue. Area, 951 km^2. The Rhone delta with large lakes and marshes, including the Nature Reserves of Etang de Vaccarès, Imperial Reserve, and Tour du Valet. Brackish lakes and pools, brackish and freshwater marshes, dunes, salt-steppes, seashores. Stone pine woods; also cultivated areas.

10. Corse. Area, 1500 km^2; altitude to 2622 m. In the centre of Corsica, with mountains and rocky coasts. An area of great beauty and diversity, with woods of Corsican, stone, and Aleppo pines, fir, and evergreen holm and cork oaks, as well as marshy lakes, maquis, and coastal grasslands. The Park has about 50 endemic species of higher plants. Scandola Nature Reserve with coastal cliffs and caves is included in the Park.

11. Luberon. Area, 1200 km^2; altitude to 1100 m. Calcareous rocks, caves, and steep-sided valleys with forests of oak and planted cedars. There are also well developed maquis, and farmlands with vineyards and orchards.

12. Pilat. Area, 600 km^2; altitude to 1432 m. Wooded granite hills with sweet chestnut, white oak, ash, and poplars, while above about 800 m are found pine, beech, and fir woods with an interesting flora.

13. Vercors. Area, 1350 km^2. Woods of pine, spruce, beech, and white oak.

14. Volcans d'Auvergne. Area, 2815 km^2; altitude, to 1885 m. Landscape of extinct volcanoes, plateaux, and many lakes. Forests of oak, sweet chestnut, beech, and fir. Extensive pastures on the plateaux, also many peat-bogs; an interesting and varied flora.

15. Vosges du Nord. Area, 1100 km^2. Rural landscape of fields, forests of oak and beech, and planted conifers. Natural grasslands on calcareous soils; peat-bogs on sandy soils.

National Reserves, Biological Reserves

16. Aiguilles Rouge National Reserve. Area, 32 km^2; altitude, 1200–2965 m. Fine mountain scenery with peaks, rocks, screes, glaciers, cliffs, lakes, and moraines. Woods of

spruce, larch, and alder, and sub-alpine and alpine vegetation on both acidic and calcareous rocks.

17. Asco (Corsica) National Reserve. Area, 29 km²; altitude, 800–1900 m. Forests of Corsican pine and maritime pine.

18. Bauges National Reserve. Area, 40 km²; altitude, 1000–2260 m. Woods; rock vegetation; grasslands.

19. Bavella Sambucco Conca (Corsica) National Reserve. Area, 39 km²; altitude, 500–1300 m. Holm oak and Corsican pine woods.

20. Belval (Ardennes) National Reserve. Area, 9 km²; altitude, 190 m. Deciduous woods of pedunculate oak and planted beech.

21. Bure Auroze (Alps) National Reserve. Area, 66 km²; altitude, 1300–2712 m. Alpine grasslands; rocks, lakes, and springs.

22. Burrus National Reserve. Area, 44 km²; altitude, 900–2800 m. Lakes and rocks. Forest of oak, beech, birch, and fir.

23. Casabianda (Corsica) National Reserve. Area, 17 km²; altitude, 4–65 m. Pine, cork oak, and eucalyptus woods; meadows, lakes, and marshes bordering the Tyrrhenian Sea.

24. Cauterets (Pyrénées) National Reserve. Area, 29 km²; altitude, 1000–2800 m. Coniferous forests; meadows; rock vegetation; lakes.

25. Chambord National Reserve. Area, 54 km²; altitude, 100 m. Oak and birch forest on sand; Scots pine woods. Lakes and marshes, with interesting wetland vegetation.

26. Chizé National Reserve. Area, 26 km²; altitude, 75 m. Woods and copses of white oak and beech.

27. Combeynot (Alps) National Reserve. Area, 47 km²; altitude, 1500–1800 m. Alpine meadows; woods.

28. Donzère–Mondragon National Reserve. Area, 18 km²; altitude, 60 m. Plantations of acacia species in the Rhone valley.

29. Etangs des Impériaux (Rhone) National Reserve. Area, 28 km²; altitude, 0–5 m. A western extension of the Camargue Regional Park.

30. Fontainbleau Forest Biological Reserve. Area, 5.5 km²; altitude, 75–150 m. Acid and calcareous beech woods; white oak and birch.

31. Markstein National Reserve. Area, 37 km²; altitude, 500–1300 m. Forests, meadows, and lakes.

32. Pelvoux (Alps) National Reserve. Area, 87 km²; altitude, 1500–4102 m. Forests and thickets of mountain pine, Arolla pine, larch, and green alder. Alpine grassland; rocks, waterfalls.

33. Petit Pierre National Reserve. Area, 43 km²; altitude, 380 m. Mixed forest of beech, Scots pine, and oak, on sandy soils.

34. Pic du Midi d'Ossau National Reserve. Area, 60 km²; altitude, 2885 m. Forests, pastures, fallow-land, rocks.

35. Pointe d'Arçay National Reserve. Area, 3 km². Important wetlands reserve, and least disturbed eco-system on the Atlantic coast, with an estuary with sand-bars, maritime pine, heathlands, and grasslands. Part of the Regional Park of Marais Poitevin, with its extensive grasslands. Has many floristically interesting drainage channels.

36. Vallée de la Grande Pierre et de Vitain National Reserve. Area, 2.9 km². Dense coppiced woodland and calcareous grasslands with an interesting flora of 159 recorded species. Woods of white oak, with juniper, box, and St. Lucie's cherry.

WEST GERMANY

National Parks

1. Bayerischer Wald. Area, 131 km²; altitude, to 1453 m. Largest forest in Central Europe—over 98 per cent of area forest-covered. Predominantly coniferous forests of pine, spruce, and silver fir; on steep slopes, also mixed coniferous and deciduous forests of maple, elm, ash, alder, aspen, willows, and bird cherry. In valleys are streams and springs with rich flora; also raised bogs with mountain pine.

2. Berchtesgaden. Area, 208 km²; altitude, to 2712 m. Beautiful alpine scenery with four mountain massifs and three main valleys with extensive high plateaux, rocks, and steep valleys. Deciduous woodlands at lower altitudes, and woods of spruce, beech, and fir above. Sub-alpine coniferous forests of spruce, larch, and Arolla pine, and above dwarf mountain pine, green alder, and alpine grasslands. The flora is rich and varied.

3. Nordfriesisches Wattenmeer (proposed). Area, 1400 km²; Islands, sand-flats, cliffs, dunes, marshes. It includes 10 coastal reserves; very important wildlife area.

Nature Parks

4. Altmühltal. Area, 2908 km²; altitude, 600 m. Beautiful landscape of Danube river valley with rocks, cliffs, and streams. Woods of both coniferous and deciduous trees, on sandstone and limestone rocks respectively.

5. Bayerischer Spessart. Area, 1307 km²; altitude, to 651 m. Hilly wooded country with extensive oak forests, also planted pines, spruce, larch, Douglas fir, Weymouth pine, beech, and oak. Also bogs and calcareous areas. Several nature reserves included within the Park.

6. Fränkische Schweiz-Veldensteiner Forst. Area, 1747 km²; altitude, 600+ m. Rocky peaks and steep-sided valleys with streams, dolomite cliffs, pinnacles, caves, and over 1000 dolines, with rich flora. Forest plantations of beech and conifers cover a quarter of the Park; also some spruce and pine. Small woodlands also occur in agricultural areas.

7. Nordeifel. Area, 1743 km²; altitude, to 698 m. Old rocks of schists, sandstones, and some limestone; numerous deep valleys with reservoirs. Half of Park wooded with conifers, and with deciduous trees of beech, oak, ash, sycamore, alder, and birch. Dry limestone grasslands with interesting flora.

8. Siebengebirge. Area, 42 km²; altitude, to 460 m. Fine landscape with wooded hills, valleys, and streams, with a rich flora under southerly influence. 80 per cent of area

covered with woods of oak, beech, and hornbeam; also alder and ash woods in valleys.

Nature Reserves and Landscape Protection Areas

9. Ammergebirge Nature Reserve. Area, 276 km²; altitude, 2000–2500 m. Alpine forests on calcareous rocks; meadows, with sub-alpine and alpine flora.

10. Augsburg Forest Nature Reserve. Area, 16 km²; altitude, 500 m. Remains of an ancient forest of silver fir, with all stages of development to the climax community.

11. Aussendeich Nordkehdingen Nature Reserve. Area, 9 km². Bank of the Elbe river, with marshes, wet pastures, mud-flats, with extensive reed and sedge communities.

12. Bodensee Nature Reserve. Area, 7769 km². On the shore of Lake Constance, with aquatic vegetation, and with an interesting flora round lake margin; also peat bogs and damp meadows.

13. Danube Reservoirs Nature Reserve and Landscape Protection Area. Area, c.175 km². Riverine woods, meadows, marshes, ponds, flood-plains, peatlands, and reed beds. Reservoirs.

14. Die Lucie Nature Reserve. Area, 18 km²; altitude, 50 m. Part of an ancient glacial valley, with wet woodland and very rich marsh vegetation.

15. Diepholzer Moorniederung Nature Reserve and Landscape Protection Area. Area, 178 km². Moors, heaths, local peat bogs, and birch woods, on glacial sands. It includes three Nature Reserves.

16. Dümmer Nature Reserve. Area, 36 km²; altitude, 100 m. Large shallow lakes with raised bogs, reed swamps, marshes, and inland dunes. An important wetland area with aquatic plants.

17. Eggstätt-Hemhofer Seenplatte Nature Reserve. Area, 10 km²; altitude, 600 m. A number of lakes on glacial deposits, with a rich marsh flora and mires.

18. Federsee Nature Reserve. Area, 14 km²; altitude, 400 m. Wetlands surrounding a lake, with reed beds and fen.

19. Graburg Nature Reserve. Area, 1.8 km². Shell limestone on side of Ringgau mountain, with very rich flora of sub-alpine and some Mediterranean species. Rock-faces and screes with scrub oak, whitebeam, lime, and rowan; also extensive beech woods with interesting ground flora.

20. Heidenhäuschen Nature Reserve. Area, 1.14 km²; altitude, c.400 m. Basaltic rocks with fine cliffs and boulder slopes, with beech woods; lower down pedunculate and sessile oaks and hazel coppice, with interesting ground flora; also planted conifers.

21. Hochkienberg Nature Reserve. Area, 95 km²; altitude, 1800–2000 m. Rich alpine and sub-alpine flora on calcareous hills.

22. Isar Valley Nature Reserve. Area, 39 km²; altitude, 600 m. A river valley with succession on sand and gravel; river-valley woods, with Scots pine.

23. Ismaninger Teichgebiet Nature Reserve. Area, 9 km². Reservoir with reed beds and river-valley woods.

24. Jadebusen and Westliche Wesermündung Nature Reserves. Area, 456 km². Maritime reserves with extensive sand, silt and mud flats, with islands with grassy foreshore; dunes.

25. Karwendel Nature Reserve. Area, 190 km²; altitude, 2000–2800 m. Mountainous country with river valleys rising to high mountains, with a rich alpine flora.

26. Kühkopf-Knoblauchsaue Nature Reserve. Area, 23 km²; altitude, 300 m. Grasslands and woodlands.

27. Lüneburger Heide Nature Reserve. Area, 197 km²; altitude, 70–169 m. Low-lying plain of moraine deposits of sands and gravels, covered by heathlands with juniper and boggy areas with interesting flora including mosses. Original woodlands of oak and birch very limited.

28. Meissner Nature Reserve. Area, 6.2 km²; altitude, 754 m. Mountain slopes with basalt boulders; small plateaux, with both acid and alkaline rocks rich in species. Montane grasslands of considerable variety; local forests rich in shrubs and herbaceous species.

29. Mohrhof Landscape Reserve. Area, 6.8 km². Many ponds with damp meadows and marshes; pine woodlands on higher ground.

30. Nordteil des Selenter Sees Nature Reserve. Area, 7.05 km². Hill country with three large lakes, with reed beds, alder carr and aquatic species.

31. Ostfriesisches Wattenmeer Nature Reserves. Area, 965 km². A chain of islands off the north coast of Germany. It includes 12 Nature Reserves, with marshes, sand dunes, and sand cliffs with a wide variety of eco-systems. Perhaps the most important bird reserve in Europe.

32. Rheinauen Bingen-Erbach Nature Reserve. Area, 4.75 km². River-valley woodlands with oak, smooth-leaved and fluttering elms, and some white poplar, Norway maple, and small-leaved lime. Interesting flora in the marshes bordering the Rhine.

33. Riddagshausen-Weddeler Teichgebiet Europa Reserve. Area, 6.5 km². Originally artificial ponds, now with extensive reedswamps, and deciduous woodland.

34. Steinhuder Meer Nature Reserve and Landscape Protection Area. Area, 58 km². The largest inland lake in northern Germany, with reedswamps and marshes, with remnants of some moorland and raised bogs.

35. Tiroler Ache Nature Reserve. Area, 5.7 km². Woods of oak, ash, and willow; reed beds and wet meadows.

36. Unterer Inn Nature Reserve. Area, 20 km². River-valley woods and river gravel communities.

37. Weltenburger Enge Nature Reserve. Area, 2.51 km². Narrow gorge with vertical cliffs and woodlands on the gentler slopes.

38. Wurzacher Ried Nature Reserve and Landscape Protection Area. Area, 13 km². Vast area of peatlands and raised bogs; also damp meadows, pools, and local calcareous marshes. Stands of dwarf mountain pines; c.500 species of flowering plants and 150 moss species recorded.

EAST GERMANY

Nature Reserves

1. Anklamer Stadtbruch. Area, 12 km². Fen, intermediate mires, raised bogs. Alder and birch woods; oak and birch woods; birch woods on raised bogs. Reeds, sedges, bog myrtle, royal fern.
2. Groszer Stechlin. Area, 18 km². Birch and oak–birch, and Scots pine woods. Raised bogs, transition mires, fens with alder carr.
3. Muritz. Area, 63 km². Scots pine, oak–birch, pine–birch oak–hornbeam woods. Fens with great fen-sedge and transition mires; raised bogs.
4. Peenemünder Haken. Area, 19 km². Coastal vegetation with saline grasslands. Also oak–birch woods.
5. Oberharz. Area, 20 km². Montane and sub-alpine grassland; mires. Spruce forests and dwarf-shrub heaths; birch woods.
6. Schwarzatal. Area, 18 km². River-valley woods.
7. Untere Mulde. Area, 13 km². River-valley woods with oak, hornbeam, elm, and ash.

Biosphere Reserves

8. Steckby-Loedderitz Forest. Area 21 km²; altitude, 50–75 m. Loams, sands, and gravels partly flooded by river Elbe. Valley woods of pedunculate oak, small-leaved elm, field maple, wild pear, and crab apple. Alder carr with ash; grasslands. Also lakes and ditches with water chestnut and the floating fern *Salvinia*.
9. Vessertal. Area, 14 km²; altitude, 430–750 m. Narrow deep-cut valley with forests of beech, sycamore, wych elm, and, locally, spruce and fir. In valley bottoms and clefts are stands of ash, alder, and Norway maple and grasslands; also stream-side communities.

GREECE

National Parks

1. Cephalonia. Area, 29 km²; altitude, to 1628 m. Beautiful limestone area with Mount Ainos in the centre. Greek fir forests, mountain grasslands, and rock flora.
2. Mount Oeti. Area, 72 km²; altitude, 600–2116 m. Mountain forests with oak, beech, sweet chestnut, black pine, and Greek fir, between 600 and 1900 m. A rich flora.
3. Mount Olympus. Area. 40 km²; altitude, to 2917 m. On lower slopes, maquis, and woods of beech and black pine; Balkan pine forms stands up to c.2500 m. Unique alpine rock and scree vegetation, with about 1500 species of flowering plants and ferns and with 19 endemic species and many Balkan endemics.
4. Mount Parnassos. Area, 35 km²; altitude, 1100–2457 m. Extensive Greek fir forests; above are limestone cliffs, alpine pastures, rock and scree vegetation; rich in Balkan species.

5. Parnes. Area, 37 km²; altitude, 300–500 m. Mountainous country with steep slopes with kermes oak, pine and Greek fir.
6. Pindus. Area, 129 km²; altitude, 1200–2177 m. Extensive forests of beech, black and Bosnian pines, fir, hornbeam, hop-hornbeam, and oak. Also horse chestnut and oriental plane, with alders and willows in the valley woods. Mediterranean alpine vegetation.
7. Lake Mikrí Prespá. Area, 195 km². Reedswamps and shallow lagoons, with very rich bird life.
8. Samaria Gorge (Crete). Area, 48 km²; altitude, 800–2200 m. Magnificent gorge with rock-walls to 600 m, streams, and springs. Woods of Italian cypress, Calabrian pine, and also Cretan maple, kermes oak, oriental plane, and olive, and an interesting herbaceous flora with a number of endemic species.
9. Vikos–Aoös Gorge. Area, 34 km². Magnificent river gorge bordering Albania. Forests of beech, fir, Bosnian pine, and black pine, and cliff vegetation.

Protected areas, Natural Monuments

10. Evros Delta protected wetland. Extensive wetlands and lagoons with bordering woods of Calabrian and black pines.
11. Gulf of Árta protected wetlands. Area, c.180 km². Extensive marshlands and reedswamps, islands, sandbars, and lagoons.
12. Rodopi Virgin Forest, National Monument. Area, 5 km². Virgin forest of beech, fir, spruce, with rich flora of both continental and Mediterranean origin.

Nature Reserves

13. Antimylos Island. Area, 20 km²; altitude, 0–686 m. Maquis, with stinking juniper, *J. foetidissima*, mastic tree; and *Cistus* species.
14. Dias Island. Area, 12 km²; altitude, 0–268 m. Mediterranean vegetation with maquis, with stinking juniper, mastic tree, and olive.
15. Guioura Island. Area, 10 km²; altitude, 0–500 m. Mediterranean vegetation with maquis, with kermes oak, box, stinking juniper, mastic tree, and olive.
16. Kerkini Lake. Area, 60 km². Wetlands with wet woodlands and marshes.

HUNGARY

National Parks and Biosphere Reserves

1. Bukk National Park. Area, 390 km²; altitude, 310–953 m. Mountain range and wide plateau, along with valleys, waterfalls, and springs; mainly limestone in the mountains but also areas of slate and sandstone. Woods of sessile and Turkey oaks and beech, to 950 m. Steppe areas with scrub forest, with some Mediterranean and Illyrian species.
2. Hortobagy National Park and Biosphere Reserve. Area, 520 km²; altitude, 86–98 m. Extensive plain (puszta) with

shallow marshes, seasonal saltmarshes and streamlets, with salt-tolerant species of both eastern and southern origin. Also aquatic sedge and rush communities. Ash and elm woods, and hills with loess-desert species, and grasslands.

3. Kiskunsag National Park and Biosphere Reserve. Area, 306 km²; altitude, 94–127 m. Wide plains between the Danube and Tisza rivers, with sand hills, hollows, salt deserts, soda lakes, and flood plains with alder, oak, and poplar woods. Plains with soda soils and characteristic saline vegetation; also shallow lakes, marshes, damp meadows, and grasslands, with distinctive species.

4. Tihany National Park. Area, 11 km²; altitude, 232 m. Volcanic region with extinct volcanoes, hot springs, and lakes, on a peninsula of Lake Balaton. A wooded area.

Nature Reserves

5. Agota Puszta. Area, 47 km². Plains with soda-salt vegetation. Seasonal saltmarshes, and rills with both natural and planted woodland.

6. Bekebarlang. Area, 6 km²; altitude, 376 m. Karst country.

7. Little Balaton (Kis-Balaton). Area, 14 km²; altitude, 105 m. Alluvial and marshy area with reedswamps and willows; alder and poplar thickets.

8. Pusztakocs. Area, 28 km². Saline puszta with plains, salt-marshes, loess-ridges; with meadows, marshes, reedswamps, and with local woods of oak, ash, and poplar.

9. Szalaika Valley. Area, 5.6 km²; altitude, 959 m. Mountainous region with streams and numerous cascades with tufa. Beech woods in the valley, oaks on the drier hills and fens at the river edge.

10. White-Water of Nagyberek. Area, 15 km². A low-lying silted-up area of Lake Balaton with marshes, peat-bogs, and meadows, with rich aquatic vegetation. Woods of willow and poplar, with some distinctive herbaceous species.

Biosphere Reserves

11. Aggtelek. Area, 192 km²; altitude, 150–600 m. Karst country, with woods of beech, sessile oak, hornbeam, and also lime–ash woods with St. Lucie's cherry and white oak. Rocky grasslands and heathlands.

12. Lake Fertö. Area, 125 km²; altitude, 114–261 m. Shallow lowland lake, rich in aquatic communities, also with reed beds, rush and sedge communities; also white oak and karst scrub.

13. Pilis. Area, 230 km²; altitude, 106–757 m. Karst region with dolomites and limestones, also volcanic formations; valleys with streams. Woods of oak and beech with sessile oak, Turkey oak, and hornbeam; steppe meadows with feather-grasses.

ICELAND

National Parks

1. Jökulsárgljúfur. Area, 151 km². Deep canyon with impressive waterfalls. In the valley, rich vegetation of sedges, downy birch, and willow; on the screes and cliffs are dwarf shrubs and an interesting flora. Also moorlands and heathlands, lakes and ponds, and sandy areas near the sea.

2. Skaftafell. Area, 200 km²; altitude, to 1430 m. Fine landscape with ravines, waterfalls, glaciers, hot springs. Birch woods with interesting ground flora; heaths, bogs, and gravel flats. Rich vegetation with c.210 species of flowering plants.

3. Thingvellir. Area, 40 km²; altitude, 103–140 m. Volcanic landscape with large lake and famous waterfalls. Dwarf birch and the moss, *Rhacomitrium lanuginosum*, covering exposed areas.

Protected Areas and Landscape Reserves

4. Mývatn og Laxá Protected Area. Area, 4400 km²; altitude, to 741 m. Lake in volcanic area with warm springs and luxuriant aquatic vegetation, particularly algae; ponds, marshes, and pastures, with surrounding fell-fields.

5. Hornstrandir Landscape Reserve. Area, 580 km²; altitude, to 600 m. Wild country in northern Iceland on the Arctic ocean, with cliffs, ravines, and valleys with streams.

IRELAND

National Parks

1. Killarney. Area, 85 km²; altitude, to 557 m. Oceanic climate, with rich fern and moss flora. Natural woods of oak and yew (with strawberry tree); upland bogs and lakes, with some Atlantic species of flowering plants. Bourn Vincent Memorial Park is included in the National Park.

2. Burren (proposed). Area, 259 km². Limestone pavement with lakelets, caves, and underground streams and low sea-cliffs. Rich and very interesting flora, with a mixture of some southern, northern, and montane species.

3. Glenveagh (proposed). Area, 100 km². Glaciated valley of granite with outcrops of schist and gneiss. Blanket bogs, heathlands, and small woods of oak and birch.

NORTHERN IRELAND

National Nature Reserves

4. Kebble. Area, 1.23 km². Western part of Rathlin Island, with lakes, grassland, and coastal cliffs.

5. Murlough. Area, 2.8 km². Sand dunes with dune-slacks and heathlands inland.

ITALY

National Parks

1. Abruzzo. Area, 400 km²; altitude, 700–2200 m. Ancient beech woods on mountain slopes with maple, black pine, and manna ash; with dwarf mountain pines and juniper at higher levels. Rocky summits with scrub vegetation and alpine grassland.

2. Calabria. Area, 170 km²; altitude, 1200–1400 m. Attractive mountain scenery with three peaks. Woods of Corsican pine, beech, fir, maple, oak, aspen, chestnut, and interesting shrub and herbaceous species.

3. Circeo. Area, 84 km²; altitude, to 541 m. Beaches, coastal dunes with Mediterranean flora. Inland—lakes and marshes with holm oak and dwarf fan palm. State Forest of Circeo, with Hungarian and pedunculate oak and Caucasian ash, alder, English elm, and some introduced tree species.

4. Etna (proposed). Area, 500 km²; altitude, 3323 m. Very large volcanic mountain crater with a rich flora. Lower slopes cultivated with olive, orange, fig, vine, and carob, the upper slopes with oak, sweet chestnut, and hazel plantations; above, forests of black pine, beech, and birch. Lava fields up to 2500 m, with interesting spiny shrub vegetation.

5. Gran Paradiso. Area, 730 km²; altitude, 1200–4061 m. Adjacent to French Vanoise National Park. Mountain valleys with larch, fir, spruce, and Arolla pine, to c.2300 m. Rocks, screes, and alpine meadows rich in species.

6. Stélvio. Area, 1370 km²; altitude, 350–3905 m. Largest Italian National Park, adjacent to the Swiss National Park. Mountainous area with forests of larch and spruce in the valleys, with Arolla pine at higher altitudes and alpine grasslands.

7. Gennargentu (proposed). Area, 1000 km². In Sardinia; a fine wilderness area of rugged landscapes, peaks, rocks, gorges, and valleys running down to the sea, with forests of holm oak and largely planted black pine. Large areas of cistus and tree heath maquis, also strawberry trees, and limestone 'islands' rich in endemic species.

8. Pollino (proposed). Area, 500 km²; altitude, to 2248 m. Region of forests, streams, and gorges. Distinctive Bosnian pines, montane pastures, rock and scree flora.

Natural Parks, Regional Parks, Forest Reserves

9. Alpe Veglia Natural Park. Area, 393 km²; Close to Simplon Pass, and surrounded by ridges and mountain summits. High-altitude woods of larch; alpine pastures with alpine flora.

10. Ademello–Brenta Natural Park. Area 436 km². Mountain area of rocks, cliffs, and steep slopes with forests of spruce, rowan, birch, and alder; and alpine flora.

11. Fusine Natural Park. Area, 0.45 km²; altitude, to 2677 m. In the Julian Alps, with two small mountain lakes. Within the Park is part of the Tarvisiano State Forest, the largest in Italy; it includes woodlands of beech, fir, larch, Scots pine, and black pine.

12. Maremma Regional Park. Area, 70 km². Unspoilt area of rocky spurs, brackish marshes, sand dunes, wet meadows, with characteristic coastal flora. Woodlands of stone pine with juniper; also other tree and shrub species.

13. Monte di Portofino Natural Park. Area, 12 km². Maquis with Aleppo and maritime pines. Forests of sweet chestnut and hornbeam.

14. Paneveggio–Pale di San Martino Natural Park. Area, 158 km². Dolomitic country with high peaks, rock cliffs, mountain lakes, and waterfalls. It includes three State Forests, largely of fir and Scots and Arolla pine with distinctive dolomitic flora. Alpine meadows rich in species.

15. Umbra Forest Nature Reserve. Area, 100 km². It includes two strict Nature Reserves. Very rich flora on the Gargano peninsula, with 2000 species recorded. Beech woods with hornbeam, field maple, flowering ash, and yew, and interesting forest species. Limited forests of oak on hillsides to 600 m, with distinctive woody species.

16. Val di Farma Forest Reserve. Area, c.45 km². Forests of pine and holm oak, with mastic and strawberry trees. Cork oaks on south-facing slopes, while sweet chestnut, holly, oak, hazel, hornbeam, and poplar grow on north-facing slopes.

Nature Reserves

17. Bolgheri. Area, 218 km². Coastal grasslands, maquis, and ash–lime woods.

18. Cossogno. Area, 10 km²; altitude, 700–2100 m. Mountain woods of coniferous and deciduous trees.

19. Orbetello, Area, 8 km². A series of tidal lagoons, with low dunes, brackish lakes, freshwater pools, fields, and woodland strips. Important migrant bird reserve.

20. Salina di Margherita. Area, 37 km². Saltmarshes, fresh and brackish lakes, rich in rushes and sedges and species of the goosefoot family, Chenopodiaceae.

LUXEMBOURG

National Park

1. Luxembourg–German Nature Park. Area, 725 km². Upland woodlands, meadows, hills and valleys, cliffs, and gorges. About one-third of the area is wooded with beech, oak, and sycamore with ash and alder in the valleys; also scattered plantations of pine, spruce, larch, and Douglas fir. Contains a rich flora, with 36 orchid species.

NETHERLANDS

National Parks

1. De Hoge Veluwe. Area, 57 km². Sandy soils with a wide variety of plant communities. Planted, sand-stabilizing, coniferous and deciduous woodlands; heathlands colonized by Scots pine. Oak–birch woods, juniper woods, wooded fens. Wet and dry heaths, dunes, and grasslands cover about half the Park.

2. De Kennemerduinen. Area, 12 km². Coastal dunes with sea-buckthorn, creeping willow, and privet scrub, and dune lakes and slacks. Corsican and Scots pine plantations.

3. Veluwezoom. Area, 45 km². Glacial moraine bank, with dunes and dry valleys. Pine and deciduous woodlands; also extensive heaths.

Nature Reserves

4. Amsterdamse Waterleidingduinen. Area, 34 km². Large dunes of calcareous sands. Scrub of sea-buckthorn, creeping willow, hawthorn, and spindle-tree. Interesting herbaceous flora, with many orchid species.
5. Biesbosch. Area, 58 km². Wetlands, reeds, wet woodlands.
6. Boschplaat (Terschelling Island). Area, 44 km². Coastal dunes, saltmarshes, dune heaths.
7. Boswachterij Schoorl. Area, 20 km². Extensive dunes and acid sands, with heaths and meadows. Inland dunes planted with Corsican pines.
8. Deelerwoud. Area, 11 km². Heathlands, pine woods, and old oak scrub.
9. De Groote Peel and Mariapeel. Area, 20 km². Peat-bogs, heaths, dry and wet areas, ponds with marginal woodlands, and marshes.
10. De Wieden and De Weerribben. Area, 72 km². Old peat cuttings showing all stages of recolonization and development, ranging from aquatic vegetation, marshes, and quaking-bogs to alder carr. Rich in species of flowering plants.
11. Dwingelosche and Kraloer Heaths. Area, 12 km². Heathlands.
12. Drentsche Aa. Area, 12 km². Rolling landscape, with alder and pine woods, heathlands and bogs, with characteristic species of acid peat soils. River with interesting aquatic flora.
13. Duinen Van Voorne. Area, 9.1 km². Calcareous dunes with a rich and wide range of plant communities, from woodlands to inland dunes, with more than 700 species recorded. Two dune lakes, with alder carr and fen.
14. Geul and Westerduinen (Texel). Area, 17 km². Coastal dunes, dune-heaths, and juniper.
15. Hulshorster Zand. Area, 2.7 km². Mobile dunes with stands of pine and juniper; rich in lichens but poor in herbaceous species.
16. Kampina Heath. Area, 11 km². Heaths, lakes, mires.
17. Kobbe. Area, 24 km². Coastal dunes, etc. on Schiermonnikoog Island.
18. Kootwijkerzand. Area, 5.9 km². Mobile sand dunes with lichen heaths; damp flats.
19. Leuvenumse Bos and Leuvenhost. Area, 19 km². Variety of types of woodland; sand hills.
20. Loonse en Drunense Duinen. Area, 15.2 km². Sandy heathlands. Plantations of Austrian and Corsican pines; small stands of oak.
21. Meijendel. Area, 13 km². Dune grassland, scrub, mixed deciduous woodlands, lakes, and marshes.
22. Naardermeer. Area, 7.5 km². Wetlands, swamps.
23. Noordhollands Duinreservaat. Area, 48 km². Rich dune flora. Highest dunes up to 45 m in height; inland dunes planted with trees.
24. Oisterwijkse Vennen. Area, 3 km². Alder, fen heathlands.
25. Oostvaardersplassen. Area, 60 km². Marshes and lakes, developed on new polder (reclaimed land).

26. Savelsbos. Area, 1.7 km². The only calcareous soils in the Netherlands. Valley woodlands with oaks, hornbeam, and birch; chalk grasslands with characteristic species.
27. Schiermonnikoog, Texel, Terschelling. Area, 102 km². Three Islands off the Dutch coast; part of an important wetland area with extensive saltmarshes, creeks, sand dunes, and mud-flats. Dunes with rich flora on Terschelling Island. Also dune heaths, dune meadows, saltmarshes, damp valleys, lakes, lagoons with a rich flora in the three reserves.
28. Schorren Achter Polder Eendracht. Area, 30 km². Coastal saltmarshes on Texel Island.
29. Strabrechtse Heath. Area, 26 km². Wetlands; heaths.
30. Zwanenwater. Area, 5.9 km². Dunes and wet valleys; two fresh-water lakes; acid heathland flora.
31. Zwarte Meer and Veluwerandmeren. Area, 55 km². Polders (reclaimed land) with reed swamps.

NORWAY

National Parks

1. Ånderdalen. Area, 68 km²; altitude, to 853 m. Mountainous granitic region with coastal forests of pine and birch on island of Senja. Two-thirds of the area lies above the tree line at c.300 m.
2. Børgefjell. Area, 1065 km²; altitude, 450–1703 m. Beautiful mountainous region, with many lakes, rivers, and waterfalls. Limited forests of birch, pine, and spruce but terrain mostly above the tree line c.600 m, with heaths and snow-bed vegetation. About 295 species of flowering plants have been recorded.
3. Dovrefjell. Area, 265 km². Rich flora on calcareous soils, with most of the Norwegian species present. Limited birch woods, with rowan and bird cherry; extensive open grasslands with many marshes, mires, and lakes.
4. Femundsmarka. Area, 386 km²; altitude, to 1415 m. Wild region with hundreds of small lakes, on a mountain plateau at 800–900 m, with bare rock and poor vegetation. Scattered forest of pine and birch.
5. Gressåmoen. Area, 180 km²; altitude, to 1009 m. Granite soils, not rich in flowering plants. Ancient spruce forest; also scattered pine and birch; lakes and marshes.
6. Gutulia. Area, 19 km²; altitude, 615–1100 m. Landscape of morainic ridges with deep depressions. To the east, flat country with open pine woods, marshes, small lakes; also spruce and birch scrub.
7. Hardangervidda. Area, 3400 km²; altitude, 1200–1719 m. High extensive plateau above the tree-line, with rich flora with many Arctic species. Pine and birch woods in valleys. Many lakes.
8. Ormtjernkampen. Area, 8 km²; altitude, to 1128 m. Spruce, pine, and birch forests with poor ground flora; bogs. At higher altitudes birch woods, and heaths with dwarf birch, relatively rich in species.
9. Øvre Anarjåkka. Area, 1120 km²; altitude, 300–500 m. Relict Scots pine and birch woods. Alpine flora, with

birch, dwarf birch, willow scrub, and lichens; also bogs and heathlands.

10. Øvre Dividal. Area, 635 km²; altitude, to 1428 m. Extensive plateaux with many lakes and bogs; broad valleys and mountain ridges with rocky peaks, with pine, birch, and alder woods. The Park is rich in flowering plants, with 315 species recorded.

11. Øvre Pasvik. Area, 60 km²; altitude, 92–202 m. Primeval Scots pine and birch forests. Arctic mires, lakes; with 192 species of vascular plants.

12. Rago. Area, 171 km²; altitude, to 1225 m. Fine rock-covered country with mountain ridges, ravines, and numerous lakes and swamps. Forests, below 400 m, of pine, with some birch, rowan, and alder; open treeless country above (fell-field).

13. Rondane. Area, 572 km²; altitude, 900–2178 m. Fine mountain scenery with 10 peaks over 2000 m, cirques, moraines, canyons, and rock-walls. Limited forests of birch and pine. Park largely covered with sparse sub-alpine and arctic-alpine vegetation.

14. Saltfjellet. Area, 1035 km². Lakes, mires, and glacier.

15. Stabbursdalen. Area, 97 km². Most northerly pine forest in the world, with birches and riverside willows. A wilderness of broad valleys and rivers flowing into the Arctic Ocean, and rocky chasms inland; wide rolling landscape with many streams and small lakes.

POLAND

National Parks and Biosphere Reserves

1. Babia Góra National Park and Biosphere Reserve. Area, 17 km²; altitude, 600–1725 m. Mountainous area of sandstones and shales with well developed and zoned forests. The lower forests are of beech, and above are spruce, fir, dwarf mountain pine, and alpine grasslands. Over 700 species of flowering plants, 200 mosses, 100 liverworts, and 100 lichens recorded.

2. Bialowieza National Park and Biosphere Reserve. Area, 51 km²; altitude, 155–172 m. Ancient forests of pine mixed with spruce, oak, hornbeam, ash, birch, alder, and aspen, with a rich shrub layer. Over 700 species of vascular plants recorded, including 23 tree and 47 shrub species. Woods of hornbeam, ash–alder, ash–elm, willows, and some conifers.

3. Bieszczady National Park. Area, 56 km²; altitude, 250–1348 m. Rich lowland forest of Scots pine and oak–hornbeam; also spruce, aspen, birch, ash and alder.

4. Gorce National Park. Area, 59 km².

5. Kampinos National Park. Area, 223 km²; altitude, 80 m. River-valley woods with alder, hornbeam, Scots pine; also fens and dunes.

6. Karkonosze National Park. Area, 56 km²; altitude, to 1604 m. Granitic mountains with cirques and lakes. Forests of beech and partly planted spruce; sub-alpine and alpine communities.

7. Ojców National Park. Area, 17 km²; altitude, 350–470 m.

Limestone karst region, with gorges, caves, and pinnacles; with beech and maple woods.

8. Pieniny National Park. Area, 27 km²; altitude, to 982 m. Calcareous mountains with river gorges and steep slopes forested with fir, beech, spruce, sycamore, ash and aspen; also larch plantations.

9. Roztocze National Park. Area, 43 km².

10. Slowiński National Park and Biosphere Reserve. Area, 181 km²; altitude, 0–110 m. Sandy coast, with fresh- and salt-water lakes, fens, sand dunes, coastal plains, heathlands, wet woodlands, raised bogs, and lake communities.

11. Swietokrzyski National Park. Area, 60 km²; altitude, to 611 m. Forested mountain range with fir, beech, and spruce woods, with larch and yew.

12. Tatra (Tatra Mountains) National Park. Area, 221 km²; altitude, 1000–2499 m. Forested mountains, with fir, beech, spruce, and sycamore to 1250 m; spruce, Arolla pine, and larch to 1650 m and dwarf mountain pine above. Alpine communities at higher altitudes.

13. Wielkopolski National Park. Area, 54 km²; altitude, 60–132 m. Moraines with many lakes, woods of oak with wild service tree, and mixed woods of Scots pine and spruce.

14. Wolin National Park. Area, 46 km²; altitude, 0–115 m. Largest Polish island in Baltic sea, with steep coastal cliffs, hills, many lakes; primitive forests of beech, oak, and Scots pine.

Nature Reserves

15. Czerwone Bagno. Area, 22 km²; altitude, c.110 m. Scots pine woods and mires.

16. Jezioro Karas. Area, 6.8 km²; altitude, c. 90 m. Lakes with mires and woodlands.

17. Jezioro Lukniany Biosphere Reserve. Area, 7.1 km²; altitude, c. 115 m. Lake with 166 species of vascular plants; sedge communities.

PORTUGAL

National Parks and Nature Parks

1. Arrábida Nature Park. Area, 108 km²; altitude, 500 m. Limestone outcrops on the coast, with steep shrub-covered slopes. The best developed maquis in Portugal, with kermes oak and characteristic Mediterranean shrubs, also woods of Portuguese oak and Phoenician juniper. Rich in western Atlantic species; over 1000 recorded in the Park.

2. Peneda-Gerês National Park. Area, 600 km²; altitude, to 1545 m. Beautiful mountain country, with peaks and deep valleys, granite escarpments, mountain streams, and three large reservoirs. Rich and varied flora in a relatively high rainfall area. Forests of oaks and pines.

3. Serra de Estrêla Nature Park. Area, 522 km²; altitude, to 1991 m. Mountains of crystalline and palaeozoic rocks, with rounded grassy summits and with glacial lakes, moraines, and deep valleys, in a high rainfall area. Forests of pedunculate oak, sweet chestnut, pine, and Pyrenean

oak, with rich shrub and herbaceous flora; also heaths with spiny dwarf shrubs and damp grasslands above.

Nature Reserves

4. Do Sapal de Castro Marim. Area, 80 km². Wetlands with lakes and salt lagoons.
5. Estuário do Tejo. Area, 228 km². Salt lagoons, mud-flats, marshes, dunes, damp hollows, and wetlands. The Reserve has a rich flora.
6. Pateira de Fermentelos. Area, 15 km². Lagoons, marshlands, reed beds, with saltmarsh vegetation.
7. Paúl do Boquilobo. Area, 3.9 km². Freshwater marshes, extensive sedge-beds, and willows and poplars on higher ground surrounding a lake in the valley of the River Tejo.
8. Ponta de Sagres. Area, c.60 km². Treeless open heathlands on exposed calcareous rocks, and also maritime limestone cliffs. Interesting heaths, with western Atlantic low shrubs and herbaceous and bulbous perennials.
9. Ria Formosa. Area, c.105 km². Area of marshes, mud-flats, lagoons, sand-bars, with some Mediterranean-type vegetation.
10. Serra de Monchique (proposed). Area, c.130 km²; altitude, to 902 m. Twin peaks with acid soils in heavy rainfall area. Planted with eucalyptus, maritime pine, cork oak, and olive groves. Interesting shrubs including rhododendron, heathers and herbaceous species.

ROMANIA

National Park

1. Retezat National Park and Biosphere Reserve. Area, 130 km²; altitude, 784–2509 m. Alpine region of mountains and lakes with 40 peaks over 2200 m. Lower slopes with sessile oak, walnut, hornbeam, beech, spruce, and fir; higher slopes with Arolla and dwarf mountain pine. Alpine meadows and dwarf shrub with *Rhododendron myrtifolium* and with arctic–alpine and Dacian species.

Nature Reserves, Forest Reserves, Biosphere reserves

2. Bucegi Nature Reserve. Area, 48 km²; altitude, 845–2509 m. Mountain massif, with fir and pine woods, and alpine meadows.
3. Ceahlau Nature Reserve. Area, 18 km²; altitude, 600–1904 m. Larch forests and characteristic montane flora.
4. Danube Delta Nature Reserve. Area, 400 km²; altitude, 0.8–2.5 m. Wetlands, reed beds, lakes, aquatic communities and islands.
5. Domogled Nature Reserve. Area, 8 km²; altitude, 168–1110 m. Deep valley, with natural vegetation including the Turkish hazel.
6. Letea Forest Reserve. Area, 7 km²; altitude, 1–3 m. Danube delta riverine forest of oak, ash, alder, white poplar, aspen, hornbeam, elm, hazel, and wild grape-vine, *Vitis vinifera* ssp. *sylvestris,* with an interesting herbaceous flora.
7. Pietrasul Mare Nature Reserve and Biosphere Reserve. Area, 27 km²; altitude, to 2303 m. The Rodna massif with

glacial cirques and lakes, with an alpine flora which includes Balkan and Central European alpine and some arctic–alpine species.

8. Rosca–Leta Nature Reserve and Biosphere Reserve. Area, 181 km²; altitude, to 15 m. Sand dunes with forest vegetation with *Quercus pedunculiflora* and pedunculate oak, white and black poplars, narrow-leaved and manna ash, and *Fraxinus pallisiae,* wild pear, silver lime, and elms, and with a rich shrub and woody climber flora. Also marshes, lakes, and canals with reed beds.
9. Slatioara Forest Reserve. Area, 6 km²; altitude, 900–1400 m. Old forest of spruce and silver fir, and alpine vegetation.
10. Snagov. Lake and Forest Reserve. Area, 18 km²; altitude, 80–100 m. Danube plain with forests of beech, oak-hornbeam and lake communities.
11. Tinovul Mare. Area, 6.7 km²; altitude, 898–969 m. Large peat-bog with bog mosses and *Vaccinium* species; also some Scots pine and birch.

SPAIN

National Parks

1. Aigües Tortes y Lago de San Mauricio. Area, 224 km²; altitude, 1600–2747 m. Two glacial valleys with many lakes, cirques, and hanging valleys. Large coniferous forests of silver fir and Scots pine, with mountain pine at higher altitudes. The Park is rich in herbaceous species.
2. Covadonga. Area, 169 km²; altitude, 140–2596 m. Mostly carboniferous limestone with karst landscape, also some sandstone and schists; scenically beautiful with glacial lakes. Forests of beech and pedunculate, sessile, and Pyrenean oaks; meadows below and heathlands and pastures above. Interesting flora rich in herbaceous species.
3. Doñana. Area, 757 km²; altitude, 0–30 m. One of the finest wetlands in Europe, with dunes, sandy heathlands, coastal marshes, shallow lakes, with stone pine woodlands; also relict cork oak forest.
4. Ordesa. Area, 157 km²; altitude 1064–3355 m. Magnificent mountain scenery, adjoining French Pyrenees National Park; four valleys with glaciers, limestone gorges, waterfalls, and high cliffs. Mountain valleys with beech, fir and Scots pine woods, with yew, aspen, birch, alder, Italian maple and willow, and with mountain pine woods above 1700 m. A rich and interesting flora, particularly in wet areas.
5. Tablas de Daimiel. Area, 18 km². Wetlands, flood-plains, and shallow lakes with many islands, with both freshwater and brackish-water species, reed-swamps, and also tamarisk scrub.

Natural Parks

6. Lago de Sanabria Natural Park. Area, 50 km²; altitude, to 2124 m. An area of lakes and mountain streams, with extensive deciduous forests of oak, sweet chestnut, Scots pine, yew, birch, poplar, alder, and ash; rich shrub flora with heathers and *Genista* species in particular.

7. Hayedo de Tejera Negra Natural Park. Area, 13.9 km²; altitude, to 2046 m. Fine beech woods on the Sierra Riaza, also oak, yew, mountain ash, and Scots pine on quartz and slate bedrock.
8. Monfrague Natural Park. Area, 178 km²; altitude, 839 m. Montane region, with virgin forests of cork oak and holm oak on the hillsides. Reafforestation with eucalyptus threatens the Park which is very rich in birdlife.

National Reserves

9. Ancares. Area, 79 km². Natural deciduous woods of oak, with beech, holly, alder, birch, and hazel on the mountain slopes, also planted Scots pines. Mountain pastures with lush vegetation; also peat-bogs locally.
10. Ebro Delta. Area, 640 km². Wetlands with salt lagoons, reed swamps, dunes.
11. Gredos. Area, 228 km²; altitude, to 2692 m. Mountains of granite, gneiss, and slate, with deep narrow valleys, waterfalls, torrents, and springs. Stands of maritime and Scots pines are found above, and remains of old cork oak and sweet chestnut woods below. An area rich in alpine species.
12. Las Batuecas. Area, 210 km²; altitude, 1723 m. Wild mountain country of deep narrow valleys with numerous springs. Forests of oaks and sweet chestnut, yew and holly in the north, and cork oak and strawberry trees in the south. Streamsides with alder, ash, and willow, and interesting shrub species.
13. Picos de Europa. Area, 76 km²; altitude, to 2648 m. Limestone mountains with extensive screes and moraine deposits, with alpine pastures rich in species. Trees in valleys include the Pyrenean oak, maples, and the large-leaved lime.
14. Saja. Area, 1800 km²; altitude, to 2538 m Near the Picos Europa, with many peaks over 2000 m. Beech and pedunculate oak woods occur at higher altitudes, while cork oak and sweet chestnut woods occur below with a rich shrub layer. Several rivers pass through the Reserve and there are many marshes and bogs. Lower levels are cultivated; above there are grazed mountain pastures.
15. Serrania de Ronda. Area, 220 km². Rugged limestone mountains, mostly covered with garigue. Relict woodlands of oak with some maple occur, while at lower altitudes the Spanish fir forms woods. Above c.1200 m is a rich mountain flora with some Iberian endemics.
16. Sierra de la Demanda. Area, 738 km²; altitude, to 2132 m. Mountain region with several lakes. Forests of Scots pine, beech, and Pyrenean oak are important timber sources.
17. Tortosa y Beceite. Area, 293 km²; altitude to 1447 m. Peaks and ravines, with shady forests below and hedgehog heaths above. Kermes and Portuguese oaks dominate the lower areas; beech and pines the higher woods; and spiny heaths dominated by species, largely of the pea family, occur above.

National Game Reserve

18. Cazorla y Segura. Area, 764 km²; altitude, to 2106 m. A game reserve with peaks, valleys, gorges, cliffs, and karst limestone. Upland coniferous woods of black, maritime, and Aleppo pines, and below woods of Portuguese and kermes oaks, with narrow-leaved ash and black poplars, and with many shrub species. Very rich flora of between 1200 and 1300 species, with some rare and endemic species.

SWEDEN

National Parks (excluding smaller National Parks)

1. Abisko. Area, 75 km²; altitude, to 1191 m. A rich floristic area, with much birch scrub in the lower altitudes and alpine meadows and arctic–alpine flora higher up.
2. Gotska Sandøn. Area, 35 km². Baltic Island in North Gotland, with dunes and pine woods.
3. Muddus. Area, 492 km²; altitude, 150–661 m. Plateaux with deep valleys, gorges and waterfalls. Forests of spruce and Scots pine, with birch and willow, cover over half the Park. Peatlands cover over 40 per cent, the remainder being stony areas and high fell-field, with tundra species.
4. Padjelanta. Area, 2040 km²; altitude, 400–1860 m. Large area of Arctic wilderness, with wide plain, mountains and scattered lakes. Alpine meadows, dwarf willow scrub, grass heaths, and some local calcareous rocks with distinctive species.
5. Peljekaise. Area, 146 km²; altitude, 470–1133 m. Area of virgin mountain birch woods, with interesting herbaceous flora. Some lakes, mires, and alpine vegetation.
6. Sarek. Area, 1940 km²; altitude, to 2090 m. Remote area with 90 rocky peaks over 1800 m, many gorges and glaciers. Sub-alpine birch woods, alpine dwarf willow scrub, and, above 1200 m, arctic–alpine species with many lichens.
7. Sonfjället. Area, 27 km²; altitude, 430–1278 m. Coniferous forests, with birch scrub above. Alpine vegetation above the tree line.
8. Stora Sjöfallet. Area, 1380 km²; altitude, 400–2015 m. Mountain region with lakes and waterfalls. In the eastern part primeval forests of spruce and Scots pine occur. Birch scrub and dwarf willow heath in the sub-alpine zone. Also lakes and marshes.
9. Töfsingdalen. Area, 14 km²; altitude, 660–892 m. Forested area intersected by ridges, with old moraines and large boulders. Coniferous forest of spruce and pine, sub-alpine birch scrub higher up, and alpine grasslands above.
10. Vadvetjåkko. Area, 24 km²; altitude, to 1250 m. Calcareous rocks, with lakes, willow thickets, and marshes with a rich flora. Dwarf birch scrub, alpine meadows, and heaths at higher altitudes.

Nature Reserves

11. Buberget. Area, 23 km²; altitude, 500–600 m. Hilly country devastated by fire in the last century; recolonized by pine, spruce, and birch. Vestiges of ancient forest remain.

12. Bullerö and Långskär. Area, 46 km². Lakes with about 900 islands, where Lake Mälaren has an outlet into the sea. The central region has spruce trees; the outer islands pine, birch, and rowan trees. Coastal wetlands, grasslands, scrub and rocky vegetation with ericaceous shrubs; also junipers.

13. Hjälstaviken. Area, 7.9 km². Coniferous forests, reed swamps, wet meadows, grasslands.

14. Kävsjön och Store Mosse. Area, 74 km². Possibly the best example of raised bog in Sweden; also two small lakes. Because of the rich bird life it is being considered as a National Park.

15. Komosse. Area, 27 km². Mires.

16. Luletjarve. Area 6 km²; altitude, 544–792 m. Forests of pine and spruce.

17. Mittådalen. Nature Reserve and Forest Reserve. Area, 1000 km². Mountainous region with lakes and rivers, with forests of birch and conifers.

18. Ottenby. Area, 9.95 km². Southern point of island of Öland. Sandy beaches, shingle, with sheltered coves, with birch, aspen and oak woods inland; also grazed meadows.

19. Reivo. Area, 87 km²; altitude, 390–640 m. Lapland plateau covered with forests of pine and spruce.

20. Rone Ytterholme. Area, 16 km². Includes islands in Laus Holmar reserve; wetlands and grasslands.

21. Sjaunja. Area, 2900 km²; altitude, 441–1720 m. The largest mires in Europe outside the USSR; coniferous forests.

22. Svaipa. Area, 494 km²; altitude, 470–1427 m. Alpine and sub-alpine region with lakes, marshes and bogs. Birch woods and willow thickets on margins of lakes; many small islands.

23. Tåkern. Area, 56 km². A shallow freshwater lake bordered by a 250-metre strip of land; primarily a bird reserve. Vast reed beds covering one–third of the area, with interesting aquatic species.

24. Tärnasjön. Area, 118 km². Lake with marshes, reed-beds, and surrounding birch woods.

25. Tjuoltavuobme. Area, 15 km²; altitude, 500–1440 m. Largest virgin sub-alpine birch forest in Europe; extensive fell-fields.

SWITZERLAND

National Park

1. Swiss National Park. Area 169 km²; altitude, to 3165 m. Superb mountain country in the Engadine, adjoining the Italian National Park of Stélvio. Dolomitic mountains with narrow deep valleys and streams. The dwarf mountain pine covers about 25 km² of the Park; much alpine grassland rich in species, with a total of over 640 recorded. Lower altitude forests of Scots pine, spruce, larch, and Arolla pine.

Nature Reserves

2. Aletschwald. Area, 3.06 km²; altitude, 2100–2300 m. Coniferous forests bordering the longest alpine glacier, with Arolla pine and larch. Interesting moraine vegetation; also small lakes.

3. Binntal. Area, 46 km². Beautiful alpine valley rich in interesting species, with alpine heathlands, bogs, and small lakes.

4. Creux du Van et Gorges de L'Areuse. Area, 11 km²; altitude, to 1335 m. Large river gorge and cirque with a rich flora. Coniferous and deciduous forests below; alpine pastures with some rare species above the tree line.

5. Fanel. Area, 4.86 km². Canals with extensive marginal reed beds with interesting marsh flora; deciduous woodlands.

6. Gelten-Iffigen Nature Reserve and Landscape Protection Area. Area, 43 km²; altitude, to 1950 m. A calcareous alpine area, with mountain lakes and waterfalls, showing good vegetation zonation, with larch and Arolla pine woods below; dwarf shrub zone and alpine grasslands above.

7. Grimsel. Area, 100 km²; altitude, to 4160 m. High alpine country with lakes, snow-beds, and glaciers. Alpine vegetation, Arolla pines.

8. Hohgant. Area, 15 km². Sub-alpine forests.

9. Hölloch Karst. Area, 92 km². Limestone cave area, with well developed karst vegetation, grasslands, and coniferous forests.

10. Lauterbrunnen. Area, 26 km². Alpine valley with high calcareous cliffs up to 500 m, many streams and waterfalls. Pine forests and an interesting and rich alpine meadow flora.

11. Monte San Giorgio. Area, 25 km²; altitude, to 1097 m. Mountain with forested slopes of oak, sweet chestnut, beech and false acacia, bordering lake Lugano. A variety of soils and mild climate results in a rich flora, with some Mediterranean species.

12. Pfynwald (proposed). Area, 10 km². Scots pine woods in low-rainfall and warm-summer area, with distinctive species of flowering plants; also river-gravel vegetation.

13. Vallée du Doubs. Area, 34 km². River valley with steep wooded slopes, lakes, waterfalls, marshes, poor pastures, and limestone screes, with a varied flora.

14. Vallée de Joux et Haut Jura Vaudois. Area, 220 km². Limestone mountain area, with forests of beech, sycamore, and whitebeam; alpine pastures and rocks, rich in species. Also shallow lakes with reed swamps.

15. Val Languard, Val dal Fain, Val Minor. Area, 17 km²; altitude, to 3166 m. High mountain area with glaciers and lakes, with a very rich flora, and forests of larch and Arolla pine up to 2300 m.

16. Vanil Noir. Area, 6.76 km²; altitude, to 2389 m. Calcareous mountains with very rich alpine and sub-alpine flora.

Landscape Protection Areas

17. Derborence. Area, 10 km². Virgin pine forest, high-altitude mountain pastures, rock walls, screes, and a lake, all with interesting and varied flora.

18. Piora. Area, 37 km²; altitude, to 2773 m. Beautiful alpine river valley with mountains, lakes, woods, and marshes; fine alpine moorland vegetation; also Arolla pine woods.

UNITED KINGDOM

National Parks

1. Dartmoor. Area, 954 km²; altitude, to 619 m. High granite plateau, with open landscape and steep-sided valleys with fast-flowing streams. Heather moors and bogs on the plateau, and in the valleys sessile oak woods with holly and rowan, and on richer soils pedunculate oak woods with ash, alder, birch, and beech locally, with a varied flora.

2. Exmoor. Area, 686 km²; altitude to 519 m. Granite hills covered with heather moors, and steep-sided valleys with streams, and also coastal rocks and beaches. Locally, mixed woods of oak, ash, wych elm, and birch, rich in lichens.

3. Lake District. Area, 2280 km²; altitude to 978 m. The largest British National Park, with rounded mountains, glacial valleys, scree slopes, hanging valleys, waterfalls and 10 major lakes. Relict woods of oak and ash, with hazel and yew. Upper slopes with grazed grasslands; also coastal dunes. Four National Nature Reserves and 115 sites of Special Scientific Interest occur within the Park.

4. The Peak. Area, 1404 km²; altitude, to 610 m. Gritstone tableland and limestone uplands, with wooded valleys and streams. Moorlands, limestone grasslands with a rich flora, and fine ash woods in the valleys; some included in a National Nature Reserve. Blanket-bog and acid grassland on Kinder Scout and Bleaklow (area 118 km²).

5. Snowdonia. Area, 2188 km²; altitude to 1085 m. A large mountainous area, with two large glacial valleys, covered with heather moors, grasslands, mires, and including 10 woodland Nature Reserves of deciduous trees (especially oak and ash), often with a rich flora of ferns, mosses, liverworts, and lichens. On the coast are two dune and saltmarsh areas which are Nature Reserves.

6. Brecon Beacons. Area, 1343 km². Upland heaths, grasslands; oak and birch woods.

7. North York Moors. Area, 1380 km². Heath, mire, limestone grassland, ash woods.

8. Pembroke Coast. Area, 585 km². Maritime heath; rocky coastal vegetation.

9. Yorkshire Dales. Area, 1760 km². Heath; grassland.

Forest Parks

10. Argyll Forest. Area, 240 km². Lakes; forests.

11. Border Forest. Area, 453 km². Spruce plantations; mires.

12. Forest of Dean. Area, 109 km². Pedunculate oak forest; grasslands.

13. Glen Trool. Area, 526 km². Lakes; forests (oak, birch).

14. New Forest. Area, 376 km². Extensive heathland, grassland, mire; oak, beech, pine, alder, and birch woods. Rich in acid-soil-tolerant plant species.

National Nature Reserves

ENGLAND

15. Braunton Burrows National Nature Reserve and Biosphere Reserve. Area, 6 km². Fine dune system, with over 375 species of flowering plants recorded; also shingle.

16. Bridgwater Bay National Nature Reserve. Area, 25 km². Saltmarshes, silt-flats, shingle, and cliffs.

17. Hickling Broad National Nature Reserve. Area, 4.8 km². The largest broad in Norfolk, with extensive reed beds and some rare aquatic plants.

18. Holkham National Nature Reserve. Area, 17 km². Coastline area of sand dunes, saltmarshes; rich in coastal plants. Planted Corsican pine.

19. Kingley Vale National Nature Reserve. Area, 1.46 km². One of the finest yew woods in Europe. Also woods of ash and beech, and extensive chalk grasslands rich in herbaceous plants, including orchids.

20. Minsmere and Walberswick National Nature Reserves. Area, 6 km². Shallow valleys with heathlands and extensive marshes and reed-beds. Woodlands of willow, birch, and alder; also some saltmarshes.

21. Moor House National Nature Reserve and Biosphere Reserve. Area, 40 km²; altitude, 882 m. Typical moorlands of Upper Teesdale, with blanket-bogs in the highest part of the Pennines. Acid grassland, montane heath, and flush communities.

22. Scolt Head Island National Nature Reserve. Area, 7 km². Coastal marshes; shingle.

23. Upper Teesdale National Nature Reserve. Area, 26 km². Heaths, blanket-bogs, acid grassland, flush communities, and tall herbaceous communities. Many rare species.

24. Wye and Crundale Downs National Nature Reserve. Area, 1.3 km². Chalk grasslands on escarpment slopes in Kent, with a rich flora. Woodlands of ash with hazel and scrub.

SCOTLAND

25. Beinn Eighe National Nature Reserve and Biosphere Reserve. Area, 48 km²; altitude, 972 m. Mountain region with remnants of ancient Scots pine forest, also northern alpine vegetation.

26. Caenlochan National Nature Reserve. Area, 36 km². Montane willow shrub (*Salix lanata* and *S. lapponum*); tall-herb communities, blanket-bogs, and snow-bed communities.

27. Caerlaverock National Nature Reserve and Biosphere Reserve. Area, 54 km². Extensive saltmarshes, saline grasslands, mud flats. Very important winter feeding grounds for wildfowl.

28. Cairngorms National Nature Reserve. Area, 235 km²; altitude, to 1309 m. Magnificent high plateau mountain area, with steep-sided glens, steep cliffs, corries, lakes and high moorlands with heaths and blanket-bogs. Typical flora of acid mountain soils; fragments of ancient pine woods lower down.

29. Hermaness (Shetland) National Nature Reserve. Area, 9.5 km². Saltmarshes, cliffs, acid grasslands, heathlands, and blanket-bogs.

30. Inchnadamph National Nature Reserve. Area, 13 km². Grasslands, flush communities, willow thickets of whortle-leaved willow, *Salix myrsinites*.

31. Invernaver National Nature Reserve. Area, 5 km². Marshes, dunes, shingles, and cliffs. Hazel–birch scrub; also flush communities. Many rare species.

32. Inverpolly National Nature Reserve. Area, 109 km². Picturesque region with wide range of habitats, including sea shores, cliffs, islands, lochs, streams, blanket-bogs, Dryas heaths, wet heaths and mountain areas. Both acidic and calcareous floras present.

33. Loch Druidibeg National Nature Reserve and Biosphere Reserve. Area, 17 km². Oligotrophic lake with water lobelia.

34. Loch Leven National Nature Reserve. Area, 16 km². Eutrophic lake.

35. Rannoch Moor National Nature Reserve. Area, 15 km²; altitude, 360 m. Blanket-bogs, poor fens.

36. Rhum National Nature Reserve and Biosphere Reserve. Area, 107 km². Island with more than 1800 plant species recorded. Woodlands recently planted, heaths, blanket–bogs, montane grassland.

37. St. Kilda National Nature Reserve and Biosphere Reserve. Area, 8 km². Grasslands, heaths, and cliff vegetation.

38. Sands of Forvie National Nature Reserve. Area, 7 km². Dunes and crowberry heath.

39. Tentsmuir Point National Nature Reserve. Area, 5 km². Fine dunes with rich dune flora; inland dunes with dune-slacks; also alder, birch, and willow scrub. Over 400 species of flowering plants recorded, largely on the dune-heaths.

WALES

40. Cors Tregaron National Nature Reserve. Area, 7.6 km². Raised bog.

41. Dyfi National Nature Reserve and Biosphere Reserve. Area, 16 km². Saltmarshes, sand-flats, dunes, and shingle.

42. Newborough Warren National Nature Reserve. Area, 6.3 km². Dunes and saltmarshes, sand-flats and shingle; also inland dune grasslands, and dune-slacks with an interesting flora.

Nature Reserves and Biosphere Reserves

43. Cairnsmore of Fleet (Scotland) Nature Reserve. Area, 13 km². Upland grasslands and heaths; blanket-bogs.

    Dunwich Heath and Marshes Nature Reserve. Area, 19 km². Coastal sand cliffs, with heath and birch. See 20.

    Horsey Mere Nature Reserve. Area, 7 km². Aquatic vegetation, reed-swamps, and fen with great fen-sedge. See 17.

44. North Norfolk Coast Nature Reserve and Biosphere Reserve. Area, 55 km². Saltmarshes, reed swamps, dunes and shingle.

45. Slindon Estate Nature Reserve. Area, 14 km². Sussex estate, with beech woods.

46. Wicken Fen National Trust Reserve. Area, 3 km². Remnant of peat fen. Carr, fen, sedge, reed, and open water.

YUGOSLAVIA

National Parks

1. Biogradska Gora. Area, 34 km²; altitude, 830–2116 m. Very fine virgin forests of mixed beech, ash, sycamore, and fir, with 64 species of trees and shrubs in the Park. Alpine meadows and five lakes.

2. Djerdap. Area, 821 km². Largest canyon in Europe, partly in Yugoslavia and partly in Romania, with rock-walls to 300 m. In the Park are 10 species of oak, also walnut, nettle tree, Turkish hazel, lilac and wig tree, and several endemic species. Five Nature Reserves are included in the Park.

3. Durmitor. Area, 320 km²; altitude, 538–2522 m. A mountain massif with alpine lakes. Forests of beech, fir, spruce, and black pine; above, dwarf mountain pines; and at higher altitudes alpine vegetation. Also includes the magnificent Tara gorge, in places 1000 m deep, with forested slopes.

4. Fruška Gora. Area, 273 km²; altitude, 150–539 m. Lowland area on border of the Pannonian Plain, with scattered forests, steppe, meadows, vineyards, and corn-fields.

5. Galičica. Area, 238 km²; altitude, 695–2285 m. The southern part of the Dinaric range situated between lakes Prespa and Ohrid. Woods of several oak species including the Macedonian oak; also Balkan pine, horse chestnut, Grecian and stinking juniper, and the local *Celtis caucasica*. The Park contains about 150 woody species of southern, northern, and Mediterranean origin.

6. Lovćen. Area, 20 km²; altitude 1200–1749 m. Limestone and dolomitic mountains close to the Adriatic coast, with karst landscape. Hop-hornbeam and white oak dominate below, while above *c.*1100 m are beech woods. The Park contains about 1200 species of flowering plants, including many exclusively Balkan species.

7. Mavrovo. Area, 790 km²; altitude, 600–2764 m. Largest Park in Yugoslavia with 50 peaks over 2000 m. Half total area forested, largely with beech and fir; elsewhere a rich Mediterranean, continental and sub-alpine flora.

8. Mljet. Area, 31 km²; Altitude 0–381 m. A limestone island in the Adriatic, with Aleppo pine woods and well developed maquis, also with holm oak, strawberry tree, tree heath, carob, myrtle, mastic tree, juniper species, and laurel.

9. Paklenica. Area, 36 km²; altitude, 0–1563 m. Karst region of the Velebit mountains, with two limestone gorges with cliffs to 400 m and caves. Rocky slopes covered with scrub of white oak, ash, hornbeam, juniper, and a rich and interesting flora, with some Yugoslav endemic species. Beech and black pine woods cover about half the Park.

10. Pelister. Area, 120 km²; altitude, 900–2601 m. Distinctive in having a large forest (1690 hectares) of Macedonian pine; also beech and oak woods. Fine mountain pastures occur above the woodlands, and there are two glacial lakes and mountain rivers.

11. Plitvice Lakes. Area, 194 km²; altitude, 417–1270 m. A series of 16 small lakes linked by waterfalls flowing over tufa cliffs. Forests of beech, fir, spruce; also alpine meadows with numerous rare species.

12. Rišnjak. Area, 30 km²; altitude 670–1528 m. Limestone and dolomite montane landscape. Lower slopes wooded with beech and fir; above 1200 m, virgin beech forests. Alpine pastures above tree line rich in interesting species. Thirty different plant communities have been described in the Park.

13. Sutjeska. Area, 172 km²; altitude, 532–2326 m. One of the best-known virgin forests of Europe, with 12 species of conifers and 143 species of broad-leaved trees and shrubs. Forests of beech, fir and oak occur below and dwarf mountain and Serbian spruce occur at higher altitudes. There is a rich alpine flora in pastures, rocky slopes, and glacial lakes.

14. Triglav. Area, 20 km²; altitude, 660–2864 m. A mountain area in the Julian Alps, with limestone karst landscape, screes, cliffs, and lakes, and a rich flora of many rare sub-alpine and alpine species. Beech woods on lower slopes—the warmer slopes having sub-Mediterranean species; at higher altitudes spruce, larch, dwarf mountain pine woods; above, extensive alpine pastures, rocks, and screes.

Nature Reserves

15. Hutovo Blato. Area, 3.6 km²; altitude, 1.5–6 m. Part of the Neretva river dela, with willow and poplar woods, marshlands, reed-swamps, mires, and lakes, with an interesting aquatic flora of over 500 species of flowering plants and many mosses, lichens and fungi.

16. Mala Pisnica. Area, 8.6 km²; altitude, 830–1940 m. Larch forests.

17. Monostorski Riitovi. Area, 6 km². Danube valley with wet woods of black and white poplar and white willow; also a rich marsh flora.

18. Obedska Bara. Area, 7.5 km². Ancient forests of oak, hornbeam, ash; grasslands and rich marsh flora.

19. Resava. Area, 100 km². Karst country.

20. Rijeka Krka. Area, 140 km². Krka river valley gorge, with lakes and tufa waterfalls. Lower part of the valley contains Mediterranean vegetation with holm oak, the upper part of the valley contains oriental hornbeam.

21. Zvijrezda. Area, 15 km². Forests of Serbian spruce, *Picea omorica*.

Selected bibliography

Vegetation: general

Braun-Blanquet, J. (1964). *Pflanzensoziologie.* Vienna.
Cain, S.A. (1971). *Foundations of plant geography.* London.
Firbas, F. (1965). *Plant geography.* In *Strasburgers textbook of botany* (28th edn.). London.
Fosberg, F.R. (1967). Classification of vegetation. In *IBP handbook No. 4.* Oxford & Edinburgh.
Good, R. (1947). *The geography of flowering plants.* London.
Greig-Smith, P. (1964). *Quantitative plant ecology.* London.
Halliday, G. and Malloch, A. (1981). *Wild flowers. Their habitats in Britain and northern Europe.* Glasgow.
Kellman, M.C. (1975). *Plant geography.* London.
Knapp, R. (1971). *Einführung in die Pflanzensoziologie.* Stuttgart.
MacArthur, R.H. (1972). *Geographical ecology. Patterns in the distribution of species.* New York.
Middlemiss, F.A. (1969). *British stratigraphy.* London.
Moore, D. (1982). *Green planet.* Cambridge.
Polunin, N. (1960). *Introduction to plant geography.* London.
Rubner, K. and Reinhold, F. (1955). *Das natürliche Waldbild Europas.* Berlin.
Schmidt, G. (1969). *Vegetationsgeographie auf Ökologisch–Soziologishe Grundlage.* Leipzig.
Shimwell, D.W. (1971). *The description and classification of vegetation.* London.
Tüxen, R. (Ed.) (1972). *Grundfragen und Methoden in der Pflanzensoziologie.* The Hague.
—— (Ed.) (1975). *Prodromus der europäischen Pflanzengesellschaften.* Vaduz.
Valentine, D.H. (Ed.) (1972). *Taxonomy, phytogeography and evolution.* London.
Walter (1968). *Die vegetation der Erde, in Ökophysiologischer Betrachtung.* Stuttgart.
—— (1979). (Translated J. Wieser). *Vegetation of the Earth.* New York.
Wilmanns, O. (1973). *Ökologische Pflanzensoziologie.* Heidelberg.

Climate and vegetation

Money, D.C. (1965). *Climate, soils and vegetation.* Cambridge.
Riley, D. and Spolton, L. (1974). *World weather and climate.* Cambridge.
Walter, H. and Lieth, H. (1960). *Atlas of climate diagrams.* Jena.

Soils and vegetation

Eyre, S.R. (1963). *Vegetation and soils* (of the world). London.
Kubiëna, W.L. (1953). *Bestimmungsbuch und Systematik der Böden Europas.* Stuttgart.

History of vegetation (recent)

Bottema, S. (1974). *Late quaternary vegetation history of northwestern Greece.* Groningen.

Godwin, H. (1975). *The history of the British flora* (2nd edn.). Cambridge.
Matthews, J. (1950). *Origin and distribution of the British flora.* London.
Pennington, W. (1969). *The history of British vegetation.* London.
Walter, H. and Straka, H. (1970). *Arealkunde. Floristisch-historische Geobotanik.* Stuttgart.

Arctic vegetation

Ronning, O.I. (1960). *The vegetation and flora north of the Arctic Circle.* Tromsö.
UNESCO (1970). *Ecology of the sub-Arctic regions.* Helsinki.

Boreal vegetation

Acta Phytogeographia Suecica (1965). The plant cover of Sweden. *Acta Phytogeog. Suec.* **50.**
Dahl, E. (1956). *Mountain vegetation of south Norway and its relation to the environment.* Oslo.
—— (1971). *Vegetation of Scandinavian mountains, forests, wetlands, sea shore, lakes.* IBP i Norden.
Einarsson, E. (1975). *Flora and vegetation in Iceland: country and population.* Reykjavik.
Göttlich, K. (Ed.) (1976). *Moor und Torfkunde.* Stuttgart.
Hansen, H.M. (1930). *Studies in the vegetation of Iceland.* Copenhagen.
Lindquist, B. (1932). *Beech forests of Sweden.* In *Die Buchenwälder Europas* (Ed. E. Rübel). Berne.
MacVean, D.N. (1955). *Notes on the vegetation of Iceland.* Edinburgh.
Moore, J.J. (1968). In *A classification of the bogs and wet heaths of northern Europe in Pflanzensoziologische Systematik* (ed. R. Tuxen). The Hague.
Rosenvinge, L.K. (1930). *The botany of Iceland.* Copenhagen.
Toivonen and Vuokko, (1972). *Finnish vegetation.*

Atlantic vegetation

Angel, H. (1981). *The natural history of Britain and Ireland.* London.
Birks, H.J.B. (1973). *Past and present vegetation of the Isle of Skye—a palaeoecological study.* Cambridge.
Braun-Blanquet, J. and Tüxen R. (1952). *Irische Pflanzengesellschaften.* Zurich.
Burnett, J.H. (1964). *The vegetation of Scotland.* Edinburgh and London.
Condry, W. (1974). *Woodlands.* London.
Géhu, J. (Ed.) (1975). *La végetation des landes d'Europe Occidental.* Vaduz.
Gimingham, C.H. (1972). *Ecology of heathlands.* London.
Halliday, G. and Malloch, A. (1981). *Wild flowers. Their habitats in Britain and N. Europe.* Glasgow.
Haslam, S.M. (1978). *River plants.* Cambridge.
Hawkesworth, D.L. (1974). *The changing flora and fauna of Britain.* London.

Selected bibliography

Lebrun, J., Noirfalise, A., Heinemann, P., and Van den Berghen, C. (1949). *Les associations végétales de Belgique. Bull. Soc. Bot. Belg.* **82,** 105–207.

Massart, J. (1910). *Geographie botanique de la Belgique.* Brussels.

McVean, D.N. and Ratcliffe, D.A. (1962). *Plant communities of the Scottish Highlands.* London.

Moore, P.D. and Bellamy, D.J. (1974). *Peatlands.* London.

Neville, P.J. (1976). *The forests of England.* London.

Perring, F.H. (Ed.) (1970). *The flora of a changing Britain.* London.

Rackham, O. (1975). *Hayley Wood: its history and ecology.* Cambridge.

Ratcliffe, D.A. (1977). *A nature conservation review* (2 vols.). Cambridge.

Roisin, P. (1969). *Le domaine phytogéographique Atlantique d'Europe.* Gembloux.

Salisbury, E.J. (1932). *The East Anglian flora: a study in comparative plant geography.* Norwich.

Tansley, A.G. (1968). *Britain's green mantle.* London.

—— (1939). *The British Islands and their vegetation.* Cambridge.

Westhoff, W. and den Held, A.J. (1969). *Plantengemeenschappen in Nederland.* Zütphen.

White, J. (1982). *Studies in Irish vegetation.* Dublin.

Wooldridge, S. W. and Goldring, F. (1953). *The weald.* London.

Central European vegetation

Browicz, K. (1972). *Atlas of distribution of trees and shrubs in Poland.* Warsaw.

Der Kaiserstuhl (1972). Baden-Würtemberg.

Ellenberg, H. (1982). *Vegetation Mitteleuropas mit den Alpen* (3rd edn.). Stuttgart.

Fischer, M. (1976). *Österreichs Pflanzenwelt.* (In *Naturgeschichte Österreichs).* Vienna.

Freitag, H. (1962). *Einführung in die Biogeographie von Mitteleuropa.* Stuttgart.

Hartmann, F.K. (1974). *Mitteleuropaische Wälder.* Stuttgart.

—— and Jahn, G. (1967). *Waldgesellschaften des Mitteleuropäishen Gebirgsraumes Nordlich der Alpen.* Stuttgart.

Horvat, I., Glavač, V., and Ellenberg, H. (1974). *Vegetation Südosteuropas.* Stuttgart.

Knapp, R. (1971). *Einführung in die Pflanzensoziologie.* Stuttgart.

Landolt, E. (1960). *Unsere Alpenflora.* Schweizer Alpine Club.

Meusel, H., Jäger, E., and Weinert, E. (1965). *Vergleichende Chorologie der zentraleuropäischen Flora.* Jena.

Oberdorfer, E. (1970). *Pflanzensoziologische Exkursionsflora für Süddeutschland und die angrenzenden Gebiete.* Stuttgart.

Rübel, E. (Ed.) (1932). *Die Buchenwälder Europas.* Berne.

Runge, F. (1973). *Die Pflanzengesellschaften Deutschlands.* Münster.

Szafer, W. (1966). *The vegetation of Poland.* London.

Tüxen, R. (1960). *Zur Systematik der west und mitteleuropäischer Buchenwälder.* Gembloux.

—— et al. (1962). *Contribution à l'unification du système phytosociologique pour l'Europe moyenne et nord-occidental.*

Walter, H. (1974). *Die vegetation Osteuropas, N. und C. Asiens.* Stuttgart.

Wolkinger, F. and Plank, S. (1981). *Dry grasslands of Europe.* Council of Europe, Strasburg.

Mediterranean vegetation

Braun-Blanquet, J., Roussine, N., Nègre, R. (1951). *Les groupements végétaux de la France Méditerranéenne.* Montpellier.

Duvigneaud, P. (1954). *Végétation et faune de la région Méditerranéenne Française,* Brussels.

Eberle, G. (1965). *Pflanzen am Mittelmeer.* Frankfurt.

Flahault, C. (1937). *La distribution géographique des végétaux dans la région Méditerranéenne Française.* Paris.

Giacomini, V. and Fenaroli, L. (1958). *Cognosci l'Italia. La flora.* Milan.

Guillèn, R.F. (1981). *La vegetacio dels päisos Catalans.* Barcelona.

Harant, H. and Jarry, D. (1973). *Guide du naturaliste dans le Midi de la France.* Paris.

Horvat, I. (1942). *Biljni Svijet Hrvatske.* Zagreb.

—— Glavač, V., and Ellenberg, H. (1974). *Vegetation Südosteuropas.* Stuttgart.

Ozenda, P. (1975). *Sur les étages de végétation dans les montagnes du bassin Méditerranéen.* Grenoble.

Moravec (Gen. editor). (1965–1974). *Végetacé ČSSR.* Prague.

Polunin, O. (1980). *Flowers of Greece and the Balkans.* Oxford.

—— and Smythies, B. (1973). *Flowers of south-west Europe.* Oxford.

Rikli, M. (1943–8). *Das Pflanzenkleid der Mittelmeerländer.* Berne.

Šilić, Č. (1973). *Atlas Drveća Grmlja.* Sarajevo.

—— (1977). *Smske Zeljaste Biljke.* Sarajevo.

Tallon, G. (1978). *La végétation de Camargue.*

Thirgood, J.V. (1981). *Man and the Mediterranean forest.* London.

Turrill, W.B. (1929). *Plant life of the Balkan Peninsula.* Oxford.

Zohary, M. and Orshan, G. (1966). *An outline of the geobotany of Crete.* Jerusalem.

Pannonic vegetation

Horvat, I., Glavač, V., and Ellenberg, H. (1974). *Vegetation Südosteuropas.* Stuttgart.

Sóo, P. (1962 and 1963). *Systematische übersicht der pannonischen Pflanzengesellschaften. Acta. bot. acad. sci. Hung.* **8,** 335–66: **9,** 123–50.

Wendelberger, G. (1954). *Steppen, Trockenrasen und Wälder des pannonischen Raumes. Angew. Pflanzensoziologie* **1,** 573–634.

Alpine vegetation

Braun-Blanquet, J. (1948). *La végétation alpine des Pyrenées orientales*. Barcelona.

Derenne, P. (1972). *Guide du naturaliste dans les Alps*. Paris.

Ern, H. (1966). *Die dreidimensionale Anordnung der Gebirgsvegetation auf der Iberischen Halbinsel*. Bonn.

Ozenda, P. and Landolt, E. (1970). *Zur Vegetation und Flora der Westalpen*. Zurich.

Pinto da Silva and Teles, A.N. (1980). *A florae a vegetacao de Serra da Estrela*. Lisbon.

Quézel, P. (1964). *Végétation des hautes montagnes de la Grèce méridionale*. The Hague.

—— (1967). *La végétation des hautes sommets du Pinde et de l'Olympe de Thessalie*. The Hague.

Schaer, J.P. (1972). *Guide du naturaliste dans les Alps*. Paris.

Wetlands vegetation

Haslam, S.M. (1978). *River plants*. Cambridge.

Macan, T.T. (1974). *Freshwater ecology*. London.

Whitton, B. (1979). *Rivers, lakes and marshes*. London.

Coastal vegetation

Barnes, R. (1979). *Coasts and estuaries*. London.

Lewis, J.R. (1976). *The ecology of rocky shores*. London.

Ranwell, D. (1972). *Ecology of salt marshes and sand dunes*. London.

Kuhnholtz-Lordat, G. (1924). *Dynamique de la végétation des dunes Méditerranéennes*. Paris.

Vegetation maps, etc.

Darlington, A. (1969). *Natural history atlas*. London.

Jalas, J. and Suominen, J. (1973). *Atlas florae Europaeae*. Helsinki.

Küchler, A.W. (1966). *International bibliography of vegetation maps*. Kansas.

Meusel, H., Jäger, E., and Weinert, E. (1965). *Vergleichende Chorologie der zentraleuropäischer Flora*. Jena.

Ozenda, P. (1979). *Vegetation map of the Council of Europe Member States*. Strasburg.

Richard, L. (1975). *Carte écologique des Alps*. Grenoble.

Tüxen, (1961). *Bemerkungen zu einer Vegetationskarte Europas*. Paris.

UNESCO (1970). *Vegetation map of the Mediterranean zone*. Paris.

National Parks and Nature Reserves

Europe—general

Duffey, E. (1982). *National Parks and Reserves in Western Europe*. London.

Dupont, P. (1976). *373 Parcs Nationaux et Réserves d'Europe*. Paris.

Handboek Nationale Parken en Natuurreservaten in Europa (1971). Amsterdam.

IUCN (1971). *United Nations list of National Parks and equivalent Reserves*. Brussels.

IUCN (1980). *A directory of western palearctic wetlands*. Gland.

IUCN Conservation Monitoring Centre (in prep.) *Plants in Danger: What do we know?*

Poore, D. and Glyn-Ambrose, P. (1980). *Nature conservation in northern and western Europe*. Gland.

Proceedings of the International Colloquium on European National Parks (1970). *Les Parks Nationaux*. Paris.

United Nations. *1980 United Nations list of National Parks and equivalent Reserves*. Gland.

Waycott, A. (1983). *National Parks of Western Europe*. Southampton.

Wirth, H. (Ed.) (1979). *Nature Reserves in Europe*. Leipzig.

Zimmermann, G. (1979). *Natur als Erlebnis: Die Nationalparke in Mitteleuropa*. Stuttgart.

Countries of Europe

Austria. Wolkinger, F. (1981). *Die Natur und Landschafsschutzgebiete Österreichs*. Vienna.

Austria. Machura, L. (1970). *Naturschutz in Osterreich*. Graz.

Belgium. Noirfalise, A. *et al.* (1970). *Les Réserves Naturelles de la Belgique*. Brussels.

Belgium. Kesteloot, E. (1962). *Parcs Nationaux et Réserves naturelles en Belgique*. Brussels.

Bulgaria. (1981). *Protected natural sites in the People's Republic of Bulgaria*. Sofia.

Denmark. (1979). *Kort over Danmark*. Copenhagen.

Finland. (1979). *Bestimmungen für die Nationalparks*. Helsinki.

Finland. (1975). *National parks of Finland*. Helsinki.

France. (1975). Derenne, P. (1972). *Guide du naturaliste dans les Alps*. Paris.

France. Harant, H. and Jarry, D. (1973). *Guide du naturaliste dans le Midi de la France*. Paris.

Germany. Ant, H. and Engelke, H. (1973). *Die Naturschutzgebiete der Bundesrepublik Deutschland*. Bonn.

Germany. Erz, W. (1979). *Katalog der Naturschutzgebiete in der Bundesrepublik Deutchsland*. Bonn.

Iceland. Einarsson, E. (1975). *Flora and Vegetation*. Reykjavik.

Italy. Pratesi, F. (1977). *Parci Nazionale e Zone Protette d'Italia*. Aosta.

Italy. Tassi, F. (1978). *Parchi Nazionali Urbanistica Informazioni* No. 1.

Italy. Farnetti, G. *et al.* (1975). *Guida alla natura d'Italia*. Verona.

Netherlands. (1980). *Handboek van Natuurreservaten en Wandelterreinen in Nederland*. Graveland.

Norway. Holt-Jensen, A. (1978). *The Norwegian wilderness. National Parks and Protected Areas*. Oslo.

Romania. Pop, E. and Sălăgeanu, N. (1965). *Nature Reserves in Romania*. Bucharest.

Spain. Aritio, L. B. (1979). *Los Parques Nacionales Españoles*. Madrid.

Selected bibliography

Sweden. Larsson, T. (1980). *Demarcation and description of Swedish wetlands of international importance.* Solna.

Switzerland. Becherer, A. (1972). *Führer durch die Flora der Schweiz.* Basle and Stuttgart.

Switzerland. Burckherdt, D. (1960). *Schweizer Naturschutz am Werk.* Berne.

Switzerland. Burckhardt, D. (1980). *Die schönsten Naturschutzgebiete der Schweiz.* Zurich and Munich.

United Kingdom. Stamp, D. (1969). *Nature conservation in Britain.* London.

United Kingdom. Nicholson, M. (1957). *Britain's Nature Reserves.* London.

United Kingdom. Ratcliffe, D.A. (1977). *A nature conservation review* (2 vols.). Cambridge.

United Kingdom. (1984). *The Macmillan Guide to Britain's Nature Reserves.* London.

Yugoslavia. Wojterski, T. (1971). *National parks of Yugoslavia.* Poznan.

Glossary of terms

Aapa mire. usually large mires, with parallel rows of hummocks—*strings*—and depressions—*flarks*—lying at right-angles to the general slope of the mire. See p. 34.

Acid soil. Soil with an acid reaction, with a pH of less than 7. Humus is the main source of acidity in soils.

Aestival. Species of flowering plants that flower in the summer and usually die down after producing seed in the dry period, or in the autumn.

Alkaline soil. Soil with an alkaline reaction, with a pH of more than 7.

Alluvium. Soils formed of fine particles of rock, washed down by rain or rivers, and deposited in a valley or estuary. Some of the most fertile soils are alluvial.

Alpine. The higher regions of a mountain system, usually above the climatic tree line. The altitude of the alpine zone differs in different mountains, depending on the latitude and exposure of each mountain range in Europe.

Annual. A plant which completes its life-cycle in one year or less—from seed germination to seed production; cf. **biennial.**

Arctic. Commonly referred to as the region north of the Arctic circle. Geographically it comprises lands north of the 10 °C July isotherm, where July is the warmest summer month, provided that the mean temperature of the coldest month is not higher than 0 °C.

Arctic–alpine. Arctic–alpine species are common to both arctic and alpine regions.

Aromatic plants. Plants producing volatile oils with fragrant or pungent odours.

Aspect. The position of a plant community in relation to its exposure to the sun or to the local climate, e.g. north-facing or south-facing aspect.

Association. The basic unit of phytosociology. Defined according to its floristic composition and containing characteristic species. Named with the suffix -etum, e.g. *Quercetum ilicis.*

Atlantic period. A post-glacial period, from about 5500 BC to 2500 BC. It had the warmest and wettest climate of the post-glacial period, and was followed by the drier sub-Boreal period and the cooler sub-Atlantic period of the present day.

Azonal. Azonal communities are those which occur in ecologically extreme habitats, e.g. aquatic, alpine, coastal, etc., where soil and other factors override the zonal climatic factors. Consequently, similar azonal communities may occur in more than one climatic zone.

Biomass. The total amount of living and dead organic matter in a community or ecosystem.

Black-earth soil. Humus-rich fertile soil, developed as a result of low rainfall in continental regions. The upper layers are somewhat leached. = **Chernozem** (Russian term).

Blanket-bog. A type of bog covering the whole surface like a blanket, occurring in cool or cold climates with high rainfall and high atmospheric humidity (see p. 61).

Bog. An acid, peat-rich soil developed in wet localities which is deficient in calcium and other basic salts. The peat accumulates as a result of partial decay of the vegetation.

Boreal. Belonging to the north: this region has short warm summers and long cold winters with snow, and the characteristic vegetation is the coniferous forest **taiga.**

Boreal period. Approximately 9500–7500 years ago, part of the post-glacial period immediately preceding the Atlantic period.

Boulder-clay. A highly calcareous glacial deposit of ground-down particles of rock, containing boulders, which have been carried considerable distances by the ice.

Brackish. Referring to water which has a mixture of both fresh water and salt water.

Brown-earth soil. Formed largely of clay and loam, and so called because of the presence of brown iron oxide and blackish humus which accumulate in the upper soil layers. A characteristic soil of central Europe.

Calcareous. Containing calcium in the form of chalk or lime.

Calcicole. Of plants: those growing exclusively or preferentially in soils with a high proportion of available calcium salts.

Calcifuge. Of plants: growing on acid soils and not on soils rich in calcium.

Canopy. The uppermost layer of vegetation formed by the branched crowns of trees or shrubs. Open-canopy refers to this layer in cases where the crowns are widely spaced and do not form a continuous layer.

Carr. A woodland developed over a fen or bog; typical trees include alders and willows.

Chernozem. See **black-earth soil.**

Climax community. A more or less stable plant community which is in equilibrium with the existing natural environmental conditions; cf. **sub-climax.**

Co-dominant. Referring to plants in communities in which more than one species is dominant.

Community. A general ecological term for any naturally occurring group of organisms inhabiting a common environment. Each community is relatively independent of other communities.

Coniferous forest. Forests dominated mainly by cone-bearing trees—as for example pines, silver firs, larches, etc.

Continental. Referring to climatic conditions typical of areas which lie inland from the sea and are exposed to prevailing winds from the interior. Continental climates have low humidity and rainfall, and extremes of temperature in summer and winter.

Coppice. Referring to vegetation in which the trees or shrubs are regularly cut, e.g. every 10–15 years. The coppice trees and shrubs in Britain are commonly hazel, ash, and sweet chestnut, and they often occur with widely spaced trees such as oak—coppice with standards—which are only felled when they are mature.

Deciduous forest. A forest in which the trees lose their leaves and become dormant during the winter season.

Detritus. Organic debris from decomposing animals and plants.

Doline. A more or less conical depression or swallow-hole, occurring in limestone or dolomitic rocks, down which water drains and causes local subsidence; cf. **polje.**

Dolomite. Limestone rocks containing more than 15 per cent of magnesium carbonate.

Dominant. A species in a plant community which has a dominating influence on other members of the community; being larger in size, occupying more space and light, contributing more organic matter, requiring more nutrients, etc.

Drift. A deposit over the earth's surface as a result of transport by water, wind, glaciers, etc.

Dune. Wind-blown sand deposits which build up, usually in mounds or ridges, at right-angles to the prevailing wind. See p. 184.

Ecosystem. All plants, animals, and micro-organisms within a community which interact with each other and with the environment. See p. 3.

Ecotone. A transitional or boundary stage between adjacent plant communities. See p. 4.

Ecotype. A group of plants within a species, which are adapted to a particular habitat and are consequently somewhat distinct, but which are able to cross-breed freely with other ecotypes of the same species.

Encrusting lichens. Lichen species which grow over the surface and are closely attached to the substrate.

Endemic. A species only found naturally in one country, region, or island.

Ensilage. The preservation of green fodder in a pit or silo without drying.

Epiphyte. A plant growing on another plant. Differing from a parasite, in obtaining only support from the associated plant; not food and water like the parasite.

Ericaceous. Plants, usually shrubs or dwarf shrubs, belonging to the heath family, Ericaceae.

Erosion. The wearing away of the land surface by such agencies as seas, rivers, rainfall, or by glaciers, frost, snow, or wind.

Escarpment. An inland cliff or steep slope.

Glossary of terms

Esker. A long narrow ridge of sand or gravel deposited by a stream running from under a glacier and left exposed when the glacier melts.

Eutrophic. Of waters or soils: rich in mineral nutrients, either from soil water or from the decomposition of animal or plant remains; cf. **oligotrophic.**

Evergreen. Plants which retain their green leaves throughout the year; cf. **deciduous.** Semi-evergreen plants shed a large proportion of their leaves in winter or early spring.

Fen. Peatlands which are alkaline to somewhat acid, and are relatively well supplied with mineral salts. ('Fen' in German = 'Niedermoor'.).

Flood-plain. A plain bordering a river and formed from the deposition of sediments carried down by the river, which is periodically flooded by the river with the addition of more sediments.

Floristics. Study of the composition of different types of vegetation, in terms of the species present in each type.

Flush community. A distinctive community of plants growing where water from springs or surface drainage flows locally over the surface.

Forest-tundra. Tundra covered by forest. A zone of often stunted trees forming forests in a transitional belt between Boreal forest and treeless tundra.

Formation. Referring to a community of plants extending over very large natural areas, and determined primarily by the climatic conditions. Tundra, heaths, deciduous forests, etc. are examples of formations. A basic unit of vegetation description.

Fruticose lichens. Lichens attached to the substrate by a single basal attachment and with free aerial branches; cf. **encrusting lichens.**

Garigue. A community of low, scattered, often spiny and aromatic shrubs found in the Mediterranean.

Geophytes. Perennial plants with their perennating buds situated in the upper layers of the soil during the period of dormancy, e.g. bulbs, corms, rhizomes, tubers, etc.

Glacial. Resulting from glacial activity.

Gley soils. Soils in which there is a reduction of ferric salts to ferrous salts, giving a grey colour to the soil; thus *gleying*.

Graminaceous. Referring to grasses.

Gravel. A coarse deposit of sand mixed with small pebbles, laid down by moving water or ice; thus river-gravels, glacier-gravels, etc.

Grey dunes. Sand dunes which have become stabilized with a close turf of mosses and lichens—the latter giving the greyish appearance to these dunes. See p. 188.

Halophytes. Plants able to grow in saline soils.

Heath. A community composed largely of dwarf evergreen shrubs commonly belonging to the heath Family, *Ericaceae,* and usually found on acid, porous, sandy or gravelly soils. Trees, particularly pines and birches, may be present.

Hedgehog-heath. Sub-alpine Mediterranean community of low, rounded, intricately-branched, largely spiny shrubs growing over the stony ground on mountain slopes. Members of the heath family are usually absent, but those of the cress, pea, and umbellifer families are frequent.

Helophytes. Herbaceous plants with perennating buds lying in mud.

Herbs. Non-woody plants, with sappy stems which die back at the end of the growing period. They can be either annuals or biennials, or perennials with over-wintering buds.

High forest. Mature woodland with fully-grown trees in close canopy.

Horizons. Distinctive layers developed in mature soils, as a result of water percolation and transport of soluble materials in the soil, and their subsequent precipitation in lower layers. See p. 7.

Humus. Organic matter resulting from the decomposition of plant and animal remains, and giving the soil a characteristically dark colour. Humus is important in maintaining the fertility of soils.

Hydrophytes. Plants adapted to moist conditions.

Hydrosere. Plant successional stages, or seres, originating in water.

Hygrophilous. Inhabiting or growing in wet or damp places.

Introduced. Referring to plants which are not native species, but which have been brought into a country or area by man, either accidentally or intentionally.

Inversion. Referring to atmospheric conditions in which layers of cooler air are trapped below layers of warmer air; as in valleys in mountains, etc. See p. 87.

Iron-pan. A hard concreted layer of soil particles and iron salts in the soil, which is usually impervious.

Karst. Areas of weathered limestone, distinguished by ridges, clefts, and caves, and formed as a result of erosion by water. Limestone pavement is a flat limestone area eroded into deep clefts.

Lagg. A marginal water-course flowing around a raised bog and receiving drainage water from the peat. See p. 63.

Leaching. The removal of soluble and fine insoluble particles by rainwater, percolating downwards through the upper layers of the soil, followed by their subsequent deposition in the lower layers. See p. 5.

Liane. A large woody climbing plant; the term usually applies to woody climbers in tropical forests.

Limestone. A sedimentary rock consisting largely of calcium carbonate (over 50 per cent); see **karst.**

Lithoseres. Plant successional stages, or seres, originating on rock.

Loam. A soil composed of sand, silt, and clay in approximately equal proportions. Hence sandy loam and heavy loam, when either sand or clay occur in larger proportions.

Loess. Fine particles of silt, sand, or clay transported by wind to form unstratified porous soils. One of the richest soils in the world; very fertile when irrigated.

Maquis. A dense, mostly evergreen shrub community 1–3 m high, characteristic of the Mediterranean region. It often occurs where forests have been destroyed.

Marble. A coarse-grained crystalline metamorphic rock derived from limestone.

Marl. In general terms a mixture of clay with calcium carbonate. Some marls are marine deposits, others are of fresh-water origin.

Marsh. A general term for usually low-lying land with a high water-table, which remains wet throughout the year and is periodically flooded. The soil has a mineral source, cf. **fen.**

Meadow. A closed stand of usually permanent grassland, traditionally maintained by grazing or cutting. Steppe-meadow is a meadow-like community largely of grasses, which occurs on the less arid margins of the true steppe.

Mesotrophic. Soils neither poor nor rich in mineral salts.

Micro-climates. Very local climatic conditions which are restricted to particular habitats; e.g. crevices of rocks, or the sheltered sides of boulders, etc.

Mire. A general term used by ecologists for soils and plant communities where there is an accumulation of peat. They may be either alkaline (fens), neutral (intermediate or transitional mires), or acid (bogs). See p. 5.

Montane. A general term referring to moderate elevations in mountains.

Moor. A general but not clearly definable term, usually relating to high ground with acid peat soils, and dominated by heather and other heath-like species, or grasses or sedges. ('Moor' in German = 'mire'.)

Mor. Acid humus, poor in mineral salts.

Moraine. A deposit of ice-worn particles of rock, carried down by glaciers and deposited when the glacier retreats.

Morphology. The external form and appearance of organisms.

Mosaic. Referring in vegetation to the patchwork-like distribution of different plant communities in an area, resulting from local differences in the environment.

Mudflat. A stretch of muddy land left uncovered at low tide.

Mull. Alkaline humus, rich in mineral salts.

Myxomatosis. A virus disease resulting in the marked decline of the rabbit population, and hence having a considerable effect on many plant communities grazed by rabbits.

Native species. A species occurring naturally in an area or region, which has not been introduced by human agency.

Natural vegetation. Vegetation which is unaffected by the activities of man or his domestic animals, either in the past or present; cf. **semi-natural.**

Naturalized species. Plants of foreign origin which have become fully established in a new area or region, and are able to reproduce and compete with the native species.

Neutral soils. Neither acidic nor alkaline (with equal numbers of alkaline hydroxyl and acid hydrogen ions); having a pH of 7.

Oceanic. Referring to climate; land areas which are influenced by the proximity of oceans and seas. They not only have a moderating influence on the temperature of the land mass but they increase the rainfall and humidity.

Oligotrophic. Poor in basic salts and thus in nutrients; cf. **eutrophic.**

Ombrogenous. Referring to mires: those in which the main source of water is from rainfall.

Ombrotrophic. Entirely dependent on rainfall as a source of nutrients.

Palsa mire. Raised mounds of peat with a frozen—*permafrost*—core. See p. 27.

Palynology. Analysis of fossil pollen grains, enabling the identification and distribution of species living during past geological periods to be established.

Pannonic. The relatively flat region of south-central Europe centred on Hungary, through which the Tisza and middle Danube rivers flow.

Parasites. Organisms living on other organisms and obtaining their nutriment solely from them; thus parasitic.

Pastures. Grasslands grazed by livestock.

Peat. Partly decomposed plant remains which accumulate in more or less water-logged soils, primarily because of lack of oxygen.

Permafrost. Permanently frozen sub-soil; only the surface layers thaw in the summer.

pH. A standard measure of acidity and alkalinity. A measurement of the negative logarithm of the hydrogen ion concentration. pH 7 is neutral; higher values up to pH 14 indicate degrees of alkalinity; values below pH 7 indicate degrees of acidity.

Phenology. The study of seasonal changes in plants and animals, e.g. the times of flowering of plants in relation to climate, or the dispersal of their seeds.

Phrygana. A Greek term denoting low scrub developed over dry stony soils in the Mediterranean region. See p. 138. It is, in general, equivalent to the term *garigue* which is used in the western Mediterranean.

Phytosociology. The study of plant communities: their origin, formation, composition, and structure.

Pioneer. Referring to plants: the first to colonize a substrate, or an area bare of vegetation.

Podsol. A type of soil in which the upper soil layers (A horizon) have been leached of mineral salts and humus, and deposited in the lower soil layers (B horizon); hence podsolization.

Polje. A large depression in limestone karst, which has steep sides and a flat bottom. The latter is often covered with soil and cultivated, and during heavy rains may become flooded; cf. **doline.**

Pollen analysis. See palynology.

Pre-Boreal. The period immediately following the last glaciation. It occurred from about 10 000 to 9500 years ago. See p. 16.

Pre-vernal. Referring to those species in a community which flower in early spring, before the development of the foliage of the deciduous trees and shrubs in the community.

Pseudomaquis. Similar to maquis, but characteristic of hill and montane country in the Mediterranean, and with a different assortment of woody species.

Puszta. The flat almost treeless steppe-grasslands of the Hungarian Plains.

Quartz. Crystalline silica (oxide of silicon) which is white and opaque, or sometimes forms colourless crystals; a constituent of granites and sandstones.

Quaternary period. The latest period of geological time, commencing 1–2 million years ago and continuing to the present day. There have probably been many 'ice-ages' in the Quaternary period of which the last (the Würm) was from about 110 000 to 10 000 years ago, with its maximum ice extent about 20 000 years ago.

Raised bog. A convex acid mire which builds up above the general level of the existing peat in climates of high rainfall where drainage is impeded. ('Raised bog' in German = 'Hochmoor'.)

Rand. The sloping margins of a raised bog.

Raw soils. Immature soils which have developed no soil horizons. See p. 9.

Refuge. A limited area within a larger region which has remained relatively unaffected by changes in climate, soil, etc., and where relict species or communities are able to survive.

Relict species. Species which have been able to survive in isolation having had a much wider distribution in earlier times, e.g. mountain species with discontinuous distribution.

Relief. Differences in elevation and slope of areas of the earth's surface.

Rendzina soils. Shallow alkaline soils developed over limestone or chalk. See p. 7.

Rich fen. An alkaline mire rich in mineral salts.

Rill. Water run-off drainage channels in raised bogs.

River-terrace. An area of flat land formed from flood plains along the margin of a river as it flows across a plain. As the river erodes deeper into the plain, a series of flat terraces at different levels is formed.

Run-off. Water which flows off the surface into rivulets, in contrast to water that percolates through the ground (ground water) and ultimately drains into streams.

Saline. Water containing common salt, sodium chloride. For example, the mean salinity of sea water is 35 parts of salt in 1000 of water.

Saline soils. Soils rich in sodium chloride or similar salts. See p. 8.

Saltmarsh. A marsh which is periodically flooded by sea water, or a marsh which is irrigated by water rich in salt.

Sand. Minute particles of weathered mineral rocks, transported by water or wind, and consisting largely of quartz grains.

Sandstone. A porous sedimentary rock, consisting largely of consolidated sand particles.

Saprophyte. A plant living on decaying organic matter.

Sclerophyllous. Referring to plants with thick leathery usually relatively small leaves, which reduce the loss of water by transpiration in dry climates.

Scree. Loose fragments of rock, stones, or boulders, which are detached from rock faces or cliffs by weathering, and which are mobile over short distances as a result of gravity.

Scrub. A general term, referring to low shrubs and bushes usually less then 5 m high.

Sedimentary rocks. Formed from material derived from existing rocks, often including, in addition, materials of organic origin. They are often laid down in distinct layers. Examples are sandstones, limestones, etc.

Semi-natural. Referring to plant communities which have been subjected to human interference or management, but which retain many of the natural species.

Seral, seres. Distinct stages in the development of plant communities in a particular

Glossary of terms

habitat as they mature to a stable climax. See **hydrosere, lithosere, xerosere**.

Serpentine. Rock containing magnesium silicate, often dull-greenish in colour.

Shale. A fine-grained sedimentary rock produced from clay; usually finely stratified and readily splitting into layers.

Shingle. Small rounded pebbles accumulated on beaches and offshore bars.

Shrub. A woody perennial with well-developed side branches arising from near its base and usually less than 10 m high. Dwarf shrubs or shrublets refer to woody perennials commonly less than 0.5 m high.

Šibljak. A Balkan term for a bush community of the sub-Mediterranean region, formed largely of deciduous shrubs. See p. 127.

Silica. A rock composed of silicon dioxide; it is insoluble and white or colourless. Common in natural rocks such as quartz, rock-crystal, flint, sand, etc.

Siliceous. Referring to rock or soil containing silica.

Silt. A deposit usually of fine sand and clay, laid down by rivers or lakes; composed of particles between 0.02 mm and 0.002 mm in diameter.

Slate. A fine-grained, layered metamorphic rock, developed from fine clay, which readily splits into smooth flakes. In comparison with shale it is a much harder and more resistant rock.

Snow-patch. A bank or patch of snow which persists in a certain area in the mountains for long periods, much longer than the surrounding snows. It occurs annually in the same position and has a profound effect on the local plant communities. See p. 160.

Soil horizons. See **horizons**.

Soil profile. Refers to distinctive layering in mature soils. See **horizons**.

Solifluction. The slow downhill movement of soil as a result of alternate freezing and thawing.

Soligenous. Referring to mires which are largely dependent on drainage water—normally relatively rich in minerals. By contrast, water supplied directly by rain is poor in minerals. cf **ombrogenous**.

Steppes. Extensive level lowlands, commonly covered with grasslands and usually tree-less; thus steppe-grasslands. Salt-steppes occur where salt is present in the upper layers of the soil; cf. **meadow**.

Strand-lines. The margins of seas, lakes, or rivers where terrestrial plants can establish themselves.

Sub-alpine. A zone below the true alpine zone in mountain ranges, and usually above the tree line. Dwarf shrubs and stunted trees are characteristic of this zone.

Sub-Atlantic. The last post-glacial period, from about 2500 years ago to the present day. See p. 17.

Sub-Boreal. A distinctive post-glacial period occurring from about 5000 to 2500 years ago. See p. 17.

Sub-climax community. A mature climax community developed under local soil or climatic conditions, and consequently differing from the main climatic climax community of the area.

Sub-Mediterranean. A climatic and vegetation zone lying between the true Mediterranean zone and the Central European or montane zones. In general terms it has a wider temperature range and higher rainfall than the Mediterranean zone, as well as a larger proportion of deciduous species.

Sub-montane. An ill-defined zone lying below the montane zone, with a distinctive forest zone differing from the forests of the plains.

Sub-oceanic. An area inland from the oceanic coastal regions, which has a climate intermediate between the true oceanic and continental climates. In general it has less rain and more contrasting winter and summer temperatures than the oceanic regions.

Subspecies. A subdivision within a species: a group with similar characteristics differing from the species in some minor respects, often as a result of geographical or ecological isolation.

Succession. Referring to the sequence of plant communities which replace each other as the vegetation becomes more integrated with the environment, ultimately reaching a more or less stable state—the climax community.

Succulent. A plant with fleshy, swollen, and juicy stems and leaves which retain water. Such plants are commonly adapted to arid or saline conditions.

Swamp. An area which is permanently saturated with water, with the water level above the soil surface; cf. **marsh** which in summer at least has the water level below the soil surface.

Symbiotic. An association of dissimilar organisms living together to their mutual advantage; e.g. nitrogen-fixing bacteria in the root-nodules of members of the pea family, *Leguminosae.*

Taiga. The northern coniferous forests of the Boreal region, lying to the south of the treeless tundra.

Taxon. A general term, referring to a named unit of any rank, in the classification of plants and animals, e.g. the taxa within a species may include subspecies, varieties or forma.

Taxonomy. The study of the classification of plants and animals.

Terra fusca. Refers to brown Mediterranean soils. See p. 7.

Terra rossa. A reddish clay-like soil probably resulting from the dissolution of limestone by water, and found primarily in karst regions. Fertile terra rossa soils are commonly found at the bottom of *poljes* and *dolines,* and are often under cultivation.

Tertiary period. The geological period immediately preceding the Quaternary period and lasting from about 65 million years ago to about 2 million years ago.

Topography. The surface configuration of an area, including its relief and the position of natural and man-made features.

Tramontana. A cool dry wind coming from the north, often over mountains, and influencing an otherwise Mediterranean climate.

Transitional mire. A mosaic of alkaline fen and acid bog covering an area, or cases where bog and fen succeed each other.

Tree-layer. The uppermost layer of vegetation in a forest or woodland. This profoundly influences the development and composition of the lower layers.

Tree line. The limit of growth of trees, either in the mountains or in the northern regions. In the mountains, for example, the tree line depends not only on altitude but on aspect, soils, and local climate.

Tufa. A deposit of insoluble calcium carbonate from spring-water rich in lime.

Tundra. A treeless zone, lying principally north of the Arctic circle, where winters are long and severe and summers short and relatively cool (mean July temperature not above 10 °C). The soil is permanently frozen below the surface layers. The main types of vegetation are moss-tundra, lichen-tundra, dwarf-heath tundra, and arctic–alpine tundra.

Tundra soil. See p. 8.

Valley bog. Acid mire developed in a valley, where drainage water comes from acid rocks or subsoils, and which is largely independent of rainfall.

Vernal aspect. Referring to species within a community, which flower in the late spring at the time when the dominant trees and shrubs are breaking into leaf; cf. **pre-vernal**.

Water-table. The uppermost level to which water saturates the soil or substrate. The height of the water-table usually varies with the seasons and with the water-holding capacity of the soil or substrate.

Weichselian glaciation. The last glacial period of Northern Europe. See p. 15.

Würm glaciation. The last glacial period, at its height about 20 000 years ago.

Xerothermic. Referring to plants: those that can withstand both drought and heat.

Xerosere. Plant communities developing in dry habitats.

Zonation. Referring to vegetation: the successive bands or zones of vegetation occurring under conditions which gradually change from place to place, e.g. the zonation of vegetation on mountains as a result of increasing altitudes, or the zonation of lake-side vegetation on progressing inland from open water to exposed soil.

Index of Latin names of illustrated species

Bold numbers refer to the colour plate section at the end of the book.

Index of Latin names of illustrated species

Index of Latin names of illustrated species

Index of Latin names of illustrated species

Index of English names of illustrated species

Index of English names of illustrated species

Index of English names of illustrated species

Index to plant communities of Europe

Bold numbers refer to illustrations in the colour plate section at the end of the book.
Italic numbers refer to pages on which the entry is illustrated.

237

Index of plant communities of Europe

Plate 61 (above). Arctic tundra. Porsanger Fjord, Norway. Dwarf-heath tundra with small lakelets surrounded by mires, with cottongrasses, *Eriophorum* sps., and sedges, *Carex* sps. In more sheltered areas, low willow thickets occur consisting largely of the woolly willow, *Salix lanata;* and *S. glauca.*

Plate 62 (left). Arctic lichen–tundra. Spannstind, Norway. The lichen species are commonly those of *Cladonia* and *Cetraria.* With them are the dwarf birch, *Betula nana,* (right) and the bearberry, *Arctostaphylos uva-ursi* (left).

Plate 63 (above). Arctic dwarf-heath tundra. Vadsö, Norway. Shrublets forming low mats, usually less than 20 cm high, are snow-covered for many months in the winter and are thus protected from extremes of climate. Present here are crowberry, *Empetrum nigrum;* dwarf birch, *Betula nana;* alpine bearberry, *Arctostaphylos alpina;* and bog bilberry, *Vaccinium uliginosum.*

Plate 64 (below). Arctic arctic–alpine tundra. Near Bolna, Norway. The le of snow-lie and aspect are very important in the development of arctic–alpine communities. This example has the blue heath, *Phyllodoce caerulea;* and *Cassiope tetragona,* with the dwarf willow, *Salix herbacea,* and encrusting lic such as *Ochrolechia* sps.

**Plate 65 (right). Arctic
birch wood.** Near
Nordkjosbotn, Norway.
The northern form of the
hairy birch, *Betula
pubescens* ssp. *tortuosa*,
forms low woods,
characteristically with the
lower trunks bare of
branches as a result of the
winter snow-cover which
may be 1–1½ m deep. The
dwarf shrub layer consists
largely of cowberry,
Vaccinium vitis-idaea;
crowberry, *Empetrum
nigrum;* and bearberry,
Arctostaphylos uva-ursi.

**Plate 66 (below). Arctic
meadows.** Porsanger
fjord, Norway. Semi-
natural grasslands are
annually cut for hay. They
occur along river verges
and inlets by the sea and
are largely dominated by
species of fescue,
Festuca; meadow-
grasses, *Poa;* bents,
Agrostis; and the tufted
hair-grass, *Deschampsia
cespitosa.* Manuring
usually increases the
number of non-grass
species in the meadows.

Plate 67 (above). Arctic palsa mire. Northern Sweden. An example in the extreme north of Sweden of a hummock which is up to 6 m high. Note grass tufts on the summit growing as a result of bird droppings.

Plate 68 (left). Arctic palsa mire. Near Utsjoki, Finland. Showing hummocks and hollows, resulting from the permafrost layers of frozen soil lying beneath the upper unfrozen layers in the summer. The hummocks typically have dwarf birch, *Betula nana;* cloudberry, *Rubus chamaemorus;* labrador-tea, *Ledum palustre;* and cowberry, *Vaccinium vitis-idaea.* The hollows have cottongrasses, *Eriophorum* sps.; sedges, *Carex* sps.; and commonly bogbean, *Menyanthes trifoliata;* and marsh cinquefoil. *Potentilla palustris.*

Plate 69. Boreal taiga—mixed Norway-spruce forest. Petkula, Finland. Here the spruce, *Picea abies,* is mixed with the downy birch, *Betula pubescens,* and the Scots pine, *Pinus sylvestris.* The dwarf-shrub layer includes ericaceous shrublets such as heather, *Calluna vulgaris;* cowberry, *Vaccinium vitis-idaea;* and bog bilberry, *Vaccinium uliginosum,* and the wavy hair-grass, *Deschampsia flexuosa;* and hairy wood-rush, *Luzula pilosa.*

Plate 70 (right). Boreal taiga—Scots pine forest. Norrköping, southern Sweden. The ground flora has such ubiquitous dwarf shrubs as bilberry, *Vaccinium myrtillus;* and cranberry, *Vaccinium oxycoccos,* with the common sedge, *Carex nigra,* and with lichens on the rocks. There are also scattered bushes of oak and birch.

Plate 71 (far right). Boreal taiga—Scots pine–lichen forest. West-central Sweden. The ground layer is largely of *Cladonia rangiferina* and some *Cladonia stellaris* (yellowish), with a weak growth of heather, *Calluna,* which has colonized the ground after a forest fire which took place a considerable time ago—note the burnt tree-stump.

Plate 72 (above). Boreal aapa or string mires. Sjöttvatn, northern Sweden. These mires occur as a series of ridges and depressions or 'flarks', at right angles to the slope of the ground. The ridges are formed largely by bog mosses, mainly *Sphagnum fuscum*, often overgrown by lichens, and also the dwarf birch, *Betula nana;* the cloudberry, *Rubus chamaemorus;* and hare's-tail cottongrass, *Eriophorum vaginatum*. The flarks have the white beaked-sedge, *Rhynchospora alba;* bogbean, *Menyanthes trifoliata;* bottle sedge, *Carex rostrata;* and other sedges, and the sundews *Drosera* sps.

Plate 73 (left). Boreal aapa or string mire. Petkula, Finland. This hummock or string is developed largely by bog mosses such as *Sphagnum magellanicum* and *S. fuscum;* through it are growing bogbean, *Menyanthes trifoliata;* bog rosemary, *Andromeda polifolia;* cranberry, *Vaccinium oxycoccous,* and the deergrass, *Scirpus cespitosus*.

Plate 74 (right). Boreal heath. Near Ivalo, Finland. Communities of largely ericaceous shrublets here including heather, *Calluna vulgaris;* crowberry, *Empetrum nigrum;* alpine bearberry, *Arctostaphylos alpina;* bog bilberry, *Vaccinium uliginosum;* and species of lichens.

Plate 75 (below). Boreal bog-pool. Sweden. At the water's edge are clumps of hare's-tail cottongrass, *Eriophorum vaginatum,* and carpets of bog mosses, *Sphagnum* sps. Inland are scattered Scots pine, *Pinus sylvestris,* and downy birch, *Betula pubescens.*

Plate 76 (above). Boreal grassland. Near Nyseter (Romsdal), Norway. A hay-meadow with tufted hair-grass, *Deschampsia cespitosa;* buttercups, *Ranunculus* sps.; and lady's-mantle, *Alchemilla vulgaris.* The walled field is surrounded by birch, *Betula* sps.; and grey alder, *Alnus incana.*

Plate 77 (left). Atlantic sessile oak wood. Devil's Bridge, Wales, UK. The sessile oak, *Quercus petraea,* forms woods on more acid soils and in more western oceanic climates. Here in the montane region of Wales, it is dominant, with frequent rowan, *Sorbus aucuparia;* birch, *Betula* sps.; hazel, *Corylus avellana;* and hawthorn, *Crataegus monogyna.* The dwarf shrub layer consists largely of bilberry, *Vaccinium myrtillus.*

Plate 78 (above). Atlantic pedunculate oak wood. Near Farnham, UK. The pedunculate oak, *Quercus robur,* favours richer neutral or alkaline and often damper soils in less oceanic climates. The holly, *Ilex aquifolium;* field maple, *Acer campestre;* and the climbing honeysuckle, *Lonicera periclymenum,* are common in the upper shrub layer, while the lower shrub layer contains blackberry, *Rubus* sps. and rose, *Rosa* sps.

Plate 79 (below). Atlantic Pyrenean oak wood. Gerês National Park, Portugal. The Pyrenean oak, *Quercus pyrenaica,* is restricted to the western part of the Iberian Peninsula and frequently forms woods on the sides of valleys in montane regions. With it commonly occur the strawberry tree, *Arbutus unedo;* ivy, *Hedera helix;* with the tree heath, *Erica arborea;* and brooms, *Cytisus* sps.

Plate 80 (above). Atlantic beech wood. North-east France. Beech, *Fagus sylvatica*, commonly grows in closed canopy and, due to selective felling for timber, it usually occurs in uniform stands of similar-aged trees. Bramble, *Rubus fruticosus*, is the only shrubby species in this example and, as a result of heavy leaf litter which is slow to decompose, there is very sparse field layer.

Plate 81 (left). Atlantic yew wood. Butser Hill, near Petersfield, UK. Woods of yew, *Taxus baccata*, are rare in Europe, though the tree is widespread. It is found largely on limestone, and here on the valley slopes it forms a dense canopy with no field layer. In the foreground are trees of beech, *Fagus sylvatica;* hawthorn, *Crataegus monogyna;* and whitebeam, *Sorbus aria*.

Plate 82 (above). Atlantic wet wood. Haute-Marne, France. Alder woods or 'carrs' are found on waterlogged soils rich in minerals, in river valleys and at lake-sides. They have a rich shrub layer including alder buckthorn, *Frangula alnus*; grey willow, *Salix cinerea*; and black currant, *Ribes nigrum*, and, in the field layer, yellow iris, *Iris pseudacorus*; common nettle, *Urtica dioica*; and the greater tussock-sedge, *Carex paniculata*.

Plate 83 (left). Atlantic maritime pine wood. Finisterre, Spain. *Pinus pinaster*, the maritime pine, forms woods on the Atlantic seaboard on sandy soils, often on the inland side of sand dunes. These woods have a shrub and field layer flora typical of southern Atlantic heaths.

Plate 84 (left). Atlantic deciduous bush community. Flatford Mill, Colchester, UK. A typical deciduous bush hedgerow community of southern England contains gorse, *Ulex europaeus;* blackthorn, *Prunus spinosa;* field rose, *Rosa arvensis;* buckthorn, *Rhamnus catharticus;* and hawthorn, *Crataegus monogyna,* along with other shrub species.

Plate 85 (below). Atlantic oceanic heath. Studland Heath, Dorset, UK. After fire, there is a rapid regrowth and recolonization by Dorset heath *Erica ciliaris;* heather, *Calluna vulgaris;* western gorse, *Ulex gallii;* and bog hair-grass, *Deschampsia setacea.*

Plate 86 (left). Atlantic south-western heath. Gerês National Park, Portugal. Dense thickets several metres high are common and are composed largely of heather species such as tree heath, *Erica arborea;* green heather, *Erica scoparia;* and *E. erigena,* with *Echinospartum lusitanicum, Genista falcata,* and *Chamaespartium tridentatum.* In many cases these heaths have replaced woods of pine and oak.

Plate 87 (below). Atlantic blanket bog. Sutherland, Scotland, UK. These bogs cover large areas over acidic rocks, largely where there is a high rainfall and constantly moist atmosphere. The commonest species—none of which can be considered as dominant—are the hare's-tail cottongrass, *Eriophorum vaginatum;* deergrass, *Scirpus cespitosus;* white and brown beak-sedges, *Rhynchospora alba* and *R. fusca;* black bog-rush, *Schoenus nigricans;* and purple moor-grass, *Molinia caerulea.* Hummocks of bog mosses, *Sphagnum* sps., occur commonly.

Plate 88 (above). Atlantic acid grassland. Plynlimon Fawr, Wales, UK. Montane ac grasslands are largely composed of bents, *Agrostis* sps.; mat-grass, *Nardus strict* which is often dominant; and the purple moor-grass, *Molini caerulea*, commonly with tormentil, *Potentilla erecta;* ar heath bedstraw, *Galium saxatile.*

Plate 89 (left). Atlantic calcareous grassland. South Downs, Sussex, UK. Grazing largely by sheep maintains suc grasslands in a sub-climax sta Common grasses include meadow oat-grass, *Avenula pratensis;* red fescue, *Festuca rubra;* sheep's-fescue, *F. ovin* quaking-grass, *Briza media;* ar crested hair-grass, *Koeleria cristata.* Many other species a present. Here occur rough hawkbit, *Leontodon hispidus;* common bird's-foot-trefoil, *Lotus corniculatus;* squinancywort, *Asperula cynanchica;* red clover, *Trifoliu pratense;* and common spotte orchid, *Dactylorhiza fuchsii.*

Plate 90 (left). Montane grassland. Picos de Europa, near Covadonga, Spain. This grassland of northern Spain has such widespread grasses as cock's-foot, *Dactylis glomerata;* and sweet vernal-grass, *Anthoxanthum odoratum,* with other species such as astrantia, *Astrantia major;* yellow rattle, *Rhinanthus* sps.; rough hawkbit, *Leontodon hispidus;* and devil's-bit scabious, *Succisa pratensis.*

Plate 91 (below). Central Europe sub-alpine beech wood. Bavarian Alps, West Germany. Commonly associated with these sub-alpine beech woods are the sycamore, *Acer pseudoplatanus,* and shrubby honeysuckles, such as the alpine honeysuckle, *Lonicera alpigena;* and the black-berried honeysuckle, *L. nigra;* also the alpine rose, *Rosa pendulina.* The shrub and field layers are often poorly developed as a result of heavy shading and leaf-fall. Spruce, *Picea abies;* and dwarf mountain pine, *Pinus mugo,* are here seen at similar altitudes.

Plates 92 (right) and 93 (far right). Central Europe oak-hornbeam wood. Wienerwald, Austria. The hornbeam, *Carpinus betulus;* and tne sessile oak, *Quercus petraea,* and/or the pedunculate oak, *Quercus robur,* form mixed woods in a variety of soils and climates, and with them are commonly associated maples, limes, ash, elm, and cherry, while the shrub layer usually contains hawthorns, *Crataegus* sps.; spindle, *Euonymus europaeus;* and dogweed, *Cornus sanguinea.* The field layer has many early-flowering herbaceous perennials such as greater stitchwort, *Stellaria holostea.*

Plate 94. Central Europe dwarf mountain pine. Raxalpe, Austria, c.2000 m. The highest woody zone in mountainous regions, particularly on calcareous soils, is often dominated by the dwarf mountain pine, *Pinus mugo,* which may form dense thickets 1–2 m high. On acid soils the dwarf juniper, *Juniperus communis* subsp. *nana,* may be dominant with the false medlar, *Sorbus chamaemespilus;* and alpenrose, *Rhododendron ferrugineum,* while the green alder, *Alnus viridis,* occurs on damper slopes.

Plate 95 (far left). Central Europe Austrian black pine wood. Kanjon Rakitnice, Visoclea, Yugoslavia. The Austrian black pine *P. nigra* ssp. *nigra* forms forests in montane areas from Austria to Greece. Commonly growing with it in the upper shrub layer are the snowy mespilus, *Amelanchier ovalis;* whitebeam, *Sorbus aria;* and *Cotoneaster* sps. Here it is mixed with beech.

Plate 96 (left). Central Europe Arolla pine woods. Tatra Mountains, Czechoslovakia. Here at the tree-line, the tall Norway spruce, *Picea abies,* is mixed with the rounded trees of arolla pine, *Pinus cembra.* On the higher slopes are thickets of dwarf mountain pine, *P. mugo.*

Plate 97. Central Europe silver fir forest. Pic du Midi, Pyrenees, France. The silver fir, *Abies alba,* commonly forms mixed forests in the mountains of Central Europe, often with beech, spruce, and sycamore. Here in the Pyrenees in its westernmost extension it is mixed with beech, *Fagus sylvatica.*

Plate 98. Central Europe larch forest. Engadine, Switzerland. Larch, *Larix decidua,* forms forests high in the central Alps.

Plate 99 (below far left). Central Europe wet woodland. Šur, Czechoslovakia. River valley woods subject to flooding are often dominated by the common alder, *Alnus glutinosa,* where the shrub layer contains black currant, *Ribes nigrum;* dewberry, *Rubus caesius;* alder buckthorn, *Frangula alnus,* and other shrubs. The field layer is rich in species, the common nettle, *Urtica dioica,* being ubiquitous, along with lady-fern, *Athyrium filix-femina;* and ground-elder, *Aegopodium podagraria.*

Plate 100 (left). Central Europe bog community. Weichselboden, Austria. A typical bog community in the cool mountain climate at altitudes of 700–1200 m with several bogmosses, *Sphagnum* sps., and, typically, cranberry, *Vaccinium oxycoccos;* bog rosemary, *Andromeda polifolia;* and the rannoch-rush, *Scheuchzeria palustris,* in the pools. The dwarf mountain pine, *P. mugo,* occupies the drier ground.

Plate 101. Central Europe dry montane grassland. Near Hainburg an der Donau, Austria. This dry steppe-like grassland has a rich flora which here includes *Dianthus pontederae; Erysimum odoratum;* dropwort, *Filipendula vulgaris;* bloody crane's-bill, *Geranium sanguineum; Galium glaucum;* and yellow woundwort, *Stachys recta.*

Plate 102 (far left). Central Europe sub-alpine grassland. Karec, Austria. Grassland on siliceous soil at about 2000 m, with the giant cat's-ear, *Hypochoeris uniflora;* fragrant orchid, *Gymnadenia conopsea;* bearded bellflower, *Campanula barbata;* wood crane's-bill, *Geranium sylvaticum;* and aristate yellow rattle, *Rhinanthus aristatus.*

Plate 103 (left). Central Europe sub-alpine grassland. Val di Fassa, Trento, Italy. A mineral-rich meadow at about 1500 m in the Dolomites, with meadow clary, *Salvia pratensis;* bird's-foot trefoil, *Lotus corniculatus;* mountain clover, *Trifolium montanum;* and sainfoin, *Onobrychis viciifolia.*

Plate 104 (above). Mediterranean holm oak wood. Aristi, Greece. The holm oak, *Quercus ilex*, rarely occurs in closed canopy, largely as a result of grazing and felling. It is usually found in open stands with a typical maquis shrub layer of *Cistus* sps., buckthorn *Rhamnus* sps., and *Phillyrea* sps. Oriental hornbeam, *Carpinus orientalis*, and strawberry tree, *Arbutus unedo*, are commonly present, as in this example.

Plate 105 (below). Mediterranean kermes oak wood. Karpenision, Greece. The kermes oak, *Quercus coccifera*, is widespread in the Mediterranean region, but rarely forms closed woods. Here it is heavily grazed by sheep and goats. Typical small trees include the almond-leaved pear, *Pyrus amygdaliformis;* and the turpentine tree *Pistacia terebinthus*, while shrubs include the thorny burnet, *Sarcopoterium spinosum;* spiny broom, *Calicotome villosa;* Jerusalem sage, *Phlomis fruticosa;* and often the olive, *Olea europea*.

Plate 106 (right). Mediterranean cork oak wood. Provence, France. The western Mediterranean cork oak, *Quercus suber*, requires a moister climate than most of the other evergreen oaks. Semi-natural woods occur in Spain and Portugal, but these oaks are usually planted in open stands for their bark, which is removed every 8 to 12 years, while beneath them either a rich shrub flora develops, or the ground is kept under cultivation.

Plate 107 (far right). Mediterranean Hungarian oak wood. Near Agia Trias, Greece. The deciduous Hungarian oak, *Quercus frainetto*, commonly forms mixed woods with the turkey oak, *Q. cerris*, in the sub-Mediterranean region of the Balkan Peninsula; often with other oak species. The field layer is poorly developed.

Plate 108. Mediterranean white oak wood. Panzano, Italy. The white oak, *Quercus pubescens*, is widespread in the sub-Mediterranean region, usually at higher altitudes than the evergreen oaks. It is often degraded to a deciduous bush community, or šibljak. Many tree species occur, including hop-hornbeam, *Ostrya carpinifolia;* nettle tree, *Celtis australis;* wild service tree, *Sorbus torminalis;* and maple, *Acer* sps., etc.

Plate 113 (left). Mediterranean Balkan pine wood. Mount Olympus, Greece. Two closely related species, the Balkan pine, *Pinus heldreichii,* and the Bosnian pine, *P. leucodermis,* form open woods in the higher mountains of the Balkan peninsula, from about 1700 to over 2300 m. Associated with these woods are the spurge-laurel, *Daphne laureola,* and the mezereon, *D. mezereum;* the box, *Buxus sempervirens; Cotoneaster integerrimus;* and juniper, *Juniperus communis.*

Plate 114 (below). Mediterranean sweet chestnut wood. Panzano, Italy. This example from central Italy has hazel, *Corylus avellana:* medlar, *Mespilus germanica;* broom, *Cytisus scoparius;* and alder buckthorn, *Frangula alnus,* in the shrub layer. The native woods further east in Europe have a greater number of species in the tree layer and a rich field layer.

Plate 115. Mediterranean oriental beech wood. Near Demirköy, Turkey-in-Europe. The oriental beech, *Fagus orientalis*, forms woods in the eastern part of the Balkan peninsula. Here it is associated with the rhododendron, *R. ponticum*.

Plates 116 (right) and 117 (far right). Mediterranean Greek fir wood. Mount Parnassus, Greece. The Greek fir, *Abies cephalonica*, forms woods in the montane regions of southern Greece, at altitudes between 800 and 1700 m. Further north, the hybrid of the Greek fir with the silver fir, *Abies borisii-regis*, forms woods in the Balkans. With the Greek fir are commonly associated the white oak, *Quercus pubescens;* the prickly juniper, *Juniperus oxycedrus;* and a sparse ground flora.

**Plate 118 (above).
Mediterranean funeral
cypress wood.** Samaria
Gorge, Crete. The funeral
cypress, *Cupressus
sempervirens,* is restricted to
the montane region of the
southern Balkans in Europe,
though it is widely planted and
naturalized in southern Europe
It rarely forms woods except
here in Crete, at altitudes of
about 800 m where it is
commonly associated with the
semi-evergreen maple, *Acer
sempervirens.* Many shrubby
species of the mint family,
Labiatae, are also common
such as *Phlomis;* sage, *Salvia*
sps.; thyme, *Thymus* sps.;
etc.

**Plate 119 (left).
Mediterranean Spanish
juniper wood.** Near Ucero,
Soria, Spain. The Spanish
juniper, *Juniperus thurifera,*
forms local thickets in the
mountains of southern and
central Spain and the French
Alps. It is commonly
associated with the holm oak,
Quercus ilex; Portuguese oak,
Q. faginea; and other juniper
species. It rarely forms closed
woods.

**Plate 120 (above).
Mediterranean olive grove.**
Delphi, Greece. The olive, *Olea europaea* var. *sylvestris*, is native in the Mediterranean region and it is commonly associated with the carob, *Ceratonia siliqua*. However olives in different forms have been widely cultivated in the Mediterranean region for their oil-producing fruits, and olive groves usually stand on cultivated ground with no shrub or field layers below. Wild olives usually occur in the maquis and garigue communities.

**Plate 121 (left).
Mediterranean oriental plane wood.** Sithonia, Greece. The oriental plane, *Platanus orientalis*, forms local woods in river valleys, on flood-plains, and on river sands and gravels in the Mediterranean and sub-Mediterranean regions, but these woods have been largely destroyed. Where a shrub layer develops it commonly consists of the oleander, *Nerium oleander;* chaste-tree, *Vitex agnus-castus;* with willows, *Salix* sps.; *Rubus* sps.; and tamarisks, *Tamarix* sps.

**Plate 122 (above).
Mediterranean maquis with
pine.** Sithonia, Greece.
Maquis is a dense shrub
community, several metres
high, composed largely of tree
heath, *Erica arborea;* Jupiter's
beard, *Anthyllis barba-jovis;*
sping broom, *Calicotome
villosa;* Eastern strawberry
tree, *Arbutus andrachne;* and
wild olive, *Olea europaea.* The
Aleppo pine, *Pinus halepensis,*
is here colonizing the maquis
and could become dominant if
undisturbed by man or fire.

**Plate 123 (left).
Mediterranean maquis in
the evergreen-oak zone.**
Kassandra, Greece. This
maquis is dominated by the
tree heath, *Erica arborea,* with
E. manipuliflora and the yellow
aromatic inula, *Dittrichia
viscosa.* The taller shrubs and
small trees include the kermes
oak, *Quercus coccifera;* Judas
tree, *Cercis siliquastrum;*
mastic tree, *Pistacia lentiscus;*
and Christ's thorn, *Paliurus
spina-christi.*

Plate 124 (above). Mediterranean cistus garigue. Vega de Tera, Spain. The gum cistus, *C. ladanifer,* is here dominant locally in the southern part of the Iberian Peninsula. Gum cistus communities, like other cistus maquis communities, are sub-climax communities resulting from the destruction of the oak woods that once dominated much of the Iberian Peninsula.

Plate 125 (above right). Mediterranean lavender garigue. Near Almonte, Spain. One of many types of garigue or dwarf shrub communities which now cover large areas of the Mediterranean region. Here the French lavender, *Lavandula stoechas,* is prominent, with scattered groups of cistus.

Plate 126 (right). Mediterranean garigue with dwarf fan palm. Southwest of Seville, southern Spain. Europe's only native palm is found in coastal garigue in the western Mediterranean. The great quaking grass, *Briza maxima,* is also clearly visible in this example.

**Plate 127 (above).
Mediterranean euphorbia
garigue.** Asine, near Nafplion,
Greece. Round the coastal
regions of the Mediterranean,
garigues of spurges are
common with rounded bushes
of the tree spurge, *Euphorbia
dendroides*, and/or the
Greek spiny spurge, *E.
acanthothamnos*, with, in the
eastern part of the
Mediterranean, Jerusalem
sage, *Phlomis fruticosa*; and
thorny burnet, *Sarcopoterium
spinosum*.

**Plate 128 (left).
Mediterranean eastern
thorny garigue or phrygana.**
Mycene, Greece. Dwarf
shrubs and shrublets in this
phrygana include jerusalem
sage, *Phlomis fruticosa*; spiny
broom, *Calicotome villosa*;
Genista acanthoclada;
Anthyllis hermanniae; *Ballota
acetabulosa*; and the thorny
burnet, *Sarcopoterium
spinosum*.

**Plate 129 (left).
Mediterranean Greek gorge
garigue or phrygana.** Imbros
Gorge, Crete. Characteristic of
such gorges is a garigue of
dwarf shrubs with
*Chamaecytisus creticus;
Anthyllis hermanniae; Genista
acanthoclada;* Greek spiny
spurge, *Euphorbia
acanthothamnos;* large
Mediterranean spurge, *E.
characias; Cistus parviflorus,
Phlomis lanata; Phlomis
cretica;* Jerusalem sage,
Phlomis fruticosa; and *Ballota
pseudodictamnus.*

**Plate 130 (below).
Mediterranean Greek island
phrygana.** Delos, Greece. The
dwarf spiny bushes of
*Sarcopoterium, Calicotome,
Genista,* and other dwarf
shrubs cover much of the
ground, while between the
shrubs in spring and early
summer there is a rich
flowering of herbaceous
annuals and bulbous and
rhizomatous perennials.

Plate 131 (left). Mediterranean rock-wall communities. Delphi, Greece. Cliffs and rock-walls in the montane region may be rich in species particularly in north-facing situations. Here above the ruins of Delphi, are Greek spiny spurge, *Euphorbia acanthothamnos;* large Mediterranean spurge, *E. characias;* Jerusalem sage, *Phlomis fruticosa;* yellow asphodel, *Asphodeline lutea;* and grass species.

Plate 132 (below). Pannonic steppe grasslands. Kaliakra, Black Sea, Bulgaria. Many grass species are widespread in the steppe grasslands, notably the feather-grasses, *Stipa* sps. which are very distinctive. Here on the plateau inland from the Black Sea, is a rich flora of spurges, *Euphorbia* sps.; knapweeds, *Centaurea* sps. and with *Androsace maxima* among many other herbaceous perennials, and bulbous species such as *Gagea, Ornithogalum, Asphodelus,* and *Asphodeline* sps.

Plate 133 (left). Pannonic salt-steppe grassland. Neusiedlersee, Austria. On certain types of saline soils (solončhak) the grass, *Puccinellia festuciformis* ssp. *intermedia,* may dominate locally, from eastern Austria to Romania. Often occurring with this grass are *Lepidium cartilagineum* ssp. *crassifolium;* annual sea-blight, *Suaeda maritima* ssp. *pannonicis;* and sea aster, *Aster tripolium* ssp. *pannonicus.*

Plate 134 (below). Pannonic salt-steppe grassland. Apetlon, Seewinkel, Austria. Grassland with *Lepidium cartilagineum* sps. *crassifolium* which is a typical saline species of Central and Eastern Europe.

Plate 135 (left). Pannonic inland saline community. Kirchsee, Seewinkel, Austria. A dried-up salt lake with the grass, *Crypsis aculeata,* and the annual sea-blight, *Suaeda maritima* (light green). Reeds fringe the lake.

Plate 136 (below left). Pannonic inland saline community. Kirchsee, Seewinkel, Austria. The eastern form of the annual sea-blight, *Suaeda maritima* ssp. *pannonica,* growing on a dried-up salt lake.

Plate 137 (below right). Pannonic inland saline community. Kirchsee, Seewinkel, Austria. An example in eastern Austria of an inland salt community with the sea club-rush *Scirpus maritimus;* oak-leaved goosefoot, *Chenopodium glaucum;* and the grass, *Crypsis aculeata.*

Plate 138 (above). Pannonic woodland. Lobau, Stadler Furt, Vienna, Austria. A riverside wood on a small branch of the Danube, dominated by grey poplar, *Populus canescens,* with shrubs mainly of hawthorn, *Crataegus monogyna;* dogwood, *Cornus sanguinea;* and white willow, *Salix alba.*

Plate 139 (right). Alpine shrub community. Niedere Tauern range, Styria, Austria. Alpine shrub communities are commonly found above the tree-line and include the alpenrose, *Rhododendron ferrugineum,* shown here, and the hairy alpenrose, *R. hirsutum.* Green alder, *Alnus viridis,* is found on humid soils, dwarf mountain pine, *P. mugo,* on dry soils; with the ericaceous dwarf shrubs, bilberry, *Vaccinium myrtillus;* cowberry, *V. vitis-idaea;* and bog bilberry, *V. uliginosum.*

Plate 140 (left). Alpine tall herb community. Lienzer Dolomites, Tyrol, Austria. At or below the tree-line there is often a rich tall herbaceous community dominated here by *Adenostyles alliariae,* often with the wolfsbane, *Aconitum vulparia;* Austrian leopard's-bane, *Doronicum austriacum;* and the hairy chervil, *Chaerophyllum hirsutum.* Typically above the tree-line such other tall species as common monkshood, *Aconitum napellus;* and spiniest thistle, *Cirsium spinosissimum,* occur.

Plate 141 (below). Alpine dwarf shrub community. Mount Parnon, Greece. A zone of dwarf shrubs and woody perennials often occurs on exposed and rocky slopes, and where the trees have been felled and grazing is heavy. Here on Mount Parnon are the white-flowered *Astragalus angustifolius, Cerastium candidissimum,* and *Achillea umbellata,* the two latter restricted to the mountains of Greece.

Plate 142 (right). Alpine hedge-hog heath. Mount Volakias, Greece. These montane shrub communities are composed largely of low cushion-like and often spiny dwarf shrubs with here, in the eastern Mediterranean, the Greek spiny spurge, *Euphorbia acanthothamnos; Astragalus angustifolius; Prunus prostrata;* and often with *Acantholimon androsaceum; Berberis cretica;* etc.

Plate 143 (far right). Alpine hedge-hog heath. Mount Ossa, Greece. These low spiny dwarf shrub communities are not strictly heaths as they contain no ericaceous species, but are commonly composed of *Astragalus, Euphorbia, Echinops* species, and here in Greece the grey herbaceous horehound, *Marrubium thessalum.*

Plate 144. Sub-alpine grasslands. Königstuhl c. 2300 m, Gurktaler Alpen, Austria. Developed as a result of deforestation and subsequent grazing, with remnants of arolla pine, *P. cembra,* and dwarf shrub communities of dwarf juniper, green alder, etc. The grassland is dominated by *Avenula versicolor; Anthoxanthum alpinum;* mat-grass, *Nardus stricta;* and species of fescue, *Festuca* sps.

Plate 145 (right). Alpine rock-wall community. Mount Olympus, Greece. On Mount Olympus, Greece there are 20 or more endemic species which are found nowhere else in the world. They include *Potentilla deorum* (right) and the grass, *Festuca olympica*. Also present is *Saxifraga scardica* which is more widespread in the mountains of the Balkan Peninsula.

Plate 146 (below left). Alpine grassland community. Velika Planina, Yugoslavia. This community is rich in species, including here the spring gentian, *Gentiana verna;* compact rockcress, *Arabis vochinensis;* and shrubby milkwort, *Polygala chamaebuxus.*

Plate 147 (below right). Alpine flush community. Oppland, c. 1400 m. southwestern Norway, A mountain flush community with starry saxifrage, *Saxifraga stellaris;* sedges, *Carex* sps.; cottongrass, *Eriophorum* sps.; and mosses.

Plate 148 (above). Alpine grassland. Steinwandeck, 2474 m, Radstädter Tauern, Austria. A typical alpine grassland on siliceous rocks, with *Carex curvula* (brownish) dominant near the top of the ridge, and with communities of *Festuca picta* and *Festuca violacea* dominating the south-facing slopes.

Plate 149 (left). Alpine calcareous rocks. Schneealpe, Styria, Austria. The characteristic community of alpine rocks in the eastern Alps is dominated by the sedge, *Carex firma.* Here at c. 1750 m are also found eastern cinquefoil, *Potentilla clusiana;* moss campion, *Silene acaulis;* and mountain avens, *Dryas octopetala.*

Plate 150 (above). Wetlands lake community. Durusu Golu, Turkey-in-Europe. Shallow lake communities have a very similar species composition throughout much of Europe. Here we can see the white water-lily, *Nymphaea alba,* with rushes *Juncus* sps., and the common reed, *Phragmites australis,* in the background.

Plate 151 (right). Wetlands fen-dyke community. Wicken Fen, Cambridge, UK. On the margin of the fen is the common reed, *Phragmites australis,* and in the shallow water the arrowhead, *Sagittaria sagittifolia;* the yellow water-lily, *Nuphar lutea;* and broad-leaved pondweed, *Potamogeton natans.*

Plate 152 (far right). Wetlands pond community. Near Suonenjoki, Finland. Still waters with the submerged long-stalked pondweed, *Potamogeton praelongus,* and the floating leaves of the amphibious bistort, *Polygonum amphibium.*

Plate 153 (right). Wetlands pond community. Near Suonenjoki, Finland. A pond verge here has the floating leaves of the yellow water-lily, *Nuphar lutea,* with duckweeds, *Lemna* sps., and emergent leaves of bog arum, *Calla palustris;* and marsh cinquefoil, *Potentilla palustris.* In the shallow margin where peat accumulates is a dense community of sedges, *Carex* sps., and also a young alder tree, *Alnus glutinosa,* colonizing the peat.

Plate 154 (below). Wetlands oligotrophic lake. Byglandsfjord, southern Norway. Oligotrophic or nutrient-poor lakes have a sparse flora. Here the emergent species is the water lobelia *L. dortmanna,* while in shallow water growing on the lake bottom commonly occur the shoreweed, *Littorella uniflora;* the awlwort, *Subularia aquatica;* and, less commonly, the quillwort, *Isoetes lacustris.*

Plate 155 (above). Wetlands swamp community. Iğneada, Turkey-in-Europe. Here, on the margins of ponds and ditches, occur the common club-rush, *Scirpus lacustris,* and the sea club-rush, *S. maritimus,* with bur-reeds, *Sparganium* sps., and bulrushes, *Typha* sps., and submerged pondweeds, *Potamogeton* sps.

Plate 156 (below). Wetlands swamp. Near Ucero, Soria, Spain. The submerged and floating species include the yellow water-lily, *Nuphar lutea;* pondweeds, *Potamogeton* sps.; and Canadian pondweed, *Elodea canadensis,* while the emergent species are water-cress, *Nasturtium officinale;* common club-rush, *Scirpus lacustris;* and bur-reed, *Sparganium* sp.

Plate 157 (right). Wetlands marsh. Tauvo, Finland. Large water dock, *Rumex aquaticus;* marsh cinquefoil, *Potentilla palustris;* tufted loosestrife, *Lysimachia thyrsiflora;* and a willowherb, *Epilobium* sp., occur here.

Plate 158 (below left). Wetlands lake community. Near Suonenjoki, Finland. The floating bur-reed, *Sparganium angustifolium,* in a lake in Finland.

Plate 159 (below right). Wetlands lake and ditch species. Durusu Golu, Turkey-in-Europe. The water chestnut, *Trapa natans,* growing in shallow water.

Plate 160. Coastal saltmarsh. Rømø Island, Denmark. A saltmarsh on the Danish coast, with the hybrid species Townsend's cord-grass, *Spartina x townsendii,* rapidly colonizing the mud-flats, along with sea aster, *Aster tripolium,* and the fleshy glassworts, *Salicornia* sps.

Plate 161 (far left). Coastal mud-flats. Epanomi, Greece. On these mud-flats, one of the shrubby glassworts, *Arthrocnemum glaucum,* forms clumps, while in the foreground are glassworts, *Salicornia* sps.

Plate 162 (left). Coastal mud-flats. Fanarion, Thrace, Greece. The first true land plants to colonize salt mud-flats are usually the glassworts, *Salicornia* sps. They tend to trap silt and mud, and consequently may bring about a gradual rise in the level of the substrate, thus enabling such grasses as the saltmarsh-grass, *Puccinellia maritima,* and the red fescue, *Festuca rubra,* to establish themselves, and cause a further rise in the level of the mud.

Plate 163 (above). Coastal sand dunes. Karacakoy, Turkey-in-Europe. These low sand dunes along the Black Sea coast have many species in common with coastal dunes elsewhere in Europe. Here on the primary, or fore-dune, is the sea-holly, *Eryngium maritimum;* cottonweed, *Otanthus maritimus;* and the grass, sand couch, *Elymus farctus,* all widespread species.

Plate 164 (right). Coastal sand dunes. Borth, Wales, UK. The marram grass, *Ammophila arenaria,* is here colonizing and stabilizing the higher white dunes on the Atlantic coast of Wales. Until these dunes become stabilized, few other plants can colonize them but where blown sand no longer accumulates, species like the red fescue, *Festuca rubra,* and the sand sedge, *Carex arenaria,* can establish themselves, with such other plants as sea bindweed, *Calystegia soldanella,* and sea pea, *Lathyrus japonicus.*

Plate 165. Coastal grey dunes. Finisterre, Spain. A rich maritime flora may develop when the sand dunes become stabilized, here on the Spanish coast. There is sea daffodil, *Pancratium maritimum;* shrubby pimpernel, *Anagallis monelli;* sea holly, *Eryngium martimum;* stonecrop, *Sedum acre;* sea bindweed *Calystegia soldanella;* and the remains of marram grass, *Ammophila arenaria*

Plate 166. Coastal grey dunes. Hansted Reserve, Denmark. Further stabilization of the dune surface by mosses and lichens occurs, and here it dominated by the grey hair grass, *Corynephorus canescens;* sand sedge, *Carex arenaria;* clumps of heather, *Calluna vulgaris,* and lichens of *Cladonia* sp. There still remains some depauperate marram grass, *Ammophila arenaria,* probably the original colonizing grass.

Plate 167 (right). Coastal dune heath. Hansted Reserve, Denmark. Dune heaths develop on the oldest brown dunes inland. Here is a well developed community of heather, *Calluna vulgaris;* crowberry, *Empetrum nigrum;* creeping willow, *Salix repens;* and sea plantain, *Plantago maritima,* and lichens of *Cladonia* sps.

Plate 168 (below). Coastal shingle. Slapton Ley, Devon, UK. Characteristic of such shingle shores are the sea-kale, *Crambe maritima,* and common scurvy-grass, *Cochlearia officinalis.* Other common species include shrubby sea-blight, *Suaeda vera;* sea campion, *Silene vulgaris* ssp. *maritima;* and yellow horned-poppy, *Glaucium flavum.*

Plate 169 (right). Coastal rocks and cliffs community. Near Llanes, northern Spain. Rocks on the coast of northern Spain in the Bay of Biscay, with sea-kale, *Crambe maritima;* wild carrot, *Daucus carota;* and oxeye daisy, *Leucanthemum* sp.

Plate 170 (below). Coastal cliff community. Cape St. Vincent, Portugal. Cliffs on one of the most exposed promontaries of Europe, with spiny thrift, *Armeria pungens; Calendula suffruticosa; Astragalus massiliensis;* sweet Alison, *Lobularia maritima;* and the Phoenician juniper, *Juniperus phoenicea.*